Creativity and Leadership in Science, Technology, and Innovation

Leadership is vital to creativity and successful innovation in groups and organizations; leadership is however seldom studied in the academic literature as a creativity driver. One reason for the lack of attention paid to leadership's effect on creativity may be the common belief that creativity cannot and should not be managed. Creative individuals and groups are regarded as, and indeed often are, autonomous and self-driving. From this belief the erroneous conclusion is drawn that there is no need for leadership in creative environments and situations. The better conclusion, proposed by this book, is that leadership not only stimulates creativity, but that such a leadership in the science, technology, and innovation fields should specifically possess at least two features: (a) expertise in the field(s) and (b) an ability to create, support, and encourage individuals, groups, and creative knowledge environments.

A number of specialist authors in this volume offer original theoretical, empirical, and applied chapters that elucidate how to better organize and lead creative efforts in science, technology, and innovation. A number of important research questions are raised and answered, including the following: What kinds of leaderships are needed at different levels of S&T organizations for a creative output? What social and cognitive abilities and skills are needed for leadership in creative environments? and How does leadership vary with different phases of the creative process? This book offers concrete analysis of how leaders and managers can facilitate, promote, and organize for creative performance in science, technology, and in innovating organizations, making it required reading for academic and industrial research leaders, scientists, and engineers.

Sven Hemlin is Professor of Psychology at Gothenburg Research Institute (GRI), School of Business, Economics and Law, the Department of Psychology, University of Gothenburg, and Department of Technology and Society, University of Skövde, Sweden. He has published in journals such as *Leadership Quarterly*, *Science, Technology & Human Values*, and *Creativity Research Journal.*

Carl Martin Allwood is Professor of Psychology in the Department of Psychology, University of Gothenburg, Sweden. He has edited several books and has published more than 70 articles in international scientific journals.

Ben R. Martin is Professor of Science and Technology Policy Studies at the University of Sussex, UK. Since 2004, he has been Editor of *Research Policy.*

Michael D. Mumford is the George Lynn Cross Distinguished Research Professor of Psychology at the University of Oklahoma where he directs the Center for Applied Social Research. Dr. Mumford has served as Senior Editor of the *Leadership Quarterly*, and he sits on the editorial boards of the *Creativity Research Journal*, the *Journal of Creative Behavior*, the *International Journal of Creative Problem Solving*, *IEEE Transactions on Engineering Management*, and *Ethics and Behavior* among other journals.

RIOT! Routledge Studies in Innovation, Organization and Technology

1 **Innovation in the U.S. Service Sector**
Michael P. Gallaher, Albert N. Link and Jeffrey E. Petrusa

2 **Information and Communications Technologies in Society**
E-Living in a Digital Europe
Edited by Ben Anderson, Malcolm Brynin and Yoel Raban

3 **The Innovative Bureaucracy**
Bureaucracy in an Age of Fluidity
Alexander Styhre

4 **Innovations and Institutions**
An Institutional Perspective on the Innovative Efforts of Banks and Insurance Companies
Patrick Vermeulen and Jorg Raab

5 **Knowledge and Innovation in Business and Industry**
The Importance of Using Others
Edited by Håkan Håkansson and Alexandra Waluszewski

6 **Knowledge and Innovation**
A Comparative Study of the USA, the UK and Japan
Helen Brown

7 **Industrial Innovation in Japan**
Edited by Takuji Hara, Norio Kambayashi and Noboru Matsushima

8 **Managing and Marketing Radical Innovations**
Marketing New Technology
Birgitta Sandberg

9 **Mobility and Technology in the Workplace**
Edited by Donald Hislop

10 **Energizing Management Through Innovation and Entrepreneurship**
European Research and Practice
Edited by Milé Terziovski

11 **Innovating for Sustainability**
Green Entrepreneurship in Personal Mobility
Luca Berchicci

12 **Organizational Capital**
Modelling, Measuring and Contextualising
Edited by Ahmed Bounfour

13 **User-Innovation**
Barriers to Democratization and IP Licensing
Victor R. G. Braun and Cornelius Herstatt

14 **Working on Innovation**
Edited by Christophe Midler, Guy Minguet and Monique Vervaeke

15 **Organization in Open Source Communities: At the Crossroads of the Gift and Market Economies**
Evangelia Berdou

16 **Theory and Practice of Triple Helix Model in Developing Countries**
Issues and Challenges
Edited by Mohammed Saad and Girma Zawdie

17 Global Innovation in Emerging Economies
Prasada Reddy

18 Creativity and Innovation in Business and Beyond
Social Science Perspectives and Policy Implications
Edited by Leon Mann and Janet Chan

19 Managing Networks of Creativity
Edited by Fiorenza Belussi and Udo Staber

20 Managing Environmentally Sustainable Innovation
Insights from the Construction Industry
Bart Bossink

21 Management and Information Technology
Challenges for the Modern Organization
Edited by Peter Dahlin and Peter Ekman

22 Managing Organizational Ecologies
Space, Management, and Organizations
Edited by Keith Alexander and Ilfryn Price

23 Digital Virtual Consumption
Edited by Mike Molesworth and Janice Denegri-Knott

24 The Video Game Industry
Formation, Present State, and Future
Edited by Peter Zackariasson and Timothy Wilson

25 Marketing Technologies
Corporate Cultures and Technological Change
Elena Simakova

26 Public Sector Transformation through E-Government
Experiences from Europe and North America
Edited by Vishanth Weerakkody and Christopher G. Reddick

27 Innovation Policy Challenges for the 21st Century
Edited by Deborah Cox and John Rigby

28 Eco-Innovation and Sustainability Management
Bart Bossink

29 Creativity and Leadership in Science, Technology, and Innovation
Edited by Sven Hemlin, Carl Martin Allwood, Ben R. Martin, and Michael D. Mumford

Creativity and Leadership in Science, Technology, and Innovation

Edited by Sven Hemlin, Carl Martin Allwood, Ben R. Martin, and Michael D. Mumford

Routledge
Taylor & Francis Group
NEW YORK AND LONDON

First published 2013
by Routledge
711 Third Avenue, New York, NY 10017

Simultaneously published in the UK
by Routledge
2 Park Square, Milton Park, Abingdon, Oxon OX14 4RN

Routledge is an imprint of the Taylor & Francis Group, an informa business

Library of Congress Cataloging-in-Publication Data

Creativity and leadership in science, technology, and innovation / edited by Sven Hemlin, Carl Martin Allwood, Ben R. Martin, and Michael D. Mumford.
pages cm—(Routledge studies in innovation, organization and technology)
Includes bibliographical references and index.
1. Research—Management. 2. Leadership. 3. Creative ability in science. 4. Organizational effectiveness. I. Hemlin, Sven, 1948– editor of compilation. II. Allwood, Carl Martin, editor of compilation. III. Martin, Ben R., editor of compilation. IV. Mumford, Michael D., editor of compilation.
Q180.55.M3C74 2013
506.8′4—dc23
2012048113

ISBN: 978-0-415-83484-1 (hbk)
ISBN: 978-0-203-49973-3 (ebk)

Typeset in Sabon
by Apex CoVantage, LLC

Printed and bound in the United States of America by Publishers Graphics, LLC on sustainably sourced paper.

Contents

Acknowledgments ix
Foreword xi

1 Introduction: Why Is Leadership Important for Creativity in Science, Technology, and Innovation 1
SVEN HEMLIN, CARL MARTIN ALLWOOD, BEN R. MARTIN, AND MICHAEL D. MUMFORD

PART I
Theoretical Section

2 Leading Scientists and Engineers: Cognition in a Socio-technical Context 29
MICHAEL D. MUMFORD, DAVID PETERSON, AND ISSAC ROBLEDO

3 What Connects Leadership and Creativity? The Mechanisms Through Which Leaders May Influence Follower and Team Creativity 58
LEIF DENTI AND SVEN HEMLIN

4 Leadership, Innovation, and Technology: The Evolution of the Creative Process 81
SAMUEL T. HUNTER, LILY CUSHENBERY, NICOLE GINTHER, AND JOSHUA FAIRCHILD

PART II
Empirical Section

5 Academic Leadership of High-Performing Research Groups 113
MAAIKE VERBREE, INGE VAN DER WEIJDEN, AND PETER VAN DEN BESSELAAR

6 Generation and Life-Cycle Effects on Academic Leadership 149

MAAIKE VERBREE, INGE VAN DER WEIJDEN, AND PETER VAN DEN BESSELAAR

7 Time to Create: Pathways to Earlier and Later Creative Discoveries in Nobel Prize Winners 184

DAWN L. EUBANKS, MICHAEL E. PALANSKI, JUANI SWART, MICHELLE M. HAMMOND, AND JOY OGUNTEBI

PART III
Implications Section

8 Succession Planning for Scientific Positions: Identifying, Developing, and Retaining Leaders for Innovation 211

GINAMARIE S. LIGON, KATE T. DEMBROSKI, ROBYN C. MAPP, AND BIANCA M. ZONGRONE

9 Leading Interdisciplinary Creative Teams: Challenges and Solutions 240

RONI REITER-PALMON, TRIPARNA DE VREEDE, AND GERT-JAN DE VREEDE

10 Leadership and Followership in Science and Technology 268

MICHAEL E. GORMAN AND PHILIP S. GORMAN

11 Creative Leadership: Its Meaning and Value for Science, Technology, and Innovation 287

GERARD PUCCIO, MARIE MANCE, AND JEFFERY ZACKO-SMITH

12 Conclusion 316

SVEN HEMLIN, CARL MARTIN ALLWOOD, BEN R. MARTIN, AND MICHAEL D. MUMFORD

Contributors 333
Index 341

Acknowledgments

Sven Hemlin would like to thank Laura Stearns, Stacy Noto and Lauren Verity of Routledge for their efficient and pleasant collaboration, VINNOVA—the Swedish Agency for Innovation—for supporting me with grants making it possible to work with this volume and Carl-Christian Trönnberg and Leif Denti for editorial assistance.

Foreword

Researchers engaged in science, technology and innovation inevitably, at some point, begin to reflect on their own thinking and actions, not least when they encounter obstacles and dead-ends in trying to solve the problems with which they are confronted. In particular, they are likely to wonder whether what they are doing could be improved, whether their research could be somehow be made more creative and productive. Similarly, research leaders are likely to think about how they and their research group can best be organized to work in ways that are likely to lead to more creative results. At a more general level, it is important to consider how science, technology and innovation can develop more creatively, how we might organize research and innovation development in more creative ways, and how research and innovation might best be led and managed to achieve the most creative results.

In spite of the evident importance of such questions, they have not previously been subject to much research, instead leaving it to university or business administrators to provide practical solutions. However, we would contend that scientists should indeed focus on and carry out research in this area because they have a better likelihood of arriving at appropriate and well-founded answers, if nothing else because of their more rigorous methodological competence and more systematic knowledge of previous research. Thus far, relatively few researchers have been thinking more deeply about, and have devoted research time to, the issue of organizing, conducting, and leading research in a more creative manner. Those people have mostly been researchers in management and organizational psychology or historians and political scientists. The latter have taken a broader and more historical perspective on issues such as the development of the university as an institution, and its aims and place in society.

For our part, we have mostly been inspired by the former tradition—management and organizational psychology—in which empirical research has been carried out at the individual, group and organizational levels of creative research and engineering. However, we also have other sources of inspiration such as cognitive psychology, social psychology, and leadership studies.

The book represents the results of our efforts to bring together chapters by some of the leading authors on the issue of how best to lead science, technology, and innovation toward more creative achievements. The book focuses on the role of leadership in fostering creativity in research and development, because science, technology, and innovation require collective effort coordinated and inspired by leaders with creative objectives firmly in mind. Without a considered, conscious approach to the promotion of creativity in research and innovation, the chances of being innovative and of coming up with creative solutions to difficult problems are substantially decreased.

One of the earliest and empirically well-founded investigations in this field is described in the book by Pelz and Andrews (1966). This work has been a source of great inspiration for some of us. When it comes to the topic of creativity, the work and approach of Simonton (2004, 2011), based in part on the prior work of Campbell in evolutionary epistemology (Campbell, 1960), has also had a fundamental influence, as indeed has previous work by Mumford and his colleagues (e.g., Mumford, Scott, Gaddis, & Strange, 2002).

The four editors have been working together over the last 2 years to bring this book to fruition. Although very busy, we have nevertheless found it easy to communicate, to prepare drafts, and to review and edit chapters, a process aided by meetings between two, three or all four of us at various places around the world. This work, and the interactions it has involved, has itself been an enjoyable and creative process of collaboration, stimulating new insights about creativity and leadership. We hope the readers of this book will feel the same.

Sven Hemlin, Carl Martin Allwood, Ben R Martin and Michael D Mumford
Gothenburg, Brighton, and Norman (Oklahoma), September 2012

REFERENCES

Campbell, D.T. (1960). Blind variation and selective retention in creative thought as in other knowledge processes. Psychological Review, 67, 380–400.

Mumford, M. D., Scott, G. M., Gaddis, B. H., & Strange, J. M. (2002). Leading creative people: Orchestrating expertise and relationships. The Leadership Quarterly, 13, 705–750.

Pelz, D.C., & Andrews, F. M. (1976). Scientists in organizations. Ann Arbor, MI: Institute for Social Research.

Simonton, D.K. (2004). Creativity in science: Chance, logic, genius, and zeitgeist. Cambridge, MA: Cambridge University Press.

——— (2011). Creativity and discovery as blind variation: Campbell's (1960) BVSR model after half the century mark. Review of General Psychology, 15, 158–174.

1 Introduction

Why Is Leadership Important for Creativity in Science, Technology, and Innovation?

Sven Hemlin, Carl Martin Allwood, Ben R. Martin, and Michael D. Mumford

This book is concerned with the relationship between creativity and leadership in science, technology and innovation. Why have we chosen to focus on this topic? There are several reasons. First, it is clear that science, technology, and innovation are crucial driving forces in the development of economies and societies. This is not just confined to the present and the emergence of what has become known as 'the knowledge economy'. Throughout history, science, technology, and innovation have been central to societal and economic development, with human curiosity and creative endeavors often being crucial to finding scientific, technological, and innovative solutions to the problems confronting humankind. Consider a few significant examples: We have seen detailed astronomical observations of stars and planets already in the earliest cultures of mankind in Mesopotamia over 5,000 years ago, Roman aqueducts for the transport of water made possible by new technologies 2,000 years ago, the invention of the printing press by Gutenberg in 1440, the adoption of electricity by Edison to light houses and streets in the 1880s, a number of effective cures for previously lethal diseases such as pneumonia through the introduction of antibiotics at the end of the 19th century, and so on. All of these accomplishments were driven by a combination of human curiosity, creative problem solving, and skills in an effort to promote societies and make life better. This drive to increase our knowledge in order to bring about the improvement of the world manifests itself in science, technology, and innovation.

But why study leadership in connection with creative efforts in science, technology, and innovation? Are not scientific discoveries the work of singular persons—of gifted scientists and inventors? As we shall see in the chapters that follow, there is strong evidence that creativity in these domains is much more dependent on good leadership today than previously. Scientific and technological process requires more coordinated actions than a hundred or even fifty years ago (Ziman, 2000). In science, team work is ever more prevalent, for example, as reflected not only in the increasing number of coauthored articles (e.g., Hicks & Katz, 1996) but also in the way that scientific research is organized in larger units, research groups and projects (Ziman, 2000). Likewise, innovation is often a result of the collective efforts of numerous individuals, teams or even organizations working in collaboration (e.g., Chesbrough, 2003). Consequently, the coordination of those engaged in creative activities in science, technology, and innovation requires more and better leadership, including the provision of cognitive and motivational support for creative individuals, teams, and organizations (Mumford, Peterson, & Robledo, this book). This requirement has increased as economies and societies have become ever more dependent on new knowledge—that is, with the emergence of the knowledge society (Stehr, 1994).

The purpose of this book is to collect, analyze and draw conclusions from recent studies of how leaders may stimulate creativity in science, technology, and innovation. This has involved bringing together a number of leading scholars in the field to present their latest results and views on this topic. Many of the chapters were first presented at a conference held at Gothenburg in July 9–10, 2010, where they were discussed, and criticisms and suggested improvements were put forward. When a number of gaps were identified in the coverage of the subject, other leading researchers were approached to prepare additional contributions. The selection of authors was guided by one of the editors (Mumford) who has a good overview of the field because he served for several years as the Editor of *Leadership Quarterly*. We do not claim to cover all dimensions of creativity and leadership in science, technology, and innovation, but the chapters in this book span a range of the more important topics and provide a number of original contributions, some being in the form of conceptual or theoretical overviews, others being primarily empirical papers, while a third set comprise contributions that are more applied in nature. Reviews from referees and from the editors of this book helped in the process of further developing and strengthening the contributions set out in the chapters that follow.

SOME DEFINITIONS

Creativity and Innovation

This book is about creativity and leadership in three domains of human enterprise, namely science, technology and innovation. We have so far attempted to argue why this is an important topic. Next, we want to offer some definitions of the central concepts we use. First, creativity is commonly seen as a novel and useful result or product of some process (Mumford & Gustafson, 1988). However, a more elaborate definition of creativity is offered by Amabile (1996, p. 35):

> A product or response will be judged as creative to the extent that (a) it is both a novel and appropriate, useful, correct or valuable response to the task at hand, and (b) the task is heuristic rather than algorithmic.

The second definition includes the proviso that tasks solved by following established rules are not creative in nature. Hence, not only is the end result important but also the process used for solving the problem. Indeed, some theorists also included the person and the 'press' (the relationship between human beings and their environment) in the definition of creativity in addition to the product and process (e.g., Rhodes, 1961). Rhodes's 'four Ps' (person, process, product, and press) of creativity subsequently became a standard view of how to approach creativity. In the early days of creativity research, most emphasis was given to the personal characteristics of creative people, but later on the problem-solving process leading to creative results received more attention, and more recently the environment promoting creativity has been the focus of attention (see Feist, 2006, for a recent review; also see Hemlin, Allwood, & Martin, 2004, 2008).

Creativity is often seen in terms of artistic accomplishment, for example, in music or art. However, it is also a characteristic of much scientific and technological endeavor. The related concept of 'innovation' is frequently the outcome of individual, team, and organizational efforts to create a new or significantly improved product, process, or service that is potentially attractive to some market or other (Denti & Hemlin, 2012).[1] Moreover, and very important, innovations are often based on creative accomplishments in science and technology. Hence, there is a strong link between creativity in science and technology, on one hand, and innovations, on the other. The latter, however, are not only dependent on creative efforts in science and technology. An innovation is the implementation of new ideas, whether scientific, technological, or otherwise, that is applied in a certain context in order to be useful. According to economists, there also needs to be some form of market demand for the innovation so that it can be sold or otherwise diffused to customers or clients and thus make the transition from an 'invention' to an 'innovation'. Therefore, innovation requires efforts and activities by individuals and organizations to implement, diffuse, organize, and market it. To return to the concept of creativity, one should view creativity rather as often being a forerunner to innovation. If no creativity were apparent, there would probably be little by way of innovation.

Innovations may be further differentiated, coming in a variety of forms. One distinction is that between incremental innovations and radical ones (e.g., Tidd, Bessant, & Pavitt, 2005). The former involve step-by-step improvements to products, processes, or services, for instance, better performance by car window-wipers in new automobile models. Radical innovations, by contrast, often involve new products, processes, or services not available previously, such as the electrical lightbulb developed by Thomas Edison.[2] Another innovation category is that of 'open innovation' denoting new products, processes, or services developed in collaboration between two or more firms or other organizations rather than in one alone (the latter corresponding to 'closed innovation'). A third distinction is that between 'exploratory innovation'—the production of new products, processes, or services based on *new* knowledge—and 'exploitative innovation' based on *already existing* knowledge within a business firm. Finally, a more recent concept is that of 'social innovation', which is distinguished from product innovation in the sense that the new phenomenon mainly concerns the social world such as Twitter. In summary, the innovations we see around us today may take many forms, and some may be based on science and technology.

Innovations are today playing a central role in driving economies but also in enabling societal development more widely. It may be helpful to the reader at this point to get a sense of where innovation is currently flourishing in the world, because this may also provide some indication of where creativity—and the leadership of creative and innovative efforts—is apparently working most effectively. The Global Innovation Index (GII) annually ranks nations in terms of their innovative capacity. In 2011, the GII rankings were dominated by Europe with six countries, whereas Asia had two as did North America. The top 10 nations 2011 are (in rank order) Switzerland, Sweden, Singapore, Hong Kong, Finland, Denmark, the United States, Canada, the Netherlands, and the United Kingdom. The GII also identified the leading nations in each of the main regions of the world, these being Switzerland, Singapore, the US, Israel (ranked 14th in the world), Chile (ranked 38th), and Mauritius (53rd) and India (62nd) (Dutta, 2011).

Leadership

The next concepts we need to scrutinize are those of 'leader' and 'leadership'. What is a leader? A common view is that a leader is someone who influences other persons for some specific purpose (Mumford, Peterson, & Robledo, this book). The leadership role may be of an informal or a formal nature. The former, for example, might mean a person influencing others to pursue some action in a work group because he or she is simply the oldest or most experienced member of that group (i.e., an informal leader), while a manager may influence others in a work group because it is his or her job (i.e., acting as a formal leader). For our purpose, it will be useful to distinguish between leaders and managers because they have different characteristics that are not always apparent or easily perceived. The seminal work on this was carried out by Zalesnik (1977/2004), who distinguishes between the characteristics of leaders and managers in terms of managerial cultures, attitudes to goals, work conceptions and relationship to others. First, Zalesnik observes that a managerial culture is one that is characterized by rationality and control, whereas he describes a leadership culture rather in terms of dramatic events. Second, whereas managers tend to have impersonal and rather passive attitudes to goals, the leader is a more personal, active, and inspiring actor. Third, managers are generally more concerned with coordinating and balancing views to limit choices in work. Leaders, on the other hand, deliberately take new approaches and open up the search for other options. Fourth, managers relate to people according to their roles in work and decision-making processes, whereas leaders are more concerned with people's ideas and relate to these in intuitive and empathetic ways. In conclusion, Zalesnik argues that managers are part of the organization that they work for, whereas leaders never truly belong to the organization, even though he or she works within it. These observations have at least two implications for leadership that we believe are particularly salient here. First, leaders are engaged in what they are doing because of their expertise in a given field (i.e., cognitive leadership). Second, leaders care and support their followers because they need the team that they lead to pursue the goals they aspire to by working effectively together (i.e., motivational leadership). However, this is not to imply that leading and coordinating research and development (R&D) or innovation does not require management skills—these are certainly needed as well.

LEADERSHIP STYLES IN DIFFERENT CULTURES

This book contain chapters dealing with leaders from a number of different countries including the US, the Netherlands, and Sweden, so it is perhaps relevant to consider whether there might be important differences among leaders and leadership styles in different countries. One reason for such differences among leaders may be that they act in social environments with different expectations as to their behavior. For example, the little research that exists suggests that leaders in Scandinavia and the Netherlands differ from those in China and the US, for example, but that those in Japan show some similarities to their counterparts in Scandinavia and the Netherlands. For example, leaders in China and to some extent the United States appear to take a more directive and hierarchical approach than leaders in the other countries mentioned previously do, at least in terms of Hofstede's power dimension.[3] On this particular dimension, China scores highest with 80, followed by Japan on 54, whereas the Scandinavian countries of Finland, Norway, and Sweden range from 31 to 33 and Denmark is only 18, compared with the United Kingdom with 35 and the US with 40.[4] In line with this is Koopman and Pool's (1991) characterization of organizational decision making in the Netherlands as typifying a 'participation model', which stresses participation by staff at levels below the top management. The same description is also applicable to, for example, Sweden and Norway. A further indication of the differences between leaders in the aforementioned countries comes from a study by Martinsons (2001) of senior business leaders in China, Japan, and the United States. The Chinese leaders scored much higher on the 'directive' decision style than did the Japanese and US leaders, whereas the Japanese leaders scored somewhat higher than did those from the US. However, the Japanese leaders also scored higher on the socially oriented 'behavioral' decision style; for example, they tended to involve more people at different levels in their decisions. Interviews with a subset of the participating leaders showed that organizational decision making tended to take longer in Japan, and this was reported as being due to attempts to achieve consensus around the suggested decision option. The interviews also showed that the implementation phase was generally shorter in Japan than in the other two countries, perhaps due to a more thorough predecision process. In this, Japan appears to be similar to Scandinavian countries and the Netherlands. For example, Allwood and Hedelin (2005), in interviews with CEOs in large organizations in Sweden on currently ongoing decision processes, found that the 'selling' of the championed decision to various parties within and outside the organization constituted a large part of the efforts in the decision processes reported on.

SCIENCE, TECHNOLOGY, AND R&D

Finally, we need to clarify what each of the terms *science*, *technology*, and *R&D* involves, because this will clearly have a considerable influence on how these activities are best led. First, science is a systematic and methodologically rigorous way of searching for new or improved knowledge (although new knowledge may also be produced by institutions other than scientific ones). One key characteristic of science is that it follows certain norms and rules (Merton, 1942), and is primarily organized by universities and academies following well established traditions.[5] However, science is continuously in a process of change, not least because new knowledge tends to have feedback effects on how subsequent scientific research is conducted. Given these characteristics, science may be viewed both as a social phenomenon (involving the organization of those involved) and as an epistemological (or knowledge) phenomenon (Merton, 1973). This is something to bear in mind when we come to discuss creativity and leadership in science, technology, and innovation.

Technology is commonly viewed as the search for practical solutions to technical problems, such as those involved in building a safe bridge or constructing a sophisticated machine like a computer. It may often involve applied science, particularly by engineers, who draw upon science as part of the basis for developing solutions to specific technical problems. To come up with technical solutions, there is a need not only for the appropriate technical skills but also for an element of creative problem solving (Dasgupta, 2011).

R&D may be viewed as an umbrella concept embracing basic and applied research and experimental development. It covers both academic and industrial research, for example, the development of new drugs in pharmaceutical companies, or problem solving in the construction of new engines in the automotive industry, or even societal research activities, such as environmental protection R&D performed by a government research institute. A commonly quoted definition of R&D is to be found in the Frascati Manual produced by Organisation for Economic and Co-operative Development (OECD; 1993, chap. 2):

> Research and experimental development (R&D) comprise creative work undertaken on a systematic basis in order to increase the stock

> of knowledge, including knowledge of man, culture and society and the use of this stock of knowledge to devise new applications.[6]

It should be noted that this definition explicitly highlights the 'creative' aspect of work involved in R&D.

The definitions of the key concepts employed in this book help to better understand our purpose. In particular, we seek to analyze the link between creativity and leadership in science, technology, and innovation (STI) in its social and cognitive context—in other words, in the creative knowledge environment (Hemlin et al., 2004, 2008). More specifically, we wish to show how leadership may facilitate, promote, and enhance the creative performance of individuals and teams with regard to science, technology and innovation. To do this, we bring together a number of new studies on the topic by leading scholars in this particular field.

A Short History of Creativity

Throughout history, humans have tried to understand phenomena relating to creativity. According to the first conceptions, creativity was seen as a gift from higher powers or, especially in the East, from the Universe (Chan, 2011). However, already Aristotle had recognized that a few select individuals are more likely to be creative than others and that creative products might be generated 'from within' these gifted individuals.

In general, thinking about creativity has developed from implicit assumptions, relying on taken-for-granted assumptions about creativity, to the explicit development and empirical testing by researchers of different conceptualizations of what creativity is. As illustrated in the following, an important tendency is that theories and thinking about creativity have over time included more factors of different kinds, while at the same time becoming more specific with respect to what aspects of the phenomenon are focused upon. Thus, creativity is no longer seen as merely residing within specific individuals but as being a function of the wider context. Moreover, a greater range of different forms of creativity have been studied. For example, from initially studying the accomplishments of great thinkers and producers, that is, creativity with a big *C*, researchers now also study everyday and more mundane creativity, in other words, creativity with a small *c*. It is also recognized that conditions for creativity may vary across different content domains (for example, the arts as opposed to the sciences).

The concept of the creative genius dominated thinking about creativity up to the 20th century; for example, Francis Galton, in the second half of the 19th century, studied the heredity of genius (Dacey, 2011). In opposition to Galton's approach, the Gestalt psychologists in the first half of the 20th century empirically demonstrated that creative thinking depends on seeing clusters of perceptually or conceptually interacting features (*gestalts*) in a new way, for example, by changing perspective. The Gestalt psychologists researched *insight problems*, a type of creative problem solved by breaking out of entrenched thought habits, and where the solution is often experienced as occurring in a flash (a sudden insight). An example is *functional fixedness*, where a solution is hindered because the solvers stick too rigidly to conventional object use (e.g., Duncker, 1945; Wertheimer, 1945).

In 1926, social psychologist and cofounder of London School of Economics Graham Wallas published the book *The Art of Thought,* in which he presented the, by now classical but much criticized, four-stage scheme of creative problem solving. This is based on the following four phases: preparation, incubation, illumination, and verification. Wallas (1926/1970) emphasized that these stages "constantly overlap each other as we explore different problems" (p. 92) and that incubation on the same problem may also continue when the thinker is consciously engaged in the preparation or the verification stage. Preparation is said to include "the whole process of intellectual education" (p. 92). In the incubation stage, two things happen; we take a break from conscious thinking about the problem, and "a series of unconscious and involuntary (or foreconscious and forevoluntary) mental events may take place during that period" (p. 94). Finally, the verification stage "closely resembles the first stage of preparation" and involves checking the proposed solution. Of these stages, the second and third, at least initially, elicited the most research, and as a result few researchers would argue today that any sophisticated unconscious processing takes place in the incubation stage (see, e.g., Smith 2011). Instead, incubation may have positive effects by providing refreshment, or the redistribution of activity in the memory (thereby getting rid of dead-end attempted solutions, also called 'set-breaking'), and by providing new impressions during the break that can then act as further cues for the solution (see, e.g., Smith, 2011). With respect to the illumination stage, research has showed that the feeling of illumination may often not be veridical (Martinsen & Kaufmann, 2011; Metcalfe, 1986).

In 1950 Guilford gave the presidential address to the American Psychological Association that is often seen as heralding the start of interest in a psychometric approach to the study of creativity, that is, the measurement of creativity and the construction of tests and scales. This research has helped to direct attention to the important criterion problem in creativity research, in particular, how should a creative product be defined (see the preceding discussion)?

Researchers during the so-called cognitive revolution (i.e., the start of cognitive psychology) in the mid-1950s were to a very large extent influenced by the ideas of the Gestalt psychologists. Newell and Simon's (1972) concept of a 'problem space' attempted to capture the components of a problem solver's experience of the problem, including the start- and goal-state, the allowed ways of changing from one problem state to another (the so-called *legal operators*) and all the problem states between the start and the goal state that can be reached by using these operators. This concept—problem space—best fits problems that are solved by successive developments from the start state but allows for a deeper specification of what it means for a problem to be ill defined (problems are ill defined when some component in the problem space is not well defined).

During the 20th century, researchers extended Wallas's four-step model to include a *problem identification stage* at the beginning of the process, and a later acceptance stage where the proposed solution was 'sold' to an audience. The importance of problem finding or problem identification is stressed by Kantorovich (1993, p. 29), who stated that "a discovery of a new problem may lead to significant progress of science, sometimes more so than the discovery of a solution to a known problem; such a new problem may open up a whole new field of investigation." The importance of problem finding for creativity is also supported by later research which found that training in problem finding is one of the more effective ways to develop creativity (Reiter-Palmon, 2011; see also Allwood, 1997, and Allwood & Bärmark, 1999, on how researchers deal with research problems as a resource in their work).

Later models of the processes in creative thinking have attempted to specify the processes involved in creative problem solving and have broadened the scope of the context seen as relevant for explaining creativity. For example, Campbell (1960) and Simonton (2003) explained creative production in two steps. They argued that the actual generation of creative ideas is a (pseudo)random process, which is then followed by a critical selection of the most valuable ideas. Later on, researchers have developed similar or expanded models of creativity to that of Wallas, for example, the creative problem-solving model (CPS) reviewed in the chapter by Puccio, Mance and Zacko-Smith in this book.

Today, research on process models for creativity is very active. An example of a current model of creativity is the dual pathway to creativity model, which has been tested on ideational problems, and which attends to, and provides empirical support for, the necessity of both flexibility and persistence in creativity (Nijstad, De Dreu, Rietzschel, & Baas, 2010). Because common associations tend to come first and more novel ones later, persistence was found to enhance creativity through allowing an exploration of content categories in greater depth. This model integrates many findings, for example, that an active positive mood may, first and foremost, help creativity through promoting a flexible and wide-ranging search for solutions, whereas an active negative mood, such as anger, may also have an impact through the increased time spent on the task.

Other research in recent decades has studied the concept of thinking styles (which are less stable than personality traits) and personality traits that characterize creative persons (Hemlin, 1996; Perkins, 1988; Simonton, 2003) For example, research has found that openness to experience, broad interests, attraction to complexity, self-confidence, aesthetic sensitivity, values of originality and independence, and rejection of the narrow and the mediocre to cherish the general and the fundamental are all positively linked to creativity. This fits in well with the random generation of ideas model presented by Simonton (2003). In addition, Simon (1965) and others have emphasized the importance of a substantial knowledge base for creativity, whereas Amabile (1999) and others have added and emphasized the importance of intrinsic motivation for creativity; a person who thinks that the activity as such is fun is likely to persevere longer. Likewise, a self-conception that includes high self-efficacy ("Yes, I can!"), especially *job self-efficacy* and *creative self-efficacy,* has been found to be advantageous for creativity (see, e.g., Tierney & Farmer, 2002).

Finally, an influential, more holistic model, put forward by Csikszentmihalyi (1999), places creativity in a broader social context, and this is especially relevant in an innovation context. This model presents creativity as a process that occurs at the meeting places of individuals, domains ('cultures', seen as the sum of domains of symbolic contents consisting of conceptions, tools, practices, and values) and fields (the social organization of domains and especially of individuals who act as *gatekeepers* with respect to the suggested creative product). An individual in suitable circumstances (for instance, with a broad knowledge in the domain) and facilitating personality traits may produce a creative variation of intellectual or artistic material from the domain. With sufficient rhetorical skills, the creator may then successfully communicate the creative product to the surrounding social field, which may or may not result in the acceptance of the product. The theory emphasizes the ability of the social field to provide supportive conditions, for example, in terms of economy, tolerance, and education for potentially creative individuals. It also emphasizes the importance of factors that determine the receiving field's reception of the suggested creative product, for example, interest, tolerance, sufficient social stability, and appropriate values.

Current Key Issues in Creativity and Leadership in STI

Given its implication for our world, it is not surprising that research on creativity and innovation has a long history—beginning at the turn of the 20th century (Albert & Runco, 1999). For many years, the emphasis of much creativity research was on idea generation, or divergent thinking, with research focusing on both the cognitive processes contributing to idea generation and the conditions under which people generate viable ideas (Plucker & Renzulli, 1999). Although this research served to establish a background for subsequent work, in recent years creativity (the production of new ideas) and innovation (the translation of new ideas into viable products or services) have begun to be understood as a form of complex performance (Welling, 2007). More specifically, creativity is held to involve the production of high-quality, original, and elegant solutions to complex, novel, ill-defined, or poorly structured problems (Mumford & Gustafson, 1988).

This reconceptualization of what we mean when we use the terms *creativity* and *innovation* has led to the emergence of a number of new, and critical, issues, and with this the emergence of a number of new lines of research. One key implication of the performance-based view of creativity and innovation is that creativity and innovation need to be seen as domain-dependent phenomena (Baer, 1998). In other words, the requirements for creativity and innovation will not be the same in poetry as in physics. One particularly powerful influence in this regard is expertise, or knowledge (both procedural and declarative). Studies by Ericsson and Charness (1994) and Rich and Weisberg (2004) have shown that expertise is, in fact, critical to creativity and innovation (Hemlin & Olsson, 2011). More centrally, this points to the importance of efforts such as those contained in the present volume, to study creativity and innovation in fields like science and technology (Carlson & Gorman, 1992).

Not only has a performance-based understanding of creativity led to a focus on creativity and innovation in specific domains; it has also led to shifts in the nature of the variables seen as critical to creativity. The performance-based view of creativity and innovation has led us to see creative thinking as a far more complex phenomenon (Hennessey & Amabile, 2010). Traditionally, creativity and innovation were understood with respect to a single cognitive process—idea generation. Today, however, creative performance is held to require effective execution of at least eight critical cognitive processes (Mumford, Medeiros, & Partlow, 2012; Mumford, Mobley, Uhlman, Reiter-Palmon, & Doares, 1991)—problem definition, information gathering, information organization, conceptual combination, idea generation, idea evaluation, implementation planning, and monitoring. Moreover, effective execution of each of these processes depends on the application of multiple strategies—for example, a search for anomalies vis-à-vis key facts in information gathering, or search, mapping, and elaboration strategies in conceptual combination (Mumford, Medeiros, & Partlow, 2012). Moreover, viable strategies for process execution vary with the type of knowledge being employed and the field in which the work is being conducted (Mumford, Antes, Caughron, Connelly, & Beehler, 2010; Scott, Lonergan, & Mumford, 2005).

However, any form of complex performance, including creativity and innovation, is not simply a matter of cognition. Thus, in recent years we have seen more attention given to studies of personality and motivation (DeDreu, Baas, & Nijstad, 2012; Hoff, Carlsson, & Smith, 2012). Indeed, personality and motivation influence attraction to creative problems and people's willingness to invest scarce cognitive resources in solving these problems. Other work has examined how creativity and innovation occurs in teams (Reiter-Palmon, Wigert, & de Vreede, 2012), how organizations identify and pursue problems calling for creative thought (Kazanjain & Drazin, 2012), how organizations should best structure themselves to promote creativity and innovation (Alencar, 2012; Damanpour & Aravind, 2012), and how people's reactions to the environment in which they are asked to generate creative problem solutions and innovative products influence their subsequent performance (West & Sacramento, 2012).

In fact, creative people, and creative thought, appear especially sensitive to climate, or to environmental perceptions. Not only have Hunter, Bedell-Avers, and Mumford (2007) shown that climate or environmental perceptions of support, openness, and mission challenges are strongly related to creativity and innovation, but it also appears that creative people will simply stop thinking when the environment is perceived not to be conducive to creative work (Oldham & Cummings, 1996). Of course, many organizational, or contextual, variables influence people's perception of climate (James, James, & Ashe, 1990). However, the actions taken by the leaders of an organization represent one of the most powerful influences on people's perception of climate—including the climate for creativity (Jaussi & Dionne, 2004).

One implication of the preceding observations is that leadership will be a key influence on people's willingness to engage in, and their capability for, creative problem solving and subsequent innovation. Indeed, the bulk of the available evidence supports this proposition (Mumford, Scott, Gaddis, & Strange, 2002). For example, more effective leadership has been shown to be strongly positively related (r = .40), to publication data and patent awards (Barrowe, 1975; Tierney, Farmer, & Graen, 1999). Leadership has been found to be critical to the successful development of innovative new products in fields such as avionics, information technology, and pharmaceuticals (Cardinal, 2001; Drazin, Glynn, & Kazanjian, 1999; Kidder, 1981). Indeed, the effective leadership of creative efforts appears to be linked to the long-term success, survival, and profitability of commercial organizations (Tushman & O'Reilly, 1997).

These findings, however, have raised a new question. Given the unusually strong positive impacts of effective leadership on creativity and innovation, more must surely be involved than simply the positive impacts of effective leadership on the climate for creativity and innovation. Indeed, leaders may exert an influence on creativity and innovation through a range of mechanisms. For example, planning is critical to creativity and innovation in many fields—resources must be acquired, research and development integrated, and new products created. Planning, however, is a critical function of organizational leaders (Marta, Leritz, & Mumford, 2005). Plans must be sold to obtain the resources needed for creative efforts (Nohria & Gulati, 1996). Moreover, it appears that leaders are critical to project sales (Howell & Boies, 2004). Furthermore, project teams for executing these plans must be established, and the normative interactions of team members must be defined (Taggar, 2002). Again, leaders play a key role in defining projects, creating project teams, and staffing projects (Mumford et al., 2002). Leaders, moreover, must direct these teams and evaluate the work, and the implications of this work, produced by the members of their project team. Indeed, the available evidence indicates that how leaders evaluate their project work and its implications is critical to the success of creative efforts in real-world settings (O'Connor, 1998).

These many different ways in which leaders exercise influence over the nature and success of creative and innovative efforts are noteworthy because they point to the importance of the present volume. Creativity and innovation are generally held to be the critical characteristics of successful work in science and technology (Mumford et al., 2005). Our traditional image of the scientist and engineer as the lone, and perhaps rather lonely, individual contributor, however, has previously led many to discount the importance of effective leadership. In the present volume, we examine the various ways in which effective leadership can contribute to the success of creative and innovative efforts in the fields of STI.

Research on Leadership in STI

Like research on creativity and innovation, studies of leadership have a long history (Bass & Bass, 2008). Early research into leadership and leader effectiveness focused on the traits, or characteristics, organizations should seek in more effective leaders. Broadly speaking, this research indicates that effective leaders are intelligent, are capable of divergent thinking, are achievement oriented, are power motivated, and are socially skilled (Mumford, Zaccaro, Harding, Jacobs, & Fleishman, 2000). Although the evidence accumulated over the years certainly points to the importance of these characteristics, in the 1950s the focus shifted to the behaviors exhibited by more effective leaders. This shift was, in part, driven by a new concern with developing leadership talent. These behavioral studies helped to identify four critical dimensions of leadership—consideration, initiating structure, participation, and change management—along with a number of key situational variables moderating the impact of these behaviors on measures of leader effectiveness—in particular, follower expertise, stress, interpersonal interactions, and the structure of these interactions (Yukl, 2010).

Although new behavioral dimensions of leadership continue to be proposed, for example, 'servant leadership' (Greenleaf, 1977) and 'authentic leadership' (Terry, 1993), in the 1980s attention began to shift to examples of outstanding leadership or leadership beyond normative expectations. Work in this vein gave rise to theories of charismatic (Shamir, House, & Arthur, 1993) and transformational (Bass, 1985) leadership. Research on charismatic and transformational leadership proved especially important, generating three major thrusts in leadership studies. First, one set of studies began to examine the sources or origins of key characteristics of transformational and charismatic leaders, such as the origins of a leader's vision (e.g., Mumford, 2006; Strange & Mumford, 2005). Second, studies began to be conducted examining how charismatic and transformational leadership operated at different levels in organizational systems (Yammarino & Dansereau, 2002)—namely, the individual, dyadic, team, and organizational levels. Third, studies began examining the conditions under which different populations or different occupational fields would accept charismatic or transformational leadership (Yukl, 1999). Significantly, one population that proved of special significance in this regard was that of scientists and engineers.

Scientists and engineers are not only engaged in creative work, but they also represent a population likely to reject charismatic or transformational leadership. As Feist and Gorman (1998) pointed out, scientists and engineers, especially creative scientists and engineers, exhibit substantial autonomy, dominance, achievement motivation, and skills in critical analysis—all characteristics likely to undermine externally imposed visions such as these articulated by charismatic and transformational leaders. Some evidence for this has been provided in a recent study by Rosing, Frese, and Bausch (2011), who found that transformational leadership, although beneficial for creativity and innovation in top management teams, was less beneficial if exhibited among those directly responsible for leading scientists and engineers. Recognition of the peculiar needs of scientists and engineers has consequently initiated an ongoing series of studies seeking to identify the nature of effective leadership for those who are asked to lead scientists and engineers (Amabile & Kramer, 2011; Howell & Boies, 2004; Kramer & Crespy, 2011; Mumford et al., 2002; Olsson, Hemlin, & Pousette, 2012; Redmond, Mumford, & Teach, 1993).

Research into the effective leadership of scientists and engineers—leadership that is expected to encourage creativity and innovation—has been informed by seven key findings emerging from earlier empirical studies. First, scientists and engineers accept, and respond to, attempts to influence them made by leaders only when these are based on expertise (Thamhain & Gemmill, 1974). Second, effective leaders of scientists and engineers possess substantial technical expertise (Andrews & Farris, 1967; Barrowe, 1975). Third, effective leaders of creative efforts appear to be able to use this technical expertise in conjunction with complex, high-level skills—in particular, they can forecast downstream implications of the work (O'Conner, 1998), define the fundamentals to be pursued over time (Hounshell, 1992), establish viable missions or technical production efforts (Mouly & Sankaran, 1999), evaluate progress on technical work in a sophisticated fashion (Pelz & Andrews, 1976), and make sense of crises emerging in the course of this work (Drazin et al. 1999). Fourth, effective leaders of creative efforts are not only skilled at establishing work teams, but they also create a work environment likely to facilitate creative work, an environment characterized by openness, safety, trust, and autonomy (Bain, Mann, & Pirola-Merlo, 2001; Ekvall & Ryhammar, 1999). Fifth, the leaders of creative and innovative efforts maintain an extensive base of contracts, and a wide range of contracts, both within the organization and externally across their professional fields (Anacona & Caldwell, 1992). Sixth, the leaders of creative or innovative efforts champion, or sell, their projects to others within and outside the organization (Howell & Boies, 2004; Markham, Green, & Basu, 1991). Seventh, the leaders of creative and innovative efforts seem to approach projects in an open, adaptive fashion, adjusting work to the local demands imposed by project development cycles and organizational expectations (Cooper & Kleinschmidt, 2000; Jelinek & Schoonhoven, 1990).

At one level, this description of the effective leader of creative or innovative efforts is internally coherent. These are exceptionally technically skilled people who can deal with others doing creative work and manage internal and external relationships as they impinge on the work being done. Nonetheless, the demands made on these leaders are clearly exceptional. They must be both technically and socially skilled. They must understand both the technical field and the organization. They must be able to exert influence with regard to technical work, and they must also be able to sell the work to other stakeholders. They must communicate with others but maintain a focus on the critical technical work, even as they reconfigure this work periodically. Lastly, they must think both with regard to technical specifics and with regard to the broader implications of the work for the people doing the work, the technical field, the organization, and the social systems in which the work is embedded.

These multiple, rather contradictory demands made on the leaders of creative efforts have in recent years initiated a search for new models that might help us to better understand what the leaders of creative and innovative efforts must do. For example, Amabile and Kramer (2011) have begun to develop a model of creative leadership that stresses incremental progress. Mumford and Hunter (2005) have proposed a model of leadership for creative and innovative efforts that stresses its inherently pragmatic orientation. Marion (2012) has proposed yet another model for understanding the leadership of creative and innovative efforts that views effective leadership as the management of socio-technical complexity. Yet these efforts, while potentially valuable, are still in their nascent stages of development. Thus, the present effort provides a basis for elaborating, extending, and revising extant models of leading those engaged in research and innovation, and thereby perhaps contributing to the development of a new integrative model of leading for creativity in the sciences and engineering.

The Structure and Content of This Book

This book is divided into the three sections focusing respectively on theoretical and empirical approaches to creativity and leadership in STI, and on the practical implications of such studies. The first section incorporates three theoretical contributions on leader cognition in a socio-technical context, the factors that link leadership and creativity, and how various technologies (particularly information and communication technologies [ICT]) influence creativity and how leaders may take advantage of this.

The second section includes three original empirical contributions. A group of Dutch authors investigated both high-performing academic research group leaders and the generational and life-cycle effects on such leaders and their performance. A third empirical study in this section examines the characteristics of the highly creative scientists awarded the Nobel Prize.

The final section is composed of four chapters aimed at stimulating specific practical applications with regard to the leadership of creative and innovative processes in STI. The topics here include succession planning for leaders in innovative work, leadership in interdisciplinary teams, the leader–follower aspect of leadership, and the value of a creative leadership.

PART I: THEORETICAL APPROACHES TO CREATIVITY AND LEADERSHIP IN STI

The first chapter, 'Leading Scientists: Cognition in a Socio-technical Context', offers a comprehensive introduction to the main theme of the book. The authors, Mumford, Peterson, and Robledo, propose a conceptual framework for the effective leadership for scientists and engineers based on extensive reading of the relevant literature. Here the reader is introduced to the typical features of work in STI, how this project-based work develops over time, and the nature of those who are involved in all this. One important lesson is that we need to look at the specific forms of leadership required in STI and their different environments. In other words, the context of working in such milieu influences how leaders should work to achieve creative outcomes. Generally, the complexity and ill-defined problems in such work make leaders dependent on creative thinking. Another highlighted feature is the dynamic work situation for leaders, because almost all work is now done as projects. The people working in this context are creative and demand autonomy, which sets certain limits on leaders trying to steer and coordinate their work. Leaders need to be experts and organizationally responsive, to have planning capacities (i.e., forecasting skills), to provide evaluative feedback to followers, and to create a positive interactional climate. The skills needed by leaders are analyzed by going through what the work, the organization, and the group need in order to achieve high creative performance. The chapter ends with a model of leaders who possess a wide variety of skills similar to the high-performing research group leaders found in the study by Verbree, van der Weijden and van den Besselaar in this book. In short, research group leaders tend to be 'all-rounders'.

Denti and Hemlin (2012) conducted a review of 22 empirical studies of leadership and creativity published in high-impact journals, focusing on the factors affecting when and how leadership impinges on creativity. Of these studies, 8 dealt with transformational leadership and two with leader–member exchange (LMX) theory, whereas the other 12 examined other leadership traits and behaviors. One major conclusion from the review is that leaders who want to stimulate creativity among their team members should adapt their behavior to their teams' creative potential. Transformational leaders may stimulate creativity by the mediating mechanisms of intrinsic motivation and of individual and team self-efficacy, the effect being greatest among followers who are not naturally creative, when the team is heterogeneous, and when there is a supportive team climate. Studies based on LMX theory suggest that leaders may stimulate individual creativity through a high-quality leader–follower work relationship, in which leaders may allow followers greater discretion, autonomy, and influence. The notions of transformational leadership and LMX theory can perhaps be linked through the concept of 'flexible leadership'. Flexible leaders are those who support followers' creativity by acting as role models for creativity, setting expectations for creativity. They recognize the importance of the creative abilities, self-efficacy, and intrinsic motivation of followers, thereby stimulating the creative performance of those followers.

The chapter by Hunter, Cushenbery, Ginther, and Fairchild examines how technology affects the creative process and how leaders might play a role in shaping technology use and application. The study adopts the eight-stage model of the creative process put forward by Mumford and colleagues. The analysis leads to five main conclusions. First, in the shorter term, technology's primary role is to speed up the activities involved in the creative process. Second, although most technologies supplement rather than replace current innovation activities, certain forms of technology may be fundamentally altering the creative process. Third, various emerging technologies such as crowd-sourcing are beginning to integrate the general public into the design process, which has implications for how organizations pursue their innovation strategy in the future. Fourth, as creative processes speed up, and as the boundaries between the consumer and the designer become blurred, questions are raised as to whether current creativity and innovation models will continue to be applicable in the future. Finally, leaders can play a significant role in the adoption and implementation of new technology in creative efforts.

PART II: EMPIRICAL APPROACHES TO CREATIVITY AND LEADERSHIP IN STI

A study of Dutch biomedical and health science researchers by Verbree, van der Weijden, and van den Besselaar set out to find the distinguishing features of high-performing leaders of research groups. Research performance was measured by the number of publications and citations of the groups relative to their size, distinguishing high-performers from other groups. A survey of group leaders generated information about resources, leadership issues and characteristics, group management, and network management. The results showed that the best leaders had a stronger research commitment in terms of providing more resources, put more effort into group management, and spent more time on network management than less successful group leaders. As the authors concluded, one may characterize creative leaders as 'all-rounders', that is, leaders who are highly efficient in all aspects of running a research group. The message to academic leaders is clear: You must be devoted to your task and spend many working hours in your group and with other colleagues. These results are clearly relevant to the reader interested in the creative performance of research groups, reinforcing earlier work that creativity in science demands total engagement by scientists (Feist, 2006).

The second article by the Dutch group reminds us that the career history of those leading creative work will interact with policy changes—in particular, older leaders may not respond to these changes in the same way as younger leaders do. Verbree, van der Weijden, and van den Besselaar investigated whether different generations of academic leaders have contrasting leadership practices in response to contextual changes in the science system as well as whether academic leaders within different life-cycle stages (less experienced, experienced, and old academic leaders) show different leadership practices. The study reveals that the generation factor and the life-cycle phase are both influential on the behavior of academic leaders. This chapter has implications for academic career policy and research groups.

Eubanks, Palanski, Swant, and Oguntebi also examine the effects of career stage on academic leaders. In this study the autobiographies of Nobel Prize winners were analyzed with respect to educational diversity, autonomy or freedom, leadership activities, and free cognitive capacity (e.g., time to do the work). The key dependent variable examined was the age at which their most cited publication was produced. In keeping with earlier studies (e.g., Mumford & Gustafson, 1998), they found that autonomy and educational diversity contributed to early career success. Later career success, however, was influenced by free cognitive capacity and, more critically, by the number and intensity of leader activities was positively related to late career productivity. This finding is of some importance because it suggests that scientific leadership, provided those leaders are given adequate thinking time, may, in fact, prove beneficial in an academic sense for those who take on these responsibilities. This study does, however, raise the question of how we should help prepare leaders to balance the competing demands made by their leadership role and their technical role.

PART III: APPROACHES TO CREATIVITY AND LEADERSHIP IN STI—SOME IMPLICATIONS

An increasing feature of contemporary science is the number of interdisciplinary teams working on complex problems. In these teams researchers with different backgrounds work together on a common goal. Of course, this makes leadership of such teams an even more demanding and challenging task than leadership of mono-disciplinary teams. The chapter by Reiter-Palmon, de Vreede, and de Vreede deals with this particular issue. They arrive at their conclusions for leaders after a thorough review of the interdisciplinary literature, identifying the specific characteristics of such work and what is distinctive about leading such teams. An important feature is the setting of team goals that enhance collaboration and lead to new ideas. The authors point to team trust and psychological safety as important drivers for successful creative collaboration, but also note that task conflicts may sometimes promote creative performance as does knowledge sharing within interdisciplinary teams. They conclude that, if creative and autonomous scientists from different disciplines are to work effectively in teams to solve complex problems, it may well be an advantage for leaders to use participative leadership techniques and to emphasize team communication.

M. Gorman and P. Gorman argue for an ethically responsible and broad approach to leader- and followership in the development of creative innovations in science and technology. They present and illustrate a number of guidelines for leaders and followers for maximizing the benefits of innovations and minimizing their harm. Because creativity demands both novelty and quality, and because an important aspect of quality is sustainability, this approach helps to secure creative innovation. The approach includes heeding not just the interests of the shareholders but also those of the various categories of stakeholders affected by the innovation. In order to improve communication between different categories of people within and between organizations, the authors suggest creating meeting places, or *trading zones*, in which 'catalysts' who possess an insight into the values and concepts of the other parties facilitate the communication to ensure that it is constructive and results in the project moving forward. This is likely to promote creative solutions since conceptual blending is advantageous for creativity. Involvement of all relevant parties, a wide spread of responsibility, and flexibility with respect to the aspects of the situation attended to by leaders and experts are seen as beneficial for the production of innovations that are creative in a broad and full sense.

Of course, one way to address this issue is to provide leaders of creative efforts with relevant training—for example, time-management training. Another is to formulate systematic career development interventions. The study by Ligon, Demborski, Mapp, and Zongrove (this volume) provides a model of how such developmental interventions might be formulated though systematic succession planning. They examine the differences between more junior (e.g., project managers) and more senior (e.g., vice presidents) scientific and engineering leadership positions with respect to 12 key dimensions: (a) environmental analysis, (b) strategy formation, (c) mission definition, (d) team building, (e) team construction, (f) climate definition, (g) idea generation, (h) idea evaluation, (i) support and resources, (j) planning, (k) project management, and (l) succession planning. The specific differences observed between more junior and more senior leaders with regard to these dimensions suggest a number of developmental interventions that might prove useful in preparing people to lead creative efforts, such as mentoring, job assignment, and rotational assignments. In this regard, however, it is important to bear in mind one particular point made by Ligon et al.—namely, that different types, intensity, and duration of developmental interventions may be required to develop creative leaders at different stages in their careers.

A new concept of leadership—the *creative leader*—is argued for by Puccio, Mance, and Zacko-Smith (this volume). Creativity is needed to solve ill-defined problems generated when the pace of change is very rapid, as it is today. The creative leader is described as 'fully relational' and as embodying change (i.e., he or she has internalized creativity). The creative leader falls somewhere between the *transformational* leader (concerned with social relationships and empowerment) and the *transcendent* leader (concerned with the impact of the relationship between the leader and his or her constituency or followers on the wider society) on a continuum that starts with the *transactional* leader (concerned with the task, efficiency, and reward). In addition, the authors explain and illustrate how creativity can be trained and successfully stimulated, for example, through programs based on cognitively oriented process models such as the creative problem-solving model. The chapter also offers some simple guidelines for how to increase creativity in complex problem-solving situations and suggestions on how to improve the creative climate in organizations.

In conclusion, the premise underpinning this book is that creativity in STI is of immense importance. If creative new ideas are not stimulated and welcomed, they are not likely to be forthcoming or to survive. As argued earlier, the current literature suggests that creativity in STI is strongly influenced by the way leaders think and act. Hence, the aim of this book is to bring together a number of scholarly papers taking new approaches to analyzing how leaders can best stimulate and organize creativity in the field of STI.

NOTES

1. The Organisation for Economic Co-operation and Development (OECD) suggests the following definition of innovation: "the implementation of a new or significantly improved product (good or service), or process, a new marketing method, or a new organizational method in business practices, workplace organization or external relations" (2005, p. 46).
2. Innovation and invention are in general language often used interchangeably and particularly for radical innovations. However, in the economic literature there is a clear distinction where innovation is a marketed invention (a discovery or scientific result).
3. Hofstede's power dimension reflects the extent to which staff lower in the hierarchy of organizations and institutions accept and indeed expect an unequal distribution of power—see, for example, Hofstede (1980).
4. Data downloaded from http://www.clearlycultural.com/geert-hofstede-cultural-dimensions/power-distance-index/ on January 26, 2012.
5. This is not to imply that universities are completely resistant to change, although they can certainly exhibit considerable inertia or resistance to change. Moreover, many academics would contend that the economic and managerial practices that have come to exert an influence on university organization in

recent years have often had a rather negative impact on the creativity of scientific researchers (e.g., Ziman, 2000).

6. The full definition from the Frascati Manual is as follows: "Research and experimental development [R&D] comprise creative work undertaken on a systematic basis in order to increase the stock of knowledge, including knowledge of man, culture and society and the use of this stock of knowledge to devise new applications. R&D is a term covering three activities: basic research, applied research and experimental development. Basic research is experimental or theoretical work undertaken primarily to acquire new knowledge of the underlying foundation of phenomena and observable facts, without any particular application or use in view. Applied research is also original investigation undertaken to acquire new knowledge. It is, however, directed primarily towards a specific practical aim or objective. Experimental development is systematic work, drawing on existing knowledge gained from research and/or practical experience, that is directed to producing new materials, products or devices, to installing new processes, systems and services, or to improving substantially those already produced or installed" (OECD, 1993, chap. 2).

REFERENCES

Albert, R. S., & Runco, M. A. (1999). A history of research on creativity. In R. J. Sternberg (Ed.), *Handbook of creativity* (pp. 16–34). Cambridge, England: Cambridge University Press.

Alencar, E. M. L. (2012). Creativity in organizations: Facilitations and inhibitions. In M. D. Mumford (Ed.), *Handbook of organizational creativity* (pp. 87–111). San Diego, CA: Elsevier.

Allwood, C. M. (1997). The function and handling of research problems in the social sciences. *Science Studies*, *10*, 50–73.

Allwood, C. M., & Bärmark, J. (1999). The role of research problems in the process of research. *Social Epistemology*, *13*, 59–83.

Allwood, C. M., & Hedelin, L. (2005). Adjusting new initiatives to the social environment: Organizational decision making as learning, commitment creating and behavior regulation. In H. Montgomery, R. Lipshitz, & B. Brehmer (Eds.), *How professionals make decisions* (pp. 223–232). Mahwah, NJ: Lawrence Erlbaum Associates.

Amabile, T. M. (1996). *Creativity in context.* Boulder, CO: Westview Press.

——— (1999). How to kill creativity. In *Harvard Business Review on Breakthrough Thinking* (pp. 1–28). Boston, MA: Harvard Business School Press.

Amabile, T. M., & Kramer, S. J. (2011). *The progress principle: Using small wins to ignite joy, engagement, and creativity at work*. Boston, MA: Harvard Business Review Press.

Anacona, D. G., & Caldwell, D. F. (1992). Bridging the boundary: External activity and performance in organizational teams. *Administrative Science Quarterly*, *37*, 634–655.

Andrews, F. M., & Farris, G. F. (1967). Supervisory practices and innovation in scientific teams. *Personnel Psychology*, *20*, 497–515.

Baer, J. (1998). The case for the domain specificity of creativity. *Creativity Research Journal*, *11*, 173–177.

Bain, P. G., Mann, C., & Pirola-Merlo, A. (2001). The innovation imperative: The relationships between team, climate, innovation, and performance in research and development teams. *Small Group Research*, *32*, 55–73.

Barrowe, T. (1975). Leadership and performance outcomes in research organizations: The supervision of scientists as a source of assistance. *Organizational Behavior and Human Performance*, *14*, 264–280.

Bass, B. M. (1985). *Leadership beyond expectations*. New York, NY: Free Press.

Bass, B. M., & Bass, R. (2008). *The Bass handbook of leadership*. New York, NY: Free Press.

Campbell, D. T. (1960). Blind variation and selective retention in creative thought as in other knowledge processes. *Psychological Review*, *67*, 380–400.

Cardinal, L. B. (2001). Technological innovation in the pharmaceutical industry: The use of organizational control in managing research and development. *Organization Science*, *12*, 19–36.

Carlson, W. B., & Gorman, M. E. (1992). A cognitive framework to understand technological creativity: Bell, Edison, and the telephone. In R. J. Weber & D. N. Perkins (Eds.), *Inventive minds: Creativity in technology* (pp. 48–79). New York, NY: Oxford University Press.

Chan, D. W. (2011) East vs. West. In M. A. Runco & S. R. Pritzker (Eds.), *Encyclopedia of creativity* (Vol. 1, pp. 415–421). New York, NY: Academic Press.

Chesbrough, H. W. (2003). *Open innovation: The new imperative for creating and profiting from technology*. Boston, MA: Harvard Business School Press.

Cooper, R. G., & Kleinschmidt, E. J. (2000). New product performance: What distinguishes star products. *Australian Journal of Management*, *25*, 17–46.

Csikszentmihalyi, M. (1999). Implications of a systems perspective for the study of creativity. In R. T. Sternberg (Ed.), *Handbook of creativity* (pp.312–338). Cambridge, England: Cambridge University Press.

Dacey, J. (2011). Historical conceptions of creativity. In M. A. Runco & S. R. Pritzker (Eds.), *Encyclopedia of creativity* (Vol. 1, pp. 608–616). New York, NY: Academic Press.

Damanpour, F., & Aravind, D. (2012). Organizational structure and innovation revisited: From organic to ambidextrous structure. In M. D. Mumford (Ed.), *Handbook of organizational creativity* (pp. 483–514). San Diego, CA: Elsevier.

Dasgupta, S. (2011). Contesting (Simonton's) blind variation, selective retention theory of creativity. *Creativity Research Journal*, *23*(2), 166–182.

DeDreu, C. K. W., Baas, M., & Nijstad, B. A., (2012). The emotive roots of creativity: Basic and applied issues on affect and motivation. In M. D. Mumford (Ed.), *Handbook of organizational creativity* (pp. 217–240). San Diego, CA: Elsevier.

Denti, L., & Hemlin, S. (2012). Leadership and innovation in organizations. A systematic review of factors that mediate or moderate that relationship. *International Journal of Innovation Management*, *16* (3), 1-20. doi:10.1142/S1363919612400075

Drazin, R., Glynn, M. A., & Kazanjian, R. K. (1999). Multilevel theorizing about creativity in organizations: A sense-making perspective. *Academy of Management Review*, *24*, 286–307.

Duncker, K. (1945). *On problem-solving*. Westport, CT: Greenwood Press.

Dutta, S. (Ed.). (2011). *The Global Innovation Index 2011. Accelerating growth and development*. City, ST: INSEAD.

Ekvall, G., & Ryhammar, L. (1999). The creative climate: Its determinants and effects at a Swedish university. *Creativity Research Journal*, *12*, 303–310.

Ericsson, K. A., & Charness, N. (1994). Expert performance: Its structure and acquisition. *American Psychologist*, *49*, 725–747.

Feist, G. J. (2006). *The psychology of science and the origins of the scientific mind*. New Haven, CT, & London, England: Yale University Press.

Feist, G. J., & Gorman, M. E. (1998). The psychology of science: Review and integration of a nascent discipline. *Review of General Psychology*, *2*, 3–47.

Greenleaf, R.K. (1977). *Servant leadership: A journey into the nature of legitimate power and greatness*. New York, NY: Paulist Press.

Guilford, J.P. (1950). Creativity. *American Psychologist, 5*, 444–454.

Hemlin, S. (1996). Research on research evaluation. *Social Epistemology, 10*, 209–250.

Hemlin, S., Allwood, C.M., & Martin, B.R. (Eds.). (2004). *Creative knowledge environments: Micro and macro Studies of R&D and higher education*. Aldershot, England, and Brookfield, VT: Edward Elgar.

——— (2008). Creative knowledge environments. *Creativity Research Journal, 20*, 196–210.

Hemlin, S., & Olsson, L. (2011). Creativity stimulating leadership: A critical incident study of leader's influence on creativity in R&D groups. *Creativity and Innovation Management, 20*(1), 49–58.

Hennessey, B. A., & Amabile, T. M. (2010). Creativity. *Annual Review of Psychology, 61*, 569–598.

Hicks, D., & Katz, S. (1996). Where is science going? *Science, Technology and Human Values, 21*, 379–406.

Hoff, E. V., Carlsson, I. M., & Smith, G. J. (2012). Personality. In M. D. Mumford (Ed.), *Handbook of organizational creativity* (pp. 241–270). San Diego, CA: Elsevier.

Hofstede, G. (1980). Motivation, leadership, and organization: Do American theories apply abroad? *Organizational Dynamics, 9*, 42–63.

Hounshell, E. A. (1992). Innovation in the industrial research laboratory: Individual or collective process. In R. J. Weber & D. N. Perkins (Eds.), *Inventive minds: Creativity in technology* (pp. 273–291). Oxford, England: Oxford University Press.

Howell, J. M., & Boies, K. (2004). Champions of technological innovation: The influences of contextual knowledge, role orientation, idea generation, and idea promotion on champion emergence. *The Leadership Quarterly, 15*, 123–143.

Hunter, S. T., Bedell-Avers, K. K., & Mumford, M. D. (2007). Climate for creativity: A qualitative review. *Creativity Research Journal, 19*, 69–90.

James, L. R., James, L. A., & Ashe, D. K., (1990). The meaning of organizations: The role of cognition and values. In B. Schneider (Ed.), *Organizational climate and culture* (pp. 40–84). San Francisco, CA: Jossey-Bass.

Jaussi, K. S., & Dionne, S. D. (2003). Leading for creativity: The role of unconventional leader behavior. *The Leadership Quarterly, 14*, 475–498.

Jelinek, M., & Schoonhoven, C. B. (1990). *The innovation marathon: Lessons learned from high-technology firms*. Oxford, England: Blackwell.

Kantorovich, A. (1993). *Scientific discovery: Logic and tinkering*. Albany: State University of New York Press.

Kazanjain, R. K., & Drazin, R. (2012). Organizational learning, knowledge management, and creativity. In M. D. Mumford (Ed.), *Handbook of organizational creativity* (pp. 547–568). San Diego, CA: Elsevier.

Kidder, T. (1981). *The soul of a new machine*. New York, NY: Avon.

Koopman, P., & Pool, J. (1991). Organizational decision making: Models, contingencies and strategies. In J. Rasmussen, B. Brehmer, & J. Leplat (Eds.), *Distributed decision making: Cognitive models for cooperative work* (pp. 19–46). New York, NY: John Wiley & Sons.

Kramer, M. W., & Crespy, D. A. (2011). Communicating collaborative leadership. *The Leadership Quarterly, 22*, 1024–1037.

Marion, R. (2012). Leadership of creativity: Entity-based, relational, and complexity perspectives. In M. D. Mumford (Ed.), *Handbook of organizational creativity* (pp. 457–482). San Diego, CA: Elsevier.

Markham, S. K., Green, S., & Basu, R. (1991). Champions and antagonists: Relationships with R&D project characteristics and management. *Journal of Engineering and Technology, 8*, 217–242.

Marta, S., Leritz, L. E., & Mumford, M. D. (2005). Leadership skills and group performance: Situational demands, behavioral requirements, and planning. *The Leadership Quarterly*, *16*, 97–120.
Martinsen, Ø. L., & Kaufmann, G. (2011). Cognitive style and creativity. In M.A. Runco & S.R. Pritzker (Eds.), *Encyclopedia of creativity* (Vol. 2, pp. 214–221). New York, NY: Academic Press.
Martinsons, M.G. (2001, August). *Comparing the decision styles of American, Chinese and Japanese business leaders.* Paper presented at the Proceedings of Academy of Management Meetings, Washington, DC. Available from SSRN: http://ssrn.com/abstract=952292
Merton, R.K. (1942). Science and technology in a democratic order. *Journal of Legal and Political Sociology*, *1*, 115–126.
——— (1973). The normative structure of science. In N.W. Storer (Ed.), *The sociology of science: Theoretical and empirical investigations.* Chicago, IL: University of Chicago Press.
Metcalfe, J. (1986). Premonitions of insight predict impending error. *Journal of Experimental Psychology: Learning, Memory, and Cognition*, *12*(4), 623–634.
Mouly, V. S., & Sankaran, J. K. (1999). The "permanent" acting leader: Lessons from a dying Indian R&D organization. *The Leadership Quarterly*, *10*, 637–651.
Mumford, M.D. (2006). *Pathways to outstanding leadership: A comparative analysis of charismatic, ideological, and pragmatic leadership.* Mahwah, NJ: Erlbaum.
Mumford, M. D., Antes, A. L., Caughron, J. J., Connelly, M. S., & Beehler, C. (2010). Cross-field differences in creative problem-solving skills: A comparison of health, biological, and social services. *Creativity Research Journal*, *22*, 14–26.
Mumford, M. D., Connelly, M. S., Scott, G. M., Espejo, J., Sohl, L., Hunter, S. T., & Bedell, K.E. (2005). Career experiences and scientific performance: A study of social, physical, life, and health sciences. *Creativity Research Journal*, *17*, 105–129.
Mumford, M. D., & Gustafson, S. B. (1988). Creativity syndrome: Integration, application, and innovation. *Psychological Bulletin*, *103*, 27–43.
Mumford, M. D., & Hunter, S. T. (2005). Innovation in organizations: A multi-level perspective on creativity. In F. Dansereau & F. J. Yammarino (Eds.), *Multi-level issues in strategy and methods* (pp. 11–73). Oxford, England: Elsevier.
Mumford, M. D., Medeiros, K. A., & Partlow, P. (2012). Creative thinking: Processes, strategies, and knowledge. *Journal of Creative Behavior*, *46*, 32–51.
Mumford, M. D., Mobley, M. I., Uhlman, C. E., Reiter-Palmon, R., & Doares, L. (1991). Process analytic models of creative capacities. *Creativity Research Journal*, *4*, 91–122.
Mumford, M. D., Scott, G. M., Gaddis, B. H., & Strange, J. M. (2002). Leading creative people: Orchestrating expertise and relationships. *The Leadership Quarterly*, *13*, 705–750.
Mumford, M. D., Zaccaro, S. J., Harding, F. D., Jacobs, T. O., & Fleishman, E. A. (2000). Leadership skills for a changing world: Solving complex social problems. *The Leadership Quarterly*, *11*, 11–35.
Newell, A., & Simon, H. A. (1972). *Human problem solving.* Englewood Cliffs, NJ: Prentice Hall.
Nijstad, B.A., De Dreu, C.K.W., Rietzschel, E.F., & Baas, M. (2010). The dual pathway to creativity model: Creative ideation as a function of flexibility and persistence. *European Review of Social Psychology*, *21*, 34–77.
Nohria, N., & Gulati, R. (1996). Is slack good or bad for innovation? *Academy of Management Journal*, *34*, 1245–1264.
O'Connor, G. C. (1998). Market learning and radical innovation: A cross case comparison of eight radical innovation projects. *Journal of Product Innovation Management*, *15*, 151–166.

Oldham, G. R., & Cummings, A. (1996). Employee creativity: Personal and contextual factors at work. *Academy of Management Journal*, *39*, 607–634.
Olsson, L., Hemlin, S., & Pousette, A. (2012). A multi-level analysis of leader-memberexchange and creative performance in research groups. *The Leadership Quarterly*, *23*, 604–619.
Organisation for Economic Co-operation and Development. (1993). *The Frascati Manual 1993: The measurement of scientific and technological activities*. Paris: Author.
Organisation for Economic Co-operation and Development. (2005). *Oslo manual. Guidelines for collecting and interpreting innovation data* (3rd ed.). Paris: Author.
Pelz, D. C., & Andrews, F. M. (1976). *Scientists in organizations*. Ann Arbor, MI: Institute for Social Research.
Perkins, D. N. (1988). Creativity and the quest for mechanism. In R. J. Sternberg & E.E. Smith (Eds.), *The psychology of human thought* (pp. 309–336). New York, NY: Cambridge University Press.
Plucker, J. A., & Renzulli, J. S. (1999). Psychometric approaches to the study of human creativity. In R. J. Sternberg (Ed.), *Handbook of creativity* (pp. 35–61). Cambridge, England: Cambridge University Press.
Redmond, M. R., Mumford, M. D., & Teach, R. J. (1993). Putting creativity to work: Leader influences on subordinate creativity. *Organizational Behavior and Human Decision Processes*, *55*, 120–151.
Reiter-Palmon, R. (2011). Problem finding. In M. A. Runco & S. R. Pritzker (Eds.), *Encyclopedia of creativity* (Vol. 2, pp. 250–253). New York, NY: Academic Press.
Reiter-Palmon, R., Wigert, B., & de Vreede, T. (2012). Team creativity and innovation: The effect of group composition, social processes, and cognition. In M. D. Mumford (Ed.), *Handbook of organizational creativity* (pp. 295–326). San Diego, CA: Elsevier.
Rhodes, M. (1961). An analysis of creativity. *Phi Delta Kappa*, *42*, 305–310.
Rich, J. D., & Weisberg, R. W. (2004). Creating all in the family: A case study in creative thinking. *Creativity Research Journal*, *16*, 247–259.
Rosing, K., Frese, M., & Bausch, A. (2011). Exploring the heterogeneity of the leadership-innovation relationship: Ambidextrous leadership. *The Leadership Quarterly*, *22*, 956–974.
Scott, G. M., Lonergan, D. C., & Mumford, M. D. (2005). Conceptual combination: Alternative knowledge structures, alternative heuristics. *Creativity Research Journal*, *17*, 79–98.
Shamir, B., House, R. J., & Arthur, M. B., (1993). The motivational effects of charismatic leadership: A self-concept based theory. *Organizational Science*, *4*, 577–594.
Simon, H. A. (1965). Understanding creativity. *Carnegie Review*, 1964/1965, No. 2 (unpaginated).
Simonton, D. K. (2003). Scientific creativity as constrained stochastic behavior: The integration of product, process, and person perspectives. *Psychological Bulletin*, *129*, 475–494.
Smith, S. M. (2011). Incubation. In M. A. Runco & S. R. Pritzker (Eds.), *Encyclopedia of creativity* (Vol. 2, pp. 653–657). New York, NY: Academic Press.
Stehr, N. (1994). *Knowledge societies*. London, England: Sage.
Strange, J. M., & Mumford, M. D. (2005). The origins of vision: Effects of reflection, models, and analysis. *The Leadership Quarterly*, *16*, 121–148.
Taggar, S. (2002). Individual creativity and group ability to utilize individual creative resources: A multi-level model. *Academy of Management Journal*, *45*, 315–330.
Terry, R. (1993). *Authentic leadership: Courage in action*. San Francisco, CA: Jossey-Bass.

Thamhain, H. J., & Gemmill, G. R. (1974). Influence styles of project managers: Some project performance correlates. *Academy of Management Journal, 17*, 216–224.

Tidd, J., Bessant, J., & Pavitt, K. (2005). *Managing innovation: Integrating technological, market and organizational change* (3rd ed.). Chichester, England: Wiley.

Tierney, P., & Farmer, S. M. (2002). Creative self-efficacy: Its potential antecedents and relationship to creative performance. *Academy of Management Journal, 45*, 1137–1148.

Tierney, P., Farmer, S. M., & Graen, G. B. (1999). An examination of leadership and employee creativity: The relevance of traits and relationships. *Personnel Psychology, 52*, 591–620.

Tushman, M. L., & O'Reilly, C. A. (2002). *Winning through innovation: A practical guide to leading organizational change and renewal.* Boston, MA: Harvard Business School Press.

Wallas, G. (1970). The art of thought. In P. E. Vernon (Ed.), *Creativity* (pp. 91–97). Harmondsworth, England: Penguin. (Original work written in 1926).

Welling, H. (2007). Four mental operations in creative cognition: The importance of abstraction. *Creativity Research Journal, 19*, 163–177.

Wertheimer, M. (1945). *Productive thinking.* New York, NY: Harper and Brothers.

West, M. A., & Sacramento, G. A. (2012). Creativity and innovation: The role of team and organizational climate. In M. D. Mumford (Ed.), *Handbook of organizational creativity* (pp. 259–285). San Diego, CA: Elsevier.

Yammarino, F. J., & Dansereau, F. (2002). *The many faces of multi-level issues.* Oxford, England: Elsevier.

Yukl, G. (1999). An evaluation of conceptual weaknesses in transformational and charismatic leadership theories. *The Leadership Quarterly, 10*, 285–305.

———. (2010). *Leadership in organizations.* Englewood Cliffs, NJ: Prentice Hall.

Zalesnik, A. (2004). Managers and leaders. Are they different? *Harvard Business Review, 82* (1), 74–81. (Original work published in 1977)

Ziman, J. (2000). *Real science: What it is and what it means.* Cambridge, England: Cambridge University Press.

Part I

Theoretical Section

2 Leading Scientists and Engineers

Cognition in a Socio-Technical Context

Michael D. Mumford, David Peterson, and Issac Robledo

> "He rarely gave orders, and instead managed to communicate his desires," as the physicist Eugene Wigner recalled, "very easily and naturally, with just his eyes, his two hands, and a half-lighted pipe." Bethe remembered that Oppie "never dictated what should be done. He brought out the best in us, like a good host with his guests." Robert Wilson felt similarly: "In his presence, I became more intelligent, more vocal, more intense, more prescient, and more poetic myself." (Bird & Sherwin, 2005, p. 218)

These quotes reflect the appraisal of physicists working on development of the first nuclear weapon concerning the capacities of their leader—Robert Oppenheimer. And, they all point to the importance of effective leadership in the research and development enterprise.

The impact of effective leadership on the performance of research and development efforts is not unique to the case of Oppenheimer. For example, Barnowe (1975) examined the productivity of 81 research and development groups, containing 963 chemists, with respect to performance criteria such as publications and patent awards. He found that leaders' skills (r = .40) were strongly related to these indices of creativity. Similarly, Tierney, Farmer, and Graen (1999) obtained correlations in the .30s between measures of leader–member exchange and measures of invention disclosure. Other work by Andrews and Farris (1967), Mouly and Sankaran (1999), and Pelz (1963) also points to a similar conclusion—effective leadership is strongly, positively, related to the performance (including objective indices such as patent awards) of research and development groups.

Although the strong, positive relationship between effective leadership and the performance of research and development groups has long been known, it has been discounted based on the cultural stereotype of the isolated genius who is presumed to be the basis of all scientific breakthroughs (Mumford, Scott, Gaddis, & Strange, 2002). This mistaken belief, however, is not without real, tangible consequences. Economic growth beyond normative expectations, based on capital and population size, is to a large extent based on the production of innovative products (Mumford, Olsen,

& James, 1989). Given the strong relationship between leadership and the performance of research and development teams, it seems clear that one way we might improve the quality of scientific production, and enhance economic growth, is by improving the leadership of scientists and engineers.

To address this issue, an issue of scientific leadership, one must of course have an understanding of the basic phenomenon at hand. In recent years, a number of models have been proposed in attempts to account for effective leadership of research and development efforts—administrative, organizational, sensegiving, team building, and engineering models. However, Engwall, Kling, and Werr (2005) found that these models are not employed in a simple or consistent fashion in the leadership of research and development efforts. Moreover, as Buijis (2007) has pointed out, these models do not seem to result in an internally consistent set of recommendations about how the leadership of scientists and engineers should occur.

Accordingly, in the present chapter, we develop an initial preliminary framework of effective leadership for scientists and engineers. The framework takes into account the context in which this unique form of leadership occurs. Specifically, three key phenomena pertaining to science and engineering influence the development of the framework: (a) the basic nature of scientific and engineering work, (b) how this work unfolds over the course of a project, and (c) the nature of the people doing this work. After a discussion of these phenomena, we then argue that effective leadership, in the context of research and development, is based on creative cognition occurring in a dynamic socio-technical context, in which technological, social, and psychological elements interact to create a unique task environment (Cooper & Foster, 1971). Finally, we will consider some of the practical implications of these observations for the development of leaders and the enhancement of leader performance. It should be noted that by 'sciences' we generally mean applied natural sciences, but the discussions herein may also pertain to other domains although further investigation is needed in support of this.

CONTEXT

The Work

When one attempts to account for effective leadership one must begin by taking into account the domain in which leadership is occurring. We do not typically develop models of effective leadership for politicians and hope they will apply to religious leaders (Mumford, 2006). And, for hundreds of years sociologists, psychologists, and philosophers have sought to understand the work being done by scientists and engineers (Kuhn, 1962; Simonton, 1990; Zuckerman, 1977). Recently, Andersen, Barker, and Chen (2006) have argued that scientists and engineers, in developing conceptual models for understanding and/or acting on natural phenomena, must establish key concepts, identify the mechanisms that give rise to relationships among these concepts, demonstrate that the model proposed accounts for the phenomenon of interest better than competing models, and develop tools, or formulate processes and procedures, for effective exploitation of this model. In each of these operations, evidence pointing to the appropriateness of the conclusions being drawn must be provided. This maps well onto the day-to-day work practices of scientists such as Einstein (Isaacson, 2007) and Faraday (Tweney, 1992) and engineers such as Edison (Hughes, 1989) and the Wright Brothers (Crouch, 1992).

Although Andersen et al.'s (2006) description of the work being conducted by scientists and engineers is straightforward, the work itself is not. The concepts, mechanisms, models, and interventions being worked with are novel and ill defined. As a result, creative thinking skills are important to the work of scientists and engineers (Mumford et al., 2005). Moreover, given the ill-defined nature of these tasks, and the need for clarity, work in the sciences and engineering requires incremental improvements over time, which depend on the bench skills of scientists and engineers (Blumberg, 1994).

The complexity of the phenomena being studied, and the need for a range of skills and relevant expertise (Dunbar, 1995), implies that work in the sciences and engineering will ultimately be collaborative (Sawyer, 2006)—not only collaborative internally, but also collaborative with regard to the broader social and technological environments. The engagement of numerous people, under conditions where observations must be obtained and observations are costly, implies that substantial resources will be required for most work in the sciences and engineering (Nohari & Gulati, 1996). The cost of the work, in turn, implies a social and technical evaluation of the worth of the investments being made. These evaluations, moreover, may change as the development of new technologies in other fields reduces the costs of conducting certain types of work. An example may be found in the impact of magnetic resonance imaging on the work being conducted in neuropsychology (Hennessey & Amabile, 2010).

The Projects

Scientists and engineers apply the processes described above in conducting a set of projects of interest to a particular discipline, field, or organization (Csikszentmihalyi, 1999; Simonton, 2009). Projects are a set of circumscribed work activities intended to provide knowledge bearing on a particular concept, mechanism, model, or intervention (Bakker, Boersma, & Oreel, 2006). Typically, projects are selected and organized around certain fundamentals which give rise to themes intended to elucidate these fundamentals (Hughes, 1989). Although science projects focus on the development and testing of substantive models and engineering projects are intended to provide usable, cost-effective solutions to practical problems, virtually all work in the sciences and engineering is project based (Mulec & Roth, 2005; Napier & Nilsson, 2006).

Projects are an inherently dynamic form of work, progressing through different stages in the course of conducting the work (Bakker et al., 2006; Basadur & Gelade, 2006). Recently, Mumford and his colleagues (Byrne, Mumford, Barrett, & Vessey, 2009; Mumford, Bedell-Avers, & Hunter, 2008) have proposed a general model of the stages occurring in project work. It holds that projects are selected to provide information bearing on themes implied by the fundamentals being addressed in a field. Figure 2.1 provides an example of how project elements (indicated by "X") from multiple themes might be integrated into a final product as a project proceeds through five stages: (1) *scanning* or initial exploration of themes; (2) *elaboration* or data gathering, concept formulation, and gap identification; (3) *development* or formation, testing, and refinement of initial models; (4) *appraisal* or assessment of limitations, impacts, and benefits; and (5) *implementation* or production and fielding of new products.

This model of project work has a number of implications for understanding the nature of work in the sciences and engineering. First, the scientists and engineers who can contribute at one stage may not be the same as those who can contribute at later stages (Mumford et al., 2008). Thus, project staffing is not necessarily fixed. Second, as work progresses through these stages, more people, drawn from a wider range of disciplines, will be required. Thus cross-functional teams will prove valuable in later stage efforts (Thamhain, 2003) and the costs of the effort and the need for organizational support will be greater in later stages (Brown & Eisenhardt, 1995). Third, learning will occur at each stage on each project (Rickards & Moger, 2006), which suggests that projects must be treated as integrated, interactive sets rather than as isolated

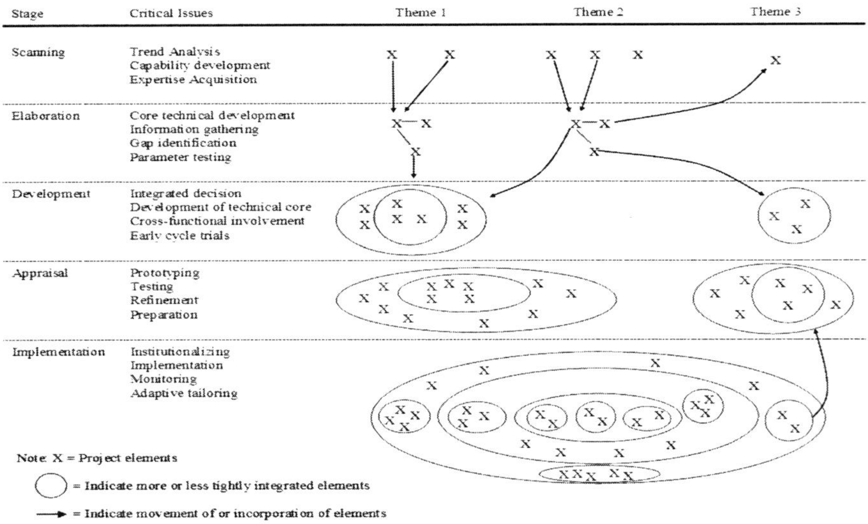

Figure 2.1 Stage model of project progression

entities (Root-Bernstein, Bernstein, & Garnier, 1996). Fourth, differences in complexity, organization, cost, problems, and criteria imply that a person who can lead effectively at one stage may not necessarily be an effective leader at other stages due to shifts in the demands made by project work (Buijis, 2007).

The People

Leadership is, of course, a phenomenon involving the exercise of direct, or indirect, social influence (Yukl, 2010). Thus, leaders must take into account the characteristics of the people doing the work. A number of studies have been conducted examining the personality of people who exhibit creativity. For example, Mumford and Gustafson (1988) found that creative people are characterized by openness, learning goals, flexibility, autonomy, and achievement motivation. More recently, Feist (1999), in a meta-analysis of the personality characteristics characterizing creative scientists, found that scientists, and presumably engineers, are open, conscientious, autonomous, ambitious/achievement oriented, and self-confident or even arrogant.

What should be recognized here, however, is that creative people wish to make their own unique contributions. Their desire for autonomy with regard to creative work (Hennessey & Amabile, 2010), in turn, may make creative people unresponsive to attempts by others to influence them—including attempts by leaders. Moreover, the locus for identification of creative people is, more often than not, the field, or profession, rather than the organization. As a result, their key concerns may lie more in professional evaluations rather than in evaluations of either the leader or the organization. More centrally, creative work is typically conducted by people working within a professional field who are experts in the area in which they are working and value feedback that pertains to their area of expertise. Thus feedback must be professionally based. However, their autonomy and self-confidence may lead them to reject even appropriate technical feedback.

What should also be recognized is that creative people may or may not accept either the leader or the projects the leader is asking them to pursue. Thus, in asking people to lead scientists and engineers, we are asking them to lead an uncertain, unfolding set of tasks, of which the outcomes are uncertain, the criteria change, and those being led must work together but may decide they have little interest in working with others, on the project, or for the leader. This is not a recipe for success. However, it is the context in which discussions of the leadership of scientists and engineers must be framed.

LEADERSHIP

As noted earlier, leadership involves the exercise of influence (Yukl, 2010). Influence, however, can be exercised by leaders through a variety of mechanisms (Yukl & Tracey, 1992)—rational persuasion, inspirational appeals, or interpersonal exchange, to mention a few. It is clear, however, that many commonly employed influence tactics do not work well when leading scientists and engineers doing creative work. For example, we commonly assume that inspirational leadership, attributed to charisma or transformational leadership (Bass, 1997; Conger & Kanungo, 1998), will allow for effective exercise of influence. However, Mumford, Scott, et al. (2002) found that inspirational influence typically produces only mixed effects in enhancing the performance of scientists and engineers (Jung, 2001; Osborn & Marion, 2009)—presumably due to the limitations that a leader's vision places on the autonomous exploration of scientists and engineers.

The one influence tactic that does appear to contribute to the effective leadership of scientists and engineers is expertise or rational persuasion. For example, Thamhain and Gemmill (1974) found that the exercise of influence through expertise was the one influence tactic that resulted in positive appraisals of the leader by scientists and engineers. Similarly, Barnowe (1975) examined the effects of five leader behaviors on the productivity of 963 chemists: (a) support, (b) participation, (c) closeness of supervision, (d) task emphasis, and (e) the leader's technical skills. It was found, consistent with earlier work by Andrews and Farris (1967), that the leader's technical skills, the basis for expertise, was the single best predictor of the productivity of these chemists. Other, more recent work by Chua and Iyengar (2008) and Li, Tan, Teo, and Tan (2006) also points to the importance of expertise in the leadership of scientists and engineers.

Expertise is not a simple matter of technical knowledge, however. As discussed earlier and in line with a socio-technical systems approach (Cooper & Foster, 1971), in the case of the leadership of scientists and engineers, the expertise needed for the effective exercise of influence may be unusually complex, requiring expertise in three distinct domains: (a) the organization, (b) the work being conducted, and (c) the interactions among the people doing this work.

The cost, and risk, entailed in the work being conducted by scientists and engineers implies that support and tangible resources must be acquired to sustain the work. And, given the uncertain nature of the outcomes of this work, such support may not be easily obtained. Dougherty and Hardy (1996) conducted interviews with key staff who had worked on 40 new product development efforts and found that sustained organizational support was critical to the success of these new product development efforts. Studies by Chandy and Tellis (2000); Klein, Conn, and Sorra (2001); and Henderson (1993) point to the same conclusion—the success of creative work in the sciences and engineering depends on sustained investment of resources.

What should be recognized here, however, is that resource investments in uncertain, high-risk activities are made on a quasi-logical basis. In other words, someone, typically the leader of a research and development group, must *sell* the research and development effort. Howell and Boies (2004) examined the sales activities of a sample of 19 product champions and 19 nonproduct champions working for organizations developing new products. Interviews were conducted with these pairs of champions, and nonchampions, and responses were coded with respect to contextual knowledge, normative knowledge, flexible role orientation, idea generation, idea promotion packaging, and idea promotion selling. It was found that product champions obtained higher scores than did nonchampions on both product packaging and product selling. Moreover, product packaging and product selling were found to depend on idea generation and contextual knowledge. These findings point to an important conclusion. The leaders of scientists and engineers must understand the needs of the organization, taking into consideration the field in which the organization operates, and must be able to formulate ideas responsive to these needs if they are to obtain the resources required to support their work.

Not only must leaders possess expertise with regard to the strategic issues confronting supporting organizations; the leaders of scientists and engineers must also possess expertise with regard to the work being conducted by scientists and engineers. To sell a project, one must have a plan (Castrogiovani, 1996). Although planning has often been held to lead to diminished performance on uncertain, dynamic, tasks such as those confronted in creative work (Mintzberg, 1991), the evidence accrued in recent years indicates that planning, by establishing frameworks for making sense of complex problems (i.e., sensemaking) and the formation of backup plans, promotes creativity and the opportunistic exploitation of emergent opportunities (Caughron & Mumford, 2008; Osburn & Mumford, 2006; Mumford et al., 2008).

Recently, Stockstrom, and Herstaat (2008) asked research and development directors in engineering firms to appraise the internal and external success of new product development efforts. It was found that the intensity of planning activities was positively related ($r = .30$) to both the internal and the external success of these new product development efforts. Additionally, Buijis (2008) found that the leaders of research and development efforts see planning as a critical component of their work, whereas Gross (2007) found that the leaders of research and development efforts do not delegate planning activities. What should be recognized here, however, is that the success of planning efforts ultimately is based on expertise and creative thinking skills such as forecasting (Byrne, Shipman, & Mumford, 2010; Mumford, 2002; Mumford, Schultz, & Van Doorn, 2001). Moreover, project planning provides a basis for defining work-group missions and structuring group tasks (Marta, Lertiz, & Mumford, 2005), selecting group members (Mumford, Eubanks, & Murphy, 2007), and evaluating project work (Mumford et al., 2008).

Although leaders must evaluate and monitor the work being conducted, overly close supervision may inhibit the performance of scientists and engineers. However, scientists and engineers do appear to actively seek evaluative feedback from leaders under certain conditions. For example, Farris

(1972) found that scientists and engineers sought contact with their leaders (a) when initially defining the problem to be pursued in the work and (b) when seeking evaluative feedback with regard to the work and its implications. What should be recognized here is that problem definition and idea evaluation skills require substantial creativity (Lonergan, Scott, & Mumford, 2004; Redmond, Mumford, & Teach, 1993). Moreover, it appears that problem definition and idea evaluation skills grow rather slowly over the course of peoples' careers. Thus leaders typically have stronger, sometimes far stronger, problem definition and idea evaluation skills than their subordinates (Mumford, Marks, Connelly, Zaccaro, & Reiter-Palmer, 2000).

Of course, evaluative feedback must be provided to others in an open, non-control-based framework (Sundgren, Selart, Ingelgard, & Bengtson, 2005). This implies that leaders must think about, and be able to *work with*, the creative scientists and engineers. They must be able to appraise and understand the concerns of others and their characteristic pattern of interactions. This social expertise may prove particularly important because leaders must get autonomous, driven, arrogant people to collaborate with each other with respect to an externally driven mission (Mumford, Scott, et al., 2002). In fact, many of the characteristics we perceive as desirable in leaders, for example, establishing perceptions of trust (Wang & Casimir, 2007) and ethical conduct (Mouly & Sankaran, 1999), may be important because they create conditions allowing for positive social interactions.

Some support for these conclusions may be found in a study by Rego, Sousa, Pina e Cunha, Correia, and Saur-Amaral (2007). They assessed the creativity of the followers of 138 leaders, who were asked to complete a survey measure that examined six dimensions of emotional intelligence, and found that leaders' scores on all six dimensions produced correlations in the .30s with the production of useful and creative ideas. One dimension, self-control with respect to criticism, however, produced a particularly strong relationship—a finding suggesting that an open and objective appraisal of others and their relationships may be critical to the leadership of scientists and engineers.

Indeed, prior research has shown that, based on interactions with followers vis-à-vis the mission, leaders act to create an interactional climate (James, James, & Ashe, 1990; Luria, 2008). What should be recognized here, however, is that in the case of creative people, the interactional climate established by leaders is a powerful variable shaping creativity and innovation (Amabile, Schatzel, Moneta, & Kramer, 2004; Bain, Mann, & Pirola-Merlo, 2001; Shipper & Davy, 2002). Accordingly, Hunter, Bedell-Avers, and Mumford (2007), in a meta-analysis of climates, perceptions, and the productivity of people working on tasks calling for creativity, found that perceptions of work climate was a powerful predictor of creative performance with commitment to the mission, perceptions of support for creative work, perceptions of autonomy or freedom, and perceptions of a collaborative work environment proving to be especially important to creative performance, especially when people were working under uncertain

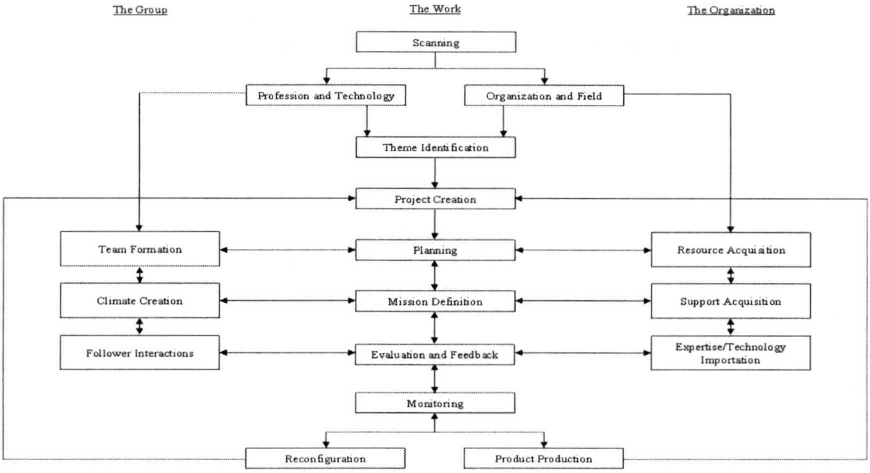

Figure 2.2 Framework of scientific leadership

conditions as is often the case of scientists and engineers. What should be recognized here, however, is that leaders must use their understanding of people, the work environment, and the mission to create these climate perceptions among the scientists and engineers being asked to do creative work.

Taken as a whole, these observations give rise to the framework of effective leadership of scientists and engineers presented in Figure 2.2. This framework assumes that leaders must acquire expertise with respect to the organization, the work, and the group. Scanning of the professional and organizational environments is used by the leader to identify themes and to define projects that would allow exploration of these themes. Project planning provides a basis for resource acquisition and team formation. Plans are then used to define the missions that scientists and engineers will be asked to pursue. Missions, in turn, provide a basis for defining the climate of the work group and acquiring support from the organization or field. As the work progresses, the leader evaluates and provides feedback, establishing a pattern of follower interactions, which indicate when the leader will need to import expertise or technology. Monitoring of the success of these efforts leads to production and, based on learning in the course of those projects, reconfiguration of the work.

LEADERSHIP SKILLS

Having provided a framework describing how leaders exercise influence, and specifically expertise-based influence, in directing the work of scientists and engineers, we now consider how each of these processes should be executed to enhance the creativity of scientists and engineers doing creative work. We have organized the discussion around the three key vectors that call for the exercise of influence—the work, the organization, and the group.

The Work

Scanning and Theme Identification

Prior research indicates that scanning of the environment, particularly in the kind of turbulent environments that characterize work in the sciences and engineering, is critical to the success of innovative efforts (Souitaris, 2001; Verhaeghe & Kfir, 2002). Scanning by leaders of creative efforts will typically be rather broad, considering two facets—(a) the profession and (b) the organization. Thus, the leaders of creative efforts may need to devote more time to scanning than is typical of other leaders because their scanning activities address two potentially distinct domains. Moreover, effective leaders of creative efforts may prove especially skilled at identifying sources, or modes of information, that point to emerging trends in both the technical profession and the organization, with leaders taking into account both the opportunities and restrictions implied by those trends.

Scanning is critical for a number of reasons. It allows leaders to identify potential staff and sources of support. More fundamentally, however, scanning provides a basis for identifying the fundamental themes to be pursued in creating the projects to be pursued by scientists and engineers. O'Connor (1998) conducted a qualitative study of eight leaders who had directed projects resulting in radical innovations. Her findings suggest that the leaders of creative efforts in the sciences and engineering are able to envision, or forecast, the downstream implications of fundamentals with respect to the production of a set of products that would serve organizational needs. Thus, Dupont's work in the development of synthetics was guided by a search for synthetics that exhibited the properties of wool (Houndshell, 1992). Although we do not know enough about how this forecasting of the implications of trends occurs among the leaders of scientists and engineers, it does seem clear that this forecasting must consider implications of fundamentals with regard to gaps, or unknowns, and the implications of these unknowns for operations of an organization operating in a given field.

Project Creation and Project Planning

With identification of fundamentals, gaps in understanding of those fundamentals, and envisioning of the products that might arise from exploration of those fundamentals, it becomes possible for leaders to create projects and formulate plans for pursuing those projects. Planning is not a rote activity. Rather, planning requires the mental simulation of future work activities (Mumford, Feldman, Hein, & Nago, 2001). Mumford, Schultz, and Osburn (2002) have proposed a model of the critical cognitive operations involved in planning. Broadly speaking, this model holds that planning involves six interconnected processes: (a) use of experiential knowledge to create frameworks for sensemaking (Drazin, Glynn, & Kazanjian, 1999); (b) identification of critical goals and causes (Hammond, 1990); (c) extended search and analysis to identify contingencies, resources, restrictions, causes, causal dependencies, and objectives (Mumford, Schultz, et al., 2002); (d) formation of an initial template plan (Mumford, Schultz, et al., 2002); (e) extensive forecasting based on this template plan in order to provide a basis for plan refinement (Shipman, Byrne, & Mumford, in press); and (f) construction of backup plans (Xiao, Milgram, & Doyle, 1997).

With regard to the leadership of scientists and engineers pursuing creative work, this description of project planning has a number of implications. First, project planning will be expertise based. Thus, leaders must have tangible experience in the field in which they are working (Weisberg, 2006). Second, the sensemaking model applied will shape the kinds of plans created by leaders. This implies that leaders of scientists and engineers must have available a number of models for understanding the kinds of problems on which scientists and engineers will be working. In addition, they must be able to integrate attributes of these models, for example, to combine multiple themes into a single project and to frame it in terms favorable to relevant stakeholders. Third, in project planning, leaders must identify critical causes and goals as well as resource contingencies and restrictions impinging on the effort. Thus, in leading research and development, one must identify ambiguities with regard to causes, resources, restrictions, and contingencies, using this analysis of ambiguities to identify critical research and development needs (Blumberg, 1994). Fourth, in plan formation, leaders must engage in extensive forecasting of likely outcomes under alternative scenarios (Shipman et al., in press)—scenarios that consider potential changes in both the technical field and the organization (Caughron & Mumford, 2008). These forecasts must, in turn, provide both work plans and frameworks for guiding adjustments in these work plans. Thus, much of the creative thinking of leaders will be based on these planning activities with project plans providing a basis for both recruiting staff and obtaining support for the project.

Mission Definition

Leaders' planning activities also provide the framework in which missions are formulated. Missions represent more than just the assignments given to individual scientists or engineers on teams of scientists and engineers. Mumford, Scott, et al. (2002) have argued that due to the importance of professional contributions and evaluations to scientists and engineers, missions also represent a critical source of both direction and motivation. Thus, the leaders of scientists and engineers must articulate the importance and significance of a mission with respect to the broader project being pursued and the significance of this mission for both the profession and the supporting organization. Finally, leaders must anticipate how this mission contributes to the work being pursued by others and the restrictions that will be imposed on the work of scientists and engineers.

The motivational and directional functions of the missions are of special importance in part because they provide a basis for the leader's attempts to define the interactional climate of the work group and acquire support and allocate resources needed by work group members for completing this mission. This implies that leaders must attend to priorities and changes in mission priorities with respect to overarching plans. Furthermore, leaders must orchestrate, or integrate, the work being pursued by scientists and engineers working on related missions (Mumford, Scott, et al., 2002b). What should be recognized here, however, is that effective integration of missions will require ongoing sensemaking by the leader as well as articulation of these sensemaking activities to those pursuing a given mission (Drazin et al., 1999).

Missions must also be defined and redefined as the work progresses through different stages of development, suggesting that the leaders of scientists and engineers must actively manage changes in the work being conducted by scientists and engineers (Nemanich & Keller, 2007), using missions to define, and articulate, the standards to be used in appraising the work of scientists and engineers (Kirchner, 1996). Thus, leaders must evidence flexibility in evaluation standards, identifying and articulating stage appropriate standards for appraising the success and failure of a mission.

Evaluation and Feedback

Traditionally, evaluation of the work of others has not been viewed as an especially creative activity—either the work was done to standard or it was not. However, in the case of people doing creative work, evaluation of this work appears to be of greater importance (Farris, 1972). A leader's evaluation of work by other researchers provides a basis for ongoing learning and progressive refinement of the work (Rickards & Moger, 2006). Moreover, it is through this progressive refinement, and ongoing learning, that creative products emerge (Huber, 1998). Thus, evaluation should not be viewed as an outcome but rather as a learning and enhancement process—a process that must be guided by the leader to facilitate both mission accomplishment and decision making with regard to pursuit of the work.

These observations imply that evaluation of scientists' and engineers' work by leaders may be an inherently creative process (Basadur, Runco, & Vega, 2000). Accordingly, Lonergan et al. (2004) asked undergraduates to assume the role of a marketing manager evaluating ideas of varying quality and originality for the production of advertising campaigns. They found that the most creative advertising campaigns were produced when managers evaluated high-quality campaigns to enhance their originality and when managers evaluated original campaigns to enhance their quality. Thus, a complex form of compensatory creativity is required of those asked to lead scientists and engineers—compensatory evaluation which calls for both divergent thinking, in forecasting the potential consequences of high-quality ideas, and convergent thinking, in evaluating original ideas against operative criteria (e.g., efficiency).

Evaluations represent a primary source of interactions of scientists and engineers with leaders where perceptions of leader support, leader trust, and positive exchange—all variables shaping creative performance (Amabile et al., 2004)—emerge as a function of these evaluations. Thus, the style of exchange in the course of evaluation will be a potentially powerful influence on leader performance. Accordingly, Sundgren et al. (2005) found that dialogue-based exchange, a form of communication characterized by informal, strategic, conversations, appears especially useful in conducting appraisals of scientists working in the pharmaceutical industry. Evaluative feedback must also provide a framework for identifying expertise or technology that must be acquired to enhance the viability of a project. Therefore, leaders must use evaluation to identify project needs, taking into account mission, stage of project development, and organizational capabilities (Mumford et al., 2008). Finally, leaders must, ultimately, use the information received in evaluation to decide when an effort should proceed, how it should proceed, who will work on the effort, and how handoffs (e.g., from one individual/team to another) in this work should occur. Notably, these handoffs require preparation of both the organization and the scientists and engineers who might be asked to work on the effort. Therefore, embedded in leaders' evaluations is an ongoing concern with regard to development of people, projects, and the organization.

Monitoring, Production, and Reconfiguration

Evaluation also provides the markers to be used in appraising the progress of the work. Tight monitoring, or overly close supervision, has been found to be negatively related to innovation among scientists and engineers (Barnowe, 1975). Therefore monitoring of work progress by leaders cannot be based on the conduct of the work per se, but rather on the identification of critical requirements in evaluation and progress in satisfying those critical requirements. Additionally, the need for and importance of monitoring will increase as projects move from initial development to field testing and product production (Mumford et al., 2008). The particular aspects being monitored by leaders will shift as a function of the stage of project development. Finally, monitoring standards should be defined as part of evaluation through identification of goals and potential impediments. Therefore monitoring standards should be defined in a consultative fashion by those leading scientists and engineers.

Monitoring should lead to the production of viable, innovative products. The products produced by scientists and engineers also generate new knowledge, which provides leaders with a basis for reconfiguration of the work being conducted and the definition of new projects. Therefore, the leaders of scientists and engineers must ultimately be curious about how the work turns out—using both successful and unsuccessful efforts as a basis for guiding future work.

The Group

Up to this point, our observations about the leadership of scientists and engineers have focused only on the work being done by the leader. However, leadership is not simply a matter of planning and evaluation. Rather, leaders must interact with their followers. Leaders can, of course, exercise influence on groups through a variety of mechanisms, ranging from group processes (Taggar, 2002) to helping followers establish knowledge networks (Christiansen & Varnes, 2007). However, three key avenues exist for the exercise of influence in leading scientists and engineers: (a) team formation, (b) climate creation, and (c) follower interactions.

Team Formation

One of the most powerful ways a leader exercises influence on creative efforts is through the formation of the project team. As noted earlier, scientists and engineers are autonomous, professionally focused individuals. Therefore, the motivation of scientists and engineers for creative work is person dependent with people selectively investing their expertise and effort. This implies that leaders must define the various forms of expertise needed in project teams and recruit individuals who have, or who are willing to develop, this expertise. They must actively recruit individuals to join this team based on the significance of the project or projects at hand for their professional development (Kidder, 1981).

Team formation also involves the acquisition of a shared mental model about the work being conducted and the processes by which group members work with each other vis-à-vis this shared mental model (Day, Gronn, & Salas, 2006). In one study along these lines, Mumford, Feldman, Hein, and Nago (2001) used training to create shared mental models among teams working on a creative project. They found that the production of higher quality and more original products occurred when team members did, as opposed to did not, share a mental model of the effort. What should be recognized here, however, is that leaders, by carefully articulating project plans, and by describing the relevance of various forms of expertise to execution of those plans, build the shared mental models critical to the success of teams of scientists and engineers working on projects that call for creativity (Howell & Boies, 2004).

Leaders also influence the processes applied by teams in executing their work. For example, Ancona and Caldwell (1992) in a study of 47 product development teams found that the extensiveness of external communication was positively related to both schedule and budget performance and the creativity of the products produced. Therefore, leaders, by encouraging external communication, can influence the creativity of scientists and engineers. However, the processes giving rise to effective teamwork by scientists and engineers may involve more than just external communication. Taggar (2002) examined the performance of groups asked to work on problem-solving tasks with respect to seven process variables: (a) team citizenship, (b) performance management, (c) effective communication, (d) involving others, (e) providing feedback, (f) articulating professional conflict, and (g) averting professional conflict. All of these process variables contributed to team creativity. They are significant because they suggest that the leaders of creative groups must articulate and help the team resolve technical conflicts while encouraging team citizenship and collaboration (Mumford & Hunter, 2005).

Climate Creation

These observations about group process are noteworthy, in part, because they suggest that the climate scientists and engineers perceive as surrounding their work will also influence creativity and innovation. Over the years, a number of studies have been conducted examining the impact of climate perceptions on the creativity of scientists and engineers (e.g., Amabile, Conti, Coon, Lazenby, & Herron, 1996; Ekvall & Ryhammer, 1999; West, 1990). In a recent meta-analysis, Hunter, Bedell-Avers, and Mumford (2007) found that perceptions of climate, or the work environment, were a powerful influence on creative production with perceptions of support, perceptions of resources, perceptions of autonomy, and perceptions of the importance of the mission all producing sizable relationships with indices of creativity and innovation.

What should be recognized in this regard, however, is that the behavior of the leader is a particularly powerful influence shaping climate perceptions (West et al., 2003). For example, Jaussi, Randel, and Dionne (2007) have shown that leaders, both by calling for creativity and by building a sense of creative self-efficacy in followers, can develop a perception that the work environment will support creativity, thereby motivating creative work. In an older series of studies, Maier and his colleagues (Maier, 1950, 1953; Maier & Hoffman, 1965; Maier & Solem, 1962) found that leaders can create positive climate perceptions by (a) requesting creative or innovative solutions, (b) framing problems in terms of broader issues and learning goals, (c) encouraging group members to consider a range of relevant information, (d) avoiding premature evaluation, (e) extending discussion and exploration of issues bearing on project work, and (f) using disagreements to frame innovative solutions. Studies of innovation in research and development by Elkins and Keller (2003) and Mossholder and Dewhirst (1980) indicate that leader behaviors that call for creativity, encourage articulation, and induce intellectual stimulation all act to create a perception of an environment in which creativity is viewed as valuable.

Follower Interactions

Our foregoing observations suggest that how leaders interact with scientists and engineers working on research and development efforts will prove of some importance. Accordingly, Tierney et al. (1999) found that the assessed creativity and invention rates of research and development chemists was related to the formation of positive interpersonal exchange relationships between leaders and followers. Moreover, these positive exchange relationships result in feelings of trust in the leader and perceptions of the leaders (Yukl, 2010), and may result in stronger processing of feedback, better collaboration, and creative contributions to a positive creative climate.

The formation of positive exchange relationships has been held to depend on perceptions of follower competence, trustworthiness, and loyalty (Graen & Uhl-Bien, 1998). Thus, to the extent that effective, competent scientists are recruited, and these scientists are asked to work in a positive environment, the positive impact of leader–follower interactions is likely to increase over time. More centrally, to the extent that evaluative feedback is provided in a developmental framework in which confidence is expressed in the individual's creative potential, leaders will find it possible to establish a positive pattern of interpersonal exchange with followers—a pattern of interaction contributing to the creativity of scientists and engineers.

The Organization

Resource Acquisition

In addition to followers, leaders must also establish a viable pattern of interaction with the organization being asked to support this work. In fact, one key attribute of project planning is that these plans provide a basis for acquiring the resources, money and equipment needed to pursue work on a project. In this regard, however, it is important to bear in mind that to obtain resource support, leaders must be able to package and sell plans in terms likely to be appealing to those organizations being asked to provide support for the initiative (Howell & Boies, 2004).

The term *sales*, however, raises another question. What must a leader know and be able to do in order to sell projects? Perhaps the first and most important requirement is that the leader must understand the problems confronting the organization and the potential contribution of the project to resolving these problems. Thus, the leaders of research and development efforts must understand the strategy of sponsoring organizations (Laugen, Boer, & Acur, 2006). Leaders must also understand how work is conducted in the organization, taking these working, day-to-day strengths and weaknesses into consideration when formulating projects (Lee & Kelley, 2008). Finally, to sell a project, one must have access to those who control the relevant resources (Jelinek & Schoonhoven, 1990), or to top management teams.

Top management teams, although responsible for resource investments, are also often unstable. In other words, people retire from, or leave, their positions. This suggests that the stable, sustained resource investments required for most research and development will require building support for the effort across various functional constituencies in the organization. As Lassen, Waehrens, and Boer (2009) have found, in a qualitative study of two cases of successful new product development, the leaders of creative efforts must be able to describe the implications of innovative efforts for the effectiveness of these functions, engaging in a process of social negotiation and helping these units to understand, or make sense of, the projects being pursued. Thus, the leaders of creative efforts must seek to build a broader organizational understanding of research and development efforts—serving as an educator, as well as a sales representative (Cohen & Levinthal, 1990).

Support Acquisition

Ultimately, these outreach activities serve to build support among other members of the organization, and the profession, for the projects being pursued by the leader and his/her scientists and engineers. Support acquisition, moreover, has been found to be critical to the success of research and development efforts (Ong, Wan, & Chang, 2003; Ramus, 2001), especially support in the form of recruiting project champions (Markham & Aiman-Smith, 2001; Markham & Griffin, 1998). Other senior managers willing to champion projects both provide resources needed to complete the work and encourage others to provide these resources. In the case of initial exploratory efforts, the acquisition of support from other members of an organization, or profession, may not prove critical. In later stage project development efforts, however, this organizational, or professional, support may prove crucial as other organizational members, for example, production units, are asked to invest staff and resources in preparing new projects for field testing, routine production, and marketing.

Acquisition of support from other members is contingent on others seeing value with regard to the growth and development of the organization, the profession, and their own functional priorities. This implies that leaders must be able to link their project to the concerns of those constituencies whose support will be required. Therefore, leaders must be skilled at stakeholder analysis, building the concerns of stakeholders into their mission definition, and articulating the importance of this support in building the climate they need to support the work of their scientists and engineers. Therefore, leaders must take stakeholder functional needs and objectives into account as they develop projects and project staff.

Expertise/Technology Importation

A leader's ability to build support among relevant stakeholder groups is noteworthy because it provides the social capital needed to import expertise and technology into research and development groups. As projects increase in complexity and move towards implementation, new forms of technology and new types of expertise must be introduced. As a result, cross-functional teams may prove beneficial to many research and development efforts (Thamhain, 2003). It should be noted that more research is clearly needed to examine the effects of diversity on creative problem solving.

With regard to leaders, however, the need to import technology and expertise has three noteworthy implications. First, the leader must recognize when this technology, or expertise, must be imported to support project work. Therefore, leaders cannot fall victim to the 'not invented here' syndrome but instead must evaluate who has the skills needed to complete the work. Second, leaders must be capable of structuring work groups and work-group activities to allow for effective importation of new techniques and new forms of expertise. Finally, leaders must develop themselves into informed consumers of the work done by others if they are to be able to use those technologies and other forms of expertise in an effective fashion.

DEVELOPING LEADERS

These observations about the nature of leadership with respect to scientists and engineers point to one inescapable conclusion—the effective leadership of scientists and engineers is unusually complex. This complexity, as well as the known impact of leader effectiveness on the success of research and development efforts (Barnowe, 1975; Tierney et al., 1999), in turn, implies that one must ask another question: How might one go about supporting the leaders of scientists and engineers? In our view, based on the framework proposed herein and on the relevant literature, at least five avenues for development should be given some consideration in this regard: (a) leadership skills training, (b) leader career development, (c) stakeholder integration, (d) survey feedback, and (e) management processes and procedures.

Leadership Skills Training

The education of scientists and engineers, including their leaders, is typically assumed to be finished once an individual has left school. However, work by Basadur and his colleagues (Basadur & Gelade, 2006; Basadur, Graen, & Green, 1982) indicates that training may prove valuable in enhancing the performance of those asked to lead scientists and engineers in creative efforts. In fact, the key causal relationships in the framework of leadership presented earlier (e.g., in the sections regarding the work, the group, and the organization) suggest that three types of training interventions might prove especially useful. First, given the importance of leader scanning, planning, and evaluation activities mentioned earlier, interventions intended to improve these key skills can be expected to have significant value. Second, training in team building might prove of value given the need for leaders to influence the group (Day et al., 2006). Finally, training programs that teach leaders how to craft project plans into viable proposals are likely to prove useful, given the need for scientists to acquire resources from the organization.

Each of these three types of training requires rather different content. For example, training intended to increase the extensiveness of forecasting should contribute to planning while training focused on providing alternative appraisal standards might prove useful in evaluation and team building

(Byrne et al., 2010; Lonergan et al., 2004). Nonetheless, the evidence provided by Scott, Leritz, and Mumford (2004) does suggest that such training interventions are most likely to prove effective if they focus on cognitive strategies for executing critical operations involved in each of these activities.

Leader Career Development

Typically, scientists and engineers are not prepared, and sometimes not expected, to lead other scientists and engineers (Mumford, 2000). This lack of systematic career development suggests the need to create programs for enhancing the leadership capabilities of scientists and engineers. For example, preparing people for leadership by having them lead a subcomponent of a longer project is likely to prove valuable. However, given the importance of technical and organizational expertise to the leadership of scientists and engineers, as discussed earlier, two forms of leadership development experiences are likely to prove especially valuable.

First, ongoing professional engagement activities (e.g., attending conferences, serving as a professional reviewer, conducting research) are likely to prove of value, especially if potential leaders are asked to engage in professional leadership activities. For example, scientists and engineers, by reviewing others' technical work, learn much about evaluation and feedback. Second, interventions must also seek to develop a broader understanding of the organization and the profession in potential leaders. For example, asking scientists and engineers to sit on marketing advisory panels will expand their understanding of what customers seek and how a company markets products. Moreover, such value-added assignments will provide the social networks needed to acquire support for various projects (Friedrich, Vessey, Schuelke, Ruark, & Mumford, 2009).

Stakeholder Integration

The fundamental dispositional characteristics of scientists and engineers, their professional orientation, autonomy, and arrogance (Ericsson & Charness, 1994; Feist, 1999) often result in a tendency to discount the views and needs of others. Therefore, interventions that provide scientists and engineers with the background needed to analyze the needs and concerns of relevant stakeholders are likely to prove of value in enhancing leadership capabilities. These interventions may improve leader scanning, by broadening and deepening the scope of that scanning, and the ability of the leader to sell projects and obtain resources and support, by providing information necessary for linking projects to the concerns of relevant constituencies.

One way this type of intervention might be accomplished is by involving high potential scientists and engineers in stakeholder studies being conducted by the organization on the profession. Another approach that might prove of value is to have scientists and engineers in project-planning meetings actively analyze the utility functions, or the goals and priorities, of different stakeholder groups. Still another approach that might prove of value is to ask scientists and engineers to consider the usability of the products being developed. Although evidence bearing on the merits of these, and other potential interventions, is lacking, it does seem reasonable to expect that such interventions might prove useful in developing the leadership of potential scientists and engineers.

Survey Feedback

One of the most common leadership-development interventions is survey feedback and coaching (Atwater, Roush, & Fischthal, 1995). Indeed, survey measures have been developed to appraise many aspects of leadership styles (e.g., Bass & Avolio, 1990; Fleishman & Harris, 1962). Given the nature of scientific and engineering leadership, it is open to question whether many of these survey feedback systems will contribute substantially to the development of technical leaders. By the same token, however, surveys are available for assessing climate perceptions (Amabile et al., 1996; West et al., 2003). Given the impact of climate perceptions on creative work, use of these surveys to appraise climate, and provide leaders with coaching as to how to adjust their behavior to improve creative climate, may prove more useful in developing effective leaders (Hunter et al., 2007).

In addition to the use of extant surveys, the leadership framework presented earlier suggests that there might be value in developing new types of survey instruments. For example, surveys examining leaders' evaluation and feedback or surveys examining team formation processes might help develop critical skills needed for leading creative scientists and engineers.

Management Processes and Procedures

A final practical implication is that one way to improve leader performance is to employ new management procedures. For example, one might ask leaders of research and development efforts to periodically provide scanning reports examining the implications of emerging techniques, or emerging technical problems, for organizational operations. Alternatively, project proposal systems might include a requirement for leaders to indicate how organizational technical capabilities would be developed by a proposed effort and to delineate the current operations that would be affected if a proposed effort was successful. Such proposal requirements would in fact help ensure the requisite support needs had been thought through more clearly. The nature of effective leadership also suggests some variations in typical project-planning procedures (Caughron & Mumford, 2008). For example, in project planning, leaders might be asked to indicate not just the time frame and budget for the overall effort; they might also be asked to delimit resources, restrictions, and contingencies likely to arise at each stage of an effort. They might be asked to indicate the evaluative criteria to be applied in appraising performance at each stage in project development. They might be asked to describe how the proposed effort would contribute to the technical and leadership development of the staff assigned.

CONCLUSION

Before turning to some of the broader conclusions flowing from the present chapter, certain limitations should be noted. To begin with, we have assumed that the leadership of creative work being conducted by scientists and engineers is a generalizable phenomenon, particularly to various subdomains of science and engineering. However, as Napier and Nilsson (2006) have pointed out, creative work, even among scientists and engineers, may occur in different ways in different organizational settings. In other words, leadership of microbiologists may be different from leadership of physicists. These cross-field differences in the nature of effective leadership of scientists and engineers need to be taken into account in future work (Csikszentmihalyi, 1999).

Along related lines, in the present chapter we have focused on the activities of the leader vis-à-vis the work, the group, and the organization. Little has been said, however, about how technical advances in the field (Wise, 1992) or organizational strategy (Sharma, 1999) act to influence the success leaders of scientists and engineers. There is a need for research on potential multi-level influences on the leadership of scientists and engineers (Mumford & Hunter, 2005).

Finally, in the present chapter we have considered the leadership of scientists and engineers involved in doing creative technical work. One implication of the statement is that this chapter should not be assumed to apply to scientists and engineers not engaged in creative work—for example, program officers or sales engineers. Similarly, given the differences observed in the nature of scientific projects and engineering projects, our assumption that these two forms of work can be equated is open to question. Unique leadership demands will likely emerge in each of these areas that need to be taken into account in discussions of leadership for technical creativity.

Even bearing these limitations in mind, we believe that the present chapter has a number of noteworthy implications. To begin with, creative work being conducted by scientists and engineers is itself complex (Feist & Gorman, 1998). However, the effective leadership of these creative efforts may be even more complex. Indeed, it may be more complex than many other forms of leadership. Effective leaders of creative efforts must exert influence, influence based on expertise, with regard to three distinct vectors—the work itself, the people doing the work, and the organization, or profession, supporting the work.

The complexity of the leadership of scientists and engineers as a phenomenon is noteworthy for three reasons. First, the leaders of scientists and engineers must possess a wide array of expertise. Second, due to the dynamic nature of creative work, they must display substantial flexibility with an ability to resolve conflicts and conflicting information arising from a variety of perspectives. Third, the multiple ways influence must be exercised suggest that multiple leaders, and shared leadership, will be required in the direction of scientists and engineers (Kamoche & Cunha, 2001).

To complicate matters further, the effective exercise of influence in the leadership of scientists and engineers involves a wide array of skills. Leaders must scan, plan, sell, create a conducive climate, evaluate, and reconfigure the work being done. In addition, leaders should have social skills in order to aid in tasks such as acquiring resources and exercising influence over unusually independent subordinates. Therefore, the leaders of creative efforts require an unusually wide array of skills.

The number and complexity of the skills required of these asked to lead the creative work of scientists and engineers suggests we need to formulate systematic leadership development programs. Unfortunately, less attention has been given by researchers to the development of leading scientists and engineers than is the case in many other areas of leadership. By the same token, we would argue that an understanding of the demands made on those leading such efforts does suggest a number of interventions that might prove of value in enhancing leader performance. Given the known impact of effective leadership on the success of research and development efforts (Barnowe, 1975; Tierney et al., 1999), we believe that such systematic developmental interventions are, in fact, warranted. We hope that the present chapter serves as an impetus for future research along these lines.

ACKNOWLEDGMENTS

We would like to thank Jamie Barrett, Sam Hunter, Eric Day, Dean Hougen, and Kim Hester for their contributions to the present chapter. Parts of this work were supported by a grant from the National Science Foundation (ITS-082589)—Michael D. Mumford, principal investigator.

REFERENCES

Amabile, T. M., Conti, R., Coon, H., Lazenby, J., & Herron, M. (1996). Assessing the work environment for creativity. *Academy of Management Journal*, *39*, 1154–1184.

Amabile, T. M., Schatzel, E. A., Moneta, G. B., & Kramer, S. J. (2004). Leader behaviors and work environment for creativity: Perceived leader support. *The Leadership Quarterly*, *15*, 5–32.

Ancona, D., & Caldwell, D. (1992). Demography and design: Predictors of new product team performance. *Organization Science*, *3*, 321–341.

Andersen, H., Barker, P., & Chen, X. (2006). *The cognitive structure of scientific revolutions*. Cambridge, MA: Cambridge University Press.

Andrews, F. M., & Farris, G. F. (1967). Supervisory practices and innovation in scientific teams. *Personnel Psychology*, *20*, 497–515.

Atwater, L. E., Roush, P., & Fischthal, A. (1995). The influence of upward feedback on self and follower ratings of leadership. *Personnel Psychology*, *48*, 35–59.

Bain, P. G., Mann, L., & Pirola-Merlo, A. (2001). The innovation imperative: The relationships between team climate, innovation, and performance in research and development teams. *Small Group Research*, *32*, 55–73.

Bakker, H., Boersma, K. V., & Oreel, S. (2006). Creativity (ideas) management in industrial R & D organizations: A crea-political process model and an empirical illustration of Corus RD & T. *Creativity and Innovation Management*, *15*, 296–309.

Barnowe, T. (1975). Leadership and performance outcomes in research organizations: The supervisor of scientists as a source of assistance. *Organizational Behavior and Human Performance*, *14*, 264–280.

Basadur, M. & Gelade, G. (2006). The role of knowledge management in the innovation process. *Creativity and Innovation Management*, *15*, 45–61.

Basadur, M. S., Graen, G. B., & Green, S. G. (1982). Training in creative problem solving. Effects on ideation and problem finding in an industrial research organization. *Organizational Behavior and Human Performance*, *30*, 41–70.

Basadur, M., Runco, M.A., & Vega, L.A. (2000). Understanding how creative thinking skills, attitudes, and behaviors work together: A causal process model. *Journal of Creative Behavior*, *34*, 77–100.

Bass, B. M. (1997). Does the transactional-transactional leadership paradigm transcend organizational and national boundaries? *American Psychologist*, *52*, 130–139.

Bass, B. M., & Avolio, B. J. (1990). *Transformational leadership development: Manual for the Multifactor Leadership Questionnaire*. Palo Alto, CA: Consulting Psychologist Press.

Bird, B., & Sherwin, M. J. (2005). *American Prometheus: The triumph and tragedy of J. Robert Oppenheimer.* New York, NY: Random House.

Blumberg, B. S. (1994). Scientific process and the hepatitis B virus. *Creativity Research Journal*, *7*, 315–325.

Brown, S., & Eisenhardt, K. (1995). Product development: Past research, present findings, and future directions. *Academy of Management Review*, *20*, 343–378.

Buijis, J. (2007). Innovation leaders should be controlled schizophrenics. *Creativity and Innovation Management*, *16*, 203–210.

——— (2008). Action planning for new product development projects. *Creativity and Innovation Management*, *17*, 319–333.

Byrne, C. L., Mumford, M. D., Barrett, J. D., & Vessey, W. B. (2009). Examining the leaders of creative efforts: What do they do, and what do they think about? *Creativity and Innovation Management*, *18*, 256–268.

Byrne, C. L. Shipman, A. S., & Mumford, M. D. (2010). The effects of forecasting on creative problem-solving: An experimental study. *Creativity Research Journal*, *22*, 119–138.

Castrogiovani, G. J. (1996). Pre-start-up planning and the survival of new small businesses: Theoretical linkages. *Journal of Management*, *22*, 801–822.

Caughron, J. J., & Mumford, M. D. (2008). Project planning: The effects of using formal planning techniques on creative problem-solving. *Creativity and Innovation Management*, *17*, 204–215.

——— (2009) Project planning: The effects of using formal planning techniques on forecasting and plan development. *Creativity and Innovation Management*, *17*, 1–12.

Chandy, R. K., & Tellis, G. J. (2000). The incumbent's curse? Incumbency, size and radical innovation. *Journal of Marketing*, *64*, 1–17.

Christiansen, J. K., & Varnes, C. J. (2007). Making decisions on innovation: Meetings or networks? *Creativity and Innovation Management*, *16*, 282–298.

Chua, R. Y. J., & Iyengar, S. S. (2008). Creativity as a matter of choice: Prior experience and task instruction as boundary conditions for the positive effect of choice on creativity. *Journal of Creative Behavior*, *42*, 164–180.

Cohen, W. M., & Levinthal, D. A. (1990). Absorptive capacity: A new perspective on learning and innovation. *Administrative Science Quarterly*, *35*, 128–152.

Conger, J. A., & Kanungo, R. S. (1998). *Charismatic leadership in organizations*. Thousand Oaks, CA: Sage.

Cooper, R., & Foster, M. (1971). Sociotechnical systems. *American Psychologist*, *26*, 467–474.

Crouch, T. D. (1992). Why Wilbur and Orville? Some thoughts on the Wright brothers and the process of invention. In R. J. Weber *&* D. N. Perkins (Eds.), *Inventive minds: Creativity in technology* (pp. 80–92). New York, NY: Oxford University Press.

Csikszentmihalyi, M. (1999). Implications of a systems perspective for the study of creativity. In R. J. Sternberg (Ed.), *Handbook of creativity* (pp. 313–338). Cambridge, England: Cambridge University Press.

Day, D. V., Gronn, P., & Salas, E. (2006). Leadership in team-based organizations: On the threshold of a new era. *The Leadership Quarterly*, *17*, 211–216.

Dougherty, D., & Hardy, C. (1996). Sustained product innovation in large, mature organizations: Overcoming innovation-to-organization problems. *Academy of Management Journal*, *39*, 1120–1153.

Drazin, R., Glynn, M. A., & Kazanjian, R. K. (1999). Multilevel theorizing about creativity in organizations: A sense-making perspective. *Academy of Management Review*, *24*, 286–329.

Dunbar, K. (1995). How scientists really reason: Scientific reasoning in real-world laboratories. In R. T. Sternberg & J. E. Davidson (Eds.), *The nature of insight* (pp. 365–396). Cambridge, MA: MIT Press.

Ekvall, G., & Ryhammer, L. (1999). The creative climate: Its determination and effects at a Swedish University. *Creativity Research Journal*, *12*, 303–310.

Elkins, T., & Keller, R. T. (2003). Leadership in research and development organizations: A literature review and conceptual framework. *The Leadership Quarterly*, *14*, 587–606.

Engwall, M., Kling, R., & Werr, A. (2005). Models in action: how management models are interpreted in new product development. *R&D Management*, *35*, 427–439.

Ericsson, K. A., & Charness, N. (1994). Expert performance: Its structure and acquisition. *American Psychologist*, *49*, 725–747.

Farris, G. F. (1972). The effect of individual roles on performance in innovative groups. *R&D Management*, *3*, 23–28.

Feist, G. J. (1999). The influence of personality on artistic and scientific creativity. In R. Sternberg (Ed.), *Handbook of creativity* (pp. 273–296). New York, NY: Cambridge University Press.

Feist, G. J., & Gorman, M. E. (1998). The psychology of science: Review and integration of a nascent discipline. *Review of General Psychology*, *2*, 3–47.

Fleishman, E. A., & Harris, E. F. (1962). Patterns of leadership behavior related to employee grievances and turnover. *Personnel Psychology*, *15*, 43–56.

Friedrich, T. L., Vessey, W. B., Schuelke, M. J., Ruark, G. A., & Mumford, M. D. (2009). A framework for understanding collective leadership: The selective utilization of leader and team expertise within networks. *The Leadership Quarterly*, *20*, 933–958.

Graen, G. B., & Uhl-Bien, M. (1998). Relationship-based approach to leadership: Development of leader-member exchange (LMX) theory of leadership over 25 years: Applying a multi-level multi-domain perspective. In F. Dansereau & F. J. Yammarino (Eds.), *Leadership: The multi-level approaches* (pp. 103–134). Stamford, CT: JAI Press.

Gross, D. (2007). Leadership in R&D projects. *Creativity and Innovation Management, 16*, 447–456.

Hammond, K. J. (1990). Case-based planning: A framework for planning from experience. *Cognitive Science, 14*, 385–443.

Henderson, R. (1993). Underinvestment and incompetence as responses to radical innovation: Evidence from the photolithographic alignment equipment industry. *RAND Journal of Economics, 24*, 248–270.

Hennessey, B. A., & Amabile, T. M. (2010). Creativity. *Annual Review of Psychology, 61*, 569–598.

Houndshell, E. A. (1992). Invention in the industrial research laboratory: Individual or collective process? In R. J. Weber, & D. N. Perkins (Eds.), *Inventive minds: Creativity in technology* (pp. 273—291). New York, NY: Oxford University Press.

Howell, J. M., & Boies, K. (2004). Champions of technological innovation: The influences of contextual knowledge, role orientation, idea generation, and idea promotion on champion emergence. *The Leadership Quarterly, 15*, 130–149.

Huber, J. C. (1998). Invention and inventivity as a special kind of creativity with implications for general creativity. *Journal of Creative Behavior, 32*, 58–72.

Hughes, T. P. (1989). *American genesis: A history of the American genius for invention.* New York, NY: Penguin.

Hunter, S. T., Bedell-Avers, K. E., & Mumford, M. D. (2007). Climate for creativity: A quantitative review. *Creativity Research Journal, 19*, 69–90.

Isaacson, W. (2007). *Einstein his life and universe.* New York, NY: Simon & Schuster.

James, L. R., James. L. A., & Ashe, D. K. (1990). The meaning of organizations: The role of cognition and values. In B. Schneider (Ed.), *Organizational Climate and Culture* (pp. 40–84). San Francisco, CA: Jossey-Bass.

Jaussi, K., Randel, A., & Dionne, S. (2007). I am, I think I can, and I do: The role of personal identity, self-efficacy, and cross-application of experiences in creativity at work. *Creativity Research Journal, 19*, 247–258.

Jelinek, M., & Schoonhoven, C. B. (1990). *The innovation marathon: Lessons learned from high technology firms.* Oxford, England: Blackwell.

Jung, D. I. (2001). Transformational and transactional leadership and their effects on creativity in groups. *Creativity Research Journal, 13*(2), 185–195.

Kamoche, K., & Cunha, M. (2001). Minimal structures: From jazz improvisation to product innovation. *Organization Studies, 22*, 733–764.

Kidder, T. (1981). *The soul of a new machine.* New York, NY: Avon.

Kirchner, P. (1996). *Oops! A stupefying survey of goofs, blunders and blotches, great and small.* Los Angeles, CA: General Publishing Group.

Klein, K. J., Conn, A. B., & Sorra, J. S. (2001). Implementing computerized technology: An organizational analysis. *Journal of Applied Psychology, 86*, 811–824.

Kuhn, T. S. (1962). *The structure of scientific revolutions.* Chicago, IL: University of Chicago Press.

Lassen, A. H., Waehrens, B. V., & Boer, H. (2009). Re-orienting the corporate entrepreneurial journey: Exploring the role of middle management. *Creativity and Innovation Management, 18*, 16–23.

Laugen, B. T., Boer, H. & Acur, N. (2006). The new product development improvement motives and practices of Miles and Snow's prospectors, analysers and defenders. *New Product Development Improvement Motives and Practices, 15*, 85–95.

Lee, H., & Kelley, D. (2008). Building dynamic capabilities for innovation: An exploratory study of key management practices. *R&D Management, 38*, 155–168.

Li, Y., Tan, C.-H., Teo, H.-H., & Tan, B.C.Y. (2006). Innovative usage of information technology in Singapore organizations: Do CIO characteristics make a difference? *IEEE Transactions on Engineering Management*, *53*, 177–190.

Lonergan, D.C., Scott, G. M., & Mumford, M. D. (2004). Evaluative aspects of creative thought: Effects of idea appraisal and revision standards. *Creativity Research Journal*, *16*, 231–246.

Luria, B. (2008) Climate strength—how leaders form consensus. *The Leadership Quarterly*, *19*, 42–53.

Maier, N. R. (1950). The quality of group discussions as influenced by the discussion leader. *Human Relations*, *3*, 155–174.

——— (1953). An experimental test of the effect of training on discussion leadership. *Human Relations*, *6*, 161–173.

Maier, N. R. F., & Hoffman, L. R. (1965). Acceptance and quality of solutions as related to leaders' attitudes toward disagreement in group problem-solving. *Journal of Applied Behavioral Science*, *1*, 373–386.

Maier, N. R. F., & Solem, A. R. (1962). Improving solutions by turning choice situations and problems. *Personnel Psychology*, *15*, 151–157.

Markham, S. K., & Aiman-Smith, L. (2001). Product champions: Truths, myths, and management. *Research Technology Management*, *44*, 44–50.

Markham, S. K., & Griffin, A. (1998). The breakfast of champions: Associations between champions and product development, environments, practices, and performance. *Journal of Product Innovation Management*, *15*, 436–454.

Marta, S., Leritz, L. E., & Mumford, M. D. (2005). Leadership skills and group performance: Situational demands, behavioral requirements, and planning. *The Leadership Quarterly*, *16*, 97–120.

Mintzberg, H. (1991). Learning 1, planning 0: Reply to Igor Ansoff. *Strategic Management Journal*, *12*, 463–466.

Mossholder, K. W., & Dewhirst, H. D. (1980). The appropriateness of management by objectives for development and research personnel. *Journal of Management*, *6*, 145–156.

Mouly, V. S., & Sankaran, J. K. (1999). The permanent acting leader: Insights from a dying Indian R&D organization. *The Leadership Quarterly*, *10*, 673–652.

Mulec, K., & Roth, J. (2005). Action, reflection, and learning—coaching in order to enhance the performance of drug development project management teams. *R&D Management*, *35*, 483–491.

Mumford, M. D. (2000). Managing creative people: Strategies and tactics for innovation. *Human Resource Management Review*, *10*, 313–351.

——— (2002). Planning in organizations: Complexity, history, and performance. In F. J. Yammarino & F. Dansereau (Eds.), *Research in multi-level issues: The many faces of multi-level issues* (pp. 89–93). Oxford, England: Elsevier.

——— (Ed.). (2006). *Pathways to outstanding leadership: A comparative analysis of charismatic, ideological, and pragmatic leadership*. Mahwah, NJ: Erlbaum.

Mumford, M. D., Bedell-Avers, K. E., & Hunter, S. T. (2008). Planning for innovation: A multi-level perspective. In M. D. Mumford, S. T. Hunter, & K. Bedell-Avers (Eds.), *Innovation in organizations: A multi-level perspective* (pp. 107–154). Oxford, England: Elsevier.

Mumford, M. D., Connelly, M. S., Scott, G. M., Espejo, J., Sohl, L.M., Hunter, S. T., & Bedell, K. E. (2005). Career experiences and scientific performance: A study of social, physical, life, and health sciences. *Creativity Research Journal*, *2*, 105–129.

Mumford, M. D., Eubanks, D.L., & Murphy, S.T. (2007). Creating the conditions for success: Best practices in leading for innovation. In J.A. Conger & R.E. Riggio (Eds.), *The practice of leadership: Developing the next generation of leaders* (pp. 129–149). San Francisco, CA: Jossey-Bass.

Mumford, M. D., Feldman, J. M., Hein, M. B., & Nago, D. J. (2001). Tradeoffs between ideas and structure: Individual versus group performance in creative-problem solving. *Journal of Creative Behavior, 35*, 1–23.

Mumford, M. D., & Gustafson, S. B. (1988). Creativity syndrome: Integration, application, and innovation. *Psychological Bulletin, 103*, 27–43.

Mumford, M. D., & Hunter, S. T. (2005). Innovation in organizations: A multi-level perspective on creativity. In F. Dansereau & F. J. Yammarino (Eds.), *Research in multi-level issues: Volume IV* (pp. 11–74). Oxford, England: Elsevier.

Mumford, M. D., Marks, M. A., Connelly, M. S., Zaccaro, S. T., & Reiter-Palmon, R. (2000). Development of leadership skills: Experiences and timing. *The Leadership Quarterly, 11*, 87–114.

Mumford, M. D., Olsen, K. A., & James, L. R. (1989). Age-related changes in the likelihood of major contributions. *International Journal of Aging and Human Development, 29*, 9–32.

Mumford, M. D., Schultz, R. A., & Osburn, H. K. (2002). Planning in organizations: Performance as a multi-level phenomenon. In F. J. Yammarino & F. Dansereau (Eds.), *Research in multi-level issues: The many faces of multi-level issues* (pp. 3–63). Oxford, England: Elsevier.

Mumford, M. D., Schultz, R. A., & Van Doorn, J. R. (2001). Performance in planning: Processes, requirements, and errors. *Review of General Psychology, 5*, 213–240.

Mumford, M. D., Scott, G. M., Gaddis, B. H., & Strange, J. M. (2002). Leading creative people: Orchestrating expertise and relationships. *The Leadership Quarterly, 13*, 705–750.

Napier, N. K., & Nilsson, M. (2006). The development of creative capabilities in and out of organizations: Three case studies. *Creativity and Innovation Management, 15*, 268–278.

Nemanich L. A., & Keller R. T. (2007). Transformational leadership in an acquisition: A field study of employees. *The Leadership Quarterly, 18*, 49–68.

Nohari, K., & Gulati, D. (1996). Is slack good or bad for innovation? *Academy of Management Journal, 39*, 799–825.

O'Connor, G. C. (1998). Market learning and radical innovation: A cross case comparison of eight radical innovation projects. *Journal of Product Innovation Management, 15*, 151–166.

Ong, C. H., Wan. D., & Chang, S. H. (2003). Factors affecting individual innovation: An examination within a Japanese subsidiary in Singapore. *Technovation, 23*, 617–633.

Osborn, R. N., & Marion, R. (2009). Contextual leadership, transformational leadership and the performance of international innovation seeking alliances. *The Leadership Quarterly, 20*, 191–206.

Osburn, H., & Mumford, M. (2006). Creativity and planning: Training interventions to develop creative problem-solving skills. *Creativity Research Journal, 18*, 173–190.

Pelz, D.C. (1963). Relationships between measures of scientific performance and other variables. In C. W. Taylor & F. Barron (Eds.), *Scientific creativity: Its recognition and development* (pp. 302–310). New York, NY: Wiley.

Ramus, C. A. (2001). Organizational support for employees: Encouraging creative ideas for environmental sustainability. *California Management Review, 43*, 85–105.

Redmond, M. R., Mumford, M. D., & Teach, R. J. (1993). Putting creativity to work: Leader influences on subordinate creativity. *Organizational Behavior and Human Decision Processes, 55*, 120–151.

Rego, A., Sousa, F., Pina e Cunha, M., Correia, A., & Saur-Amaral, I. (2007). Leader self-reported emotional intelligence and perceived employee creativity: An exploratory study. *Creativity and Innovation Management, 16*, 229–238.

Rickards, T., & Moger, S. (2006). Creative leaders: A decade of contributions from *Creativity* and *Innovation Management Journal. Creativity and Management Journal, 15*, 4–18.

Root-Bernstein, R. S., Bernstein, M., & Garnier, H. (1995). Correlations between avocations, scientific style, work habits, and professional impact of scientists. *Creativity Research Journal, 8*, 115–137.

Sawyer, R. K. (2006). *Explaining creativity: The science of human innovation.* New York, NY: Oxford University Press.

Scott, G. M., Leritz, L. E., & Mumford, M. D. (2004). The effectiveness of creativity training: A meta-analysis. *Creativity Research Journal, 16*, 361–388.

Sharma, A. (1999). Central dilemmas of managing innovation in large firms. *California Management Review, 41*, 65–85.

Shipman, A. L., Byrne, C. L., & Mumford, M. D. (in press). Vision formation and forecasting: The effects of forecasting extent, resources, and time frame. *The Leadership Quarterly.*

Shipper, F., & Davy, J. (2002). A model and investigation of managerial skills, employees' attitudes, and managerial performance. *The Leadership Quarterly, 13*, 95–120.

Simonton, D.K. (1990). *Psychology, science, and history: An introduction to historiometry.* New Haven, CT: Yale University Press.

——— (2009). Varieties of (scientific) creativity: A hierarchical model of domain-specific disposition, development, and achievement. *Perspectives on Psychological Science, 4*, 441–452.

Souitaris, V. (2001). External communication determinants of innovation in the context of a newly industrialised country: A comparison of objective and perceptual results from Greece. *Technovation, 21*, 25–34.

Stockstrom, C., & Herstatt, C. (2008). Planning and uncertainty in new product development. *R&D Management, 38*, 480–490.

Sundgren, M., Selart, M., Ingelgard, A., & Bengtson, C. (2005). Dialogue based evaluation as a creative climate indicator: Evidence from the pharmaceutical industry. *Creativity and Innovation Management, 14*, 84–98.

Taggar, S. (2002). Individual creativity and group ability to utilize individual creative resources: A multilevel model. *Academy of Management Journal, 45*, 315–330.

Thamhain, H. J. (2003). Managing innovative R&D teams. *R&D Management, 33*, 297–311.

Thamhain H. J., & Gemmill, G. R. (1974). Influence styles of project managers: Some project performance correlates. *Academy of Management Journal, 17*, 216–224.

Tierney, P., Farmer, S. M., & Graen, G. B. (1999). An examination of leadership and employee creativity: The relevance of traits and relationships. *Personnel Psychology, 52*, 591–620.

Tweney, R. D. (1992). Inventing the field: Michael Faraday and the creative "engineering" of electromagnetic field theory. In R. J. Weber & D. N. Perkins (Eds.), *Inventive Creativity in Technology* (pp. 31–47). New York, NY: Oxford University Press.

Verhaeghe, A. & Kfir, R. (2002). Managing innovation in a knowledge intensive technology organization (KITO). *R&D Management, 32*, 409–417.

Wang, K. Y., & Casimir, G. (2007). How attitudes of leaders may enhance organizational creativity: Evidence from a Chinese study. *Creativity and Innovation Management, 16*, 229–238.

Weisberg, R. W. (2006). Modes of expertise in creative thinking: Evidence from case studies. In K. A. Ericsson, N. Charness, P.J. Feltovich, & R. R. Hoffman (Eds.), *The Cambridge handbook of expertise and expert performance* (pp. 761–787). New York, NY: Cambridge University Press.

West, M. A. (1990). The social psychology of innovation in groups. Innovation and creativity at work. In M. A. West & J. L. Farr (Eds.), *Psychological and organizational strategies* (pp. 309–333). Oxford England: John Wiley & Sons.

West, M. A., Borrill, C. S., Dawson, J. F., Brodbeck, F., Shapiro, D. A., & Haward, B. (2003). Leadership clarity and team innovation in health care. *The Leadership Quarterly, 14,* 393–410.

Wise, G. (1992). Inventions and corporations in the maturing electrical industry. In R. J. Weber & D. N. Perkins (Eds.), *Inventive minds: Creativity in technology* (pp. 291–310). New York, NY: Oxford University Press.

Xiao, Y., Milgram, P., & Doyle, D.J. (1997). Planning behavior and its functional role in interactions with complex systems. *IEEE Transactions on Systems, Man, and Cybernetics, 27,* 313–325.

Yukl, G. (2010). *Leadership in organizations* (7th ed.). Englewood Cliffs, NJ: Prentice Hall.

Yukl, G., & Tracey, B. (1992). Consequences of influence tactics used with subordinates, peers, and superiors. *Journal of Applied Psychology, 77,* 525–535.

Zuckerman, H. (1977). *The scientific elite.* New York, NY: Free Press.

3 What Connects Leadership and Creativity?

The Mechanisms Through Which Leaders May Influence Follower and Team Creativity

Leif Denti and Sven Hemlin

Despite the enormous interest in leadership generally, little attention has been paid to the effect of leadership on creativity (Mumford, Scott, Gaddis, & Strange, 2002). One reason may be the "romantic conception of the creative act" that idealizes creativity as mostly an individual trait, more likely to be repressed than stimulated by supervision (Mumford et al., 2002, p. 706). However, as we demonstrate in this chapter, empirical research points in another direction. This research shows that leaders who construct suitable environments for creativity are essential to the promotion of creativity among those they supervise (Hemlin, Allwood, & Martin, 2008).

A concept related to creativity is innovation. The two concepts are sometimes used interchangeably in the leadership and other management literature (Basadur, 2004; Csikszentmihalyi, 1999). However, a distinction between the two is also made. A common view is that creativity relates to the generation of ideas, while innovation relates to the implementation of ideas that lead to new and improved business products or processes (Anderson, De Dreu, & Nijstad, 2004; Scott & Bruce, 1994).

Basing our literature review on this distinction, in this chapter we focus on the relationship between leadership and creativity in work life. Therefore, (a) we look at the mechanisms influential leaders use to promote follower and team creativity, and (b) we review only the creativity literature and exclude the innovation literature.[1] Our goal is to learn which contexts and mechanisms are important as leaders encourage creativity among their followers and teams.

We believe that leaders are crucial for the creative performance of followers and teams because, to a great extent, leaders create the environments that stimulate and enhance creativity (Hemlin et al., 2008; Shalley & Gilson, 2004). Much of the research describes leader actions that contribute to the construction of contexts and opportunities that such environments require. Successful leaders encourage intrinsic motivation (Avolio, Bass, & Jung, 1999); facilitate problem-solving (Hemlin, 2006a; Puccio, Murdock, & Mance, 2007; Tierney, Farmer, & Graen, 1999); support the generation, structuring, and promotion of ideas (Mumford et al., 2002); foster a positive team climate (Anderson & West, 1998), and establish and maintain high-quality work relationships with team members (Tierney et al., 1999). Thus, we argue that leaders promote creativity in their followers and teams indirectly through such mechanisms.

In this chapter we examine these mechanisms in order to learn *when* and *how* leaders influence creativity. However, we exclude typical tasks associated with a leader's organizational role (e.g., defining missions, setting goals, and selling projects) that may also influence creativity. Other researchers (e.g., Mumford et al., 2002; Shalley & Gilson, 2004) have already investigated those tasks and their influence on creativity (see also Mumford, Peterson, & Robledo, this volume).

This chapter reviews 22 empirical studies that were published in high-impact journals in the last 30 years. All the studies deal with the relationship between leadership and creativity at the individual and/or team level in the workplace. Although various scholars have addressed this relationship (e.g., Mumford et al., 2002; Oke, Munshi, & Walumbwa, 2009; Shalley & Gilson, 2004), there are few systematic compilations and analyses of the empirical research. Our review takes a broader view of this research than, for example, Elkins and Keller (2003) who looked only at the effects of leadership on various outcomes in research and development (R&D). However, our review takes a narrower view than do Mumford et al. (2002), who reviewed both the creativity and innovation literature on leadership. We extend the analyses of Ford (1996) and Woodman, Sawyer, and Griffin (1993) by reviewing the more recent literature. In addition, we examine leadership and creativity at the individual level and at the team level in organizations settings other than R&D.

Other researchers have pointed to the importance of the creative context and to the factors that interact with creativity (Bass & Riggio, 2006; Mumford et al., 2002; Shalley & Gilson, 2004). In this chapter we analyze the factors that moderate or mediate the effect of leadership on creativity. By examining the moderating factors, we may learn *when* leadership relates to creativity (i.e., the circumstances, situations, and human interactions in this relationship). By examining the mediating factors, we may learn *how* leadership relates to creativity (i.e., the influence of leadership at both the individual level and the team level).

PROCEDURE

We performed our literature search in several steps. We began with a computer search for journal articles using the following Internet databases: PsycINFO, PsycARTICLES, Social Services Abstracts, Sociological Abstracts, IBSS (International Bibliography of the Social Sciences), ASSIA (Applied Social Sciences Index and Abstracts), Business Source Premier, Econlit, and Regional Business News. We applied broad range descriptors using the main keywords *leadership* and *creativity.* Although the focus of

Table 3.1 Studies Reviewed in This Chapter

Study	Sample	Leadership measure	Creativity measure	Results
Redmond, Mumford, and Teach (1993)	96 university students, experimental study	Leader behavior designed to increase (1) self-efficacy, (2) learning and performance	Originality and quality of generated products	Leadership aimed at increasing problem construction and self-efficacy led to higher quality and novelty of the generated products.
Oldham and Cummings (1996)	171 and 53 employees from two manufacturing facilities	Supportive or controlling supervision	(1) Employee creativity, supervisor rated, (2) patent disclosures, (3) contributions to a suggestion program	Supportive and noncontrolling supervision was positively related to creativity. Moderators were individual creative predisposition and job complexity.
Sosik, Avolio, and Kahai (1997) Sosik, Avolio, Kahai, and Jung (1998)	159 undergraduate students, experimental study	Transformational and transactional leader behavior	Idea generation effectiveness, products' originality	Transformational leadership influenced group potency and effectiveness. This relationship was moderated by task interdependence and anonymity.
Fodor and Roffe-Steinrotter (1998)	108 participants, experimental study	Three leadership styles: (1) Rogerian, (2) considerate, (3) structuring	Ratings of creativity and complexity	Rogerian leadership produced the highest ratings of creativity in the produced solutions.
Jung and Avolio (1999)	347 asian and caucasian students, experimental study	Transformational and transactional leader behavior	Quantity and quality of generated ideas	Individualists produced more ideas when working with a transactional leader. Collectivists produced more ideas when working with a transformational leader.

Tierney, Farmer, and Graen (1999)	40 supervisors and 191 employees in a U.S. chemical company	Leader-member exchange (LMX) quality	Employee creativity: (1) supervisor ratings, (2) invention disclosures, (3) research reports.	High-quality LMX relationships were positively related to employee creativity. This relationship was moderated by employee cognitive style.
Jaussi and Dionne (2003)	364 participants in 79 groups, experimental study	Transformational leadership, leader unconventional behavior	Individual creative performance, judges' assessments	Leader role modeling creativity had a positive relationship on the individual level. Transformational leadership was negatively related to creative performance at the group level.
Shin and Zhou (2003)	290 employees in 46 Korean companies	Supervisor transformational leadership	Employee creativity, supervisor rated	Transformational leadership was positively related to employee creativity and moderated by employee conservation. Intrinsic motivation acted as a mediator.
Amabile, Schatzel, Moneta, and Kramer (2004)	238 employees from 26 teams in 7 organizations	Supportive leadership behaviors	Employee creativity, peer rated	Leaders' supportive behaviors were positively related to employees' creative behaviors.
Tierney and Farmer (2004)	140 R&D employees	Supervisor expectations for creativity, supervisor supportive behavior	Employee creativity, supervisor rated	Supervisors' creative expectations and supportive behavior was positively related to employee creativity. This relationship was mediated by employee creative self-efficacy.
Unsworth, Wall, and Carter (2005)	1,083 health service employees	Supervisors' degree of supportive leadership	Employee creativity, self-rated	Leader support was positively related to employee creativity, and leader expectations for creativity mediated this relationship.

(*Continued*)

Table 3.1 (Continued)

Study	Sample	Leadership measure	Creativity measure	Results
Carmeli and Schaubroeck (2007)	155 employees from two Israeli organizations	Perceived leader expectations for creativity	Employee creative work involvement, self-rated	Leaders' expectations of creativity was positively related to employee creative work involvement.
Shin and Zhou (2007)	75 teams from 55 Korean R&D companies	Supervisor transformational leadership	Team creativity, supervisor rated	Transformational leadership moderated the relationship between team heterogeneity and team creativity.
Kratzer, Leenders, and Van Engelen (2008)	39 engineering design teams	Leader position in different flows of communication	Team creativity, self and supervisor rated	Leaders who were too close or too peripheral in the work-flow process hampered team creativity. Leaders facilitate creativity when they stay central in the external information network.
Atwater and Carmeli (2009)	193 employees in Israeli organizations	Leader-member exchange (LMX) quality	Employee creativity, self-rated	High LMX quality was positively related to employee creativity, feelings of energy mediated this relationship.
Choi, Anderson, and Veilette (2009)	123 self-peer pairs in Canadian organizations	Aversive leader behavior and close monitoring by leaders	Employee creativity, peer rated	Aversive leadership was negatively related to employee creativity while close monitoring was positively related.
Hirst, Van Dick, and Van Knippenberg (2009)	115 supervisor employee pairs in R&D	Leader inspirational motivation and leader team prototypicality	Employee creativity, supervisor rated	When leaders were high in inspirational motivation and team prototypicality, the relationship between member team identification and creative effort was highest. This was positively related to creative performance.

Gong, Huang, and Farh (2009)	277 insurance agents in Taiwan	Supervisor transformational leadership	Employee creativity, supervisor rated	Transformational leadership and employee learning orientation were positively related to employee creativity. These relationships were mediated by employee creative self-efficacy.
Gumusluoglu and Ilsev (2009)	163 employees in 43 Turkish organizations	Supervisor transformational leadership	Employee creativity, supervisor rated	Transformational leadership was positively related to employee creativity, and this relationship was mediated by psychological empowerment.
Wang and Rode (2010)	212 employees in 71 work groups in 55 US organizations	Supervisor transformational leadership	Employee creativity, supervisor rated	Transformational leadership was not significantly related with employee creativity. A three-way interaction of transformational leadership, employee identification with leader, and innovative climate was found to be related to employee creativity.
Zhang and Bartol (2010)	367 employees and 164 supervisors	Empowering leadership	Employee creativity, supervisor rated	Empowering leadership was positively related to employee creativity through the mediating variables psychological empowerment, intrinsic motivation, and creative process engagement.

our literature review was creativity, we also used *innovation* as a key word because the terms *creativity* and *innovation* are sometimes used interchangeably in the literature. (We distinguished between the terms by following the definitions given earlier and by looking at the empirical data used in the studies).

After assembling a list of candidate articles, we analyzed each article's abstract. Our aim was to identify those articles that (a) were based on empirical studies and (b) treated leadership as an independent variable and creativity or innovation as a dependent variable. We eliminated unpublished articles and dissertations from further consideration because they had not been peer-reviewed. At this point, we had 99 articles. We then applied our three selection criteria. First, we eliminated articles published in or before 1980 as we were looking for studies that (a) used advanced methodologies capable of mediator and moderator analyses and (b) reflected leaders' influence on followers with contemporary work attitudes and values. Second, we eliminated articles that had not been published and indexed in the Web of Science. Third, we eliminated articles that had not been published in journals with a journal impact factor of >1.0 (the ratio of citations over citable articles in the journal during the preceding 2-year period).[2] We applied the second and third criteria in order to identify high quality articles published by prestigious journals. Following this elimination process, we coded the remaining articles as independent (i.e., leadership) measures, dependent (i.e., innovation or creativity) measures, and results. Finally, we excluded innovation articles that did not measure creativity (i.e., for our purposes, the study had to measure some aspect of ideation).[3] The results of this article selection process are shown in Table 3.1.

RESULTS

In our final sample of 22 studies on leadership and creativity, 8 studies measured transformational/transactional leadership, 2 studies measured leader–member exchange (LMX), and 12 studies measured other leadership traits or behaviors. In the measurement of dependent variables, 17 studies measured creativity at the individual level, and 5 studies measured creativity at the team level (see Table 3.1 for specification of the independent and dependent measures). Thus, the majority of those leader and creativity studies follow in the tradition that Feist (2006) describes in which the focus is on individual creativity rather than on team creativity.

LEADERSHIP BEHAVIOR

In this section, we discuss the results of our analysis of leadership and creativity at the individual and team levels.

Leader Support

Employees engaged in creative work often encounter poorly defined and multifaceted problems. When leaders, using their expertise, support employees, the likelihood of solving such problems increases (Reiter-Palmon & Ilies, 2004). In a longitudinal study of 238 knowledge workers in 26 teams, Amabile, Schatzel, Moneta, and Kramer (2004), using both quantitative and qualitative methodological approaches, found that leaders who support employees on a day-to-day basis positively influence employee creativity. Employees feel their leaders are supportive when leaders recognize their good work, support them emotionally, involve them in important decisions, and monitor their progress fairly. Less supportive leaders give employees ambiguous task assignments, fail to resolve important problems, and fail to monitor progress adequately. These findings are consistent with other studies that measure leader support (e.g., Amabile, Conti, Coon, Lazenby, & Herron, 1996; Fodor & Roffe-Steinrotter, 1998; George & Zhou, 2001; Krause, 2004). More recently, Choi, Anderson, and Veilette (2009) took a rarely used approach in their study of creativity and innovation by examining negative predictors of these outcomes, in particular of leaders' negative behaviors (i.e., intimidating followers and dispensing punishment). These aversive leadership behaviors were negatively related to follower creativity. Choi et al. also found that when leaders engaged in aversive behaviors and also monitored their followers too closely, creativity was further impaired. In summary, leaders who exhibit continuous support, but do not closely monitor behaviors, positively influence their followers' creativity.

Creative Expectations

The Pygmalion Effect in work settings (Eden, 1984) refers to the idea that leaders' positive expectations of followers can enhance their performance. Tierney and Farmer (2004), in their examination of the Pygmalion effect in relationship to creativity, found that employees tend to perceive leaders with high (although not overly high) expectations for creative performance as more supportive of their (the employees') creativity. As a result, the employees' creative performance was enhanced. Unsworth, Wall, and Carter (2005) obtained similar results in their survey of 1,083 health-care professionals in which they found that creative requirement (i.e., the expectation of employee creativity) was a determinant of employee creativity. Moreover, leaders' expectations may have a greater influence on employees' self-expectations than do the external expectations of customers and family (Carmeli & Schaubroeck, 2007). We propose that leaders who have expectations of creativity positively influence their followers' creativity.

Role Modeling

Jaussi and Dionne (2003) studied the effect of leaders' unconventional behavior on creativity. They used group exercises in their experiments (e.g., printing instructions on T-shirts). Their study revealed that such behavior positively influenced individual and group creativity. This study suggests that the leader can act as a role model for creative performance. Unconventional behavior by the leader can stimulate similar behavior in followers.

LEADERSHIP STYLE

Transformational Leadership

A number of studies apply transformational leadership theory (e.g., Bass, 1985) in their examination of the link between leadership and creativity. Dvir, Eden, Avolio, and Shamir (2002, p. 735) describe transformational leadership as leadership that exerts an influence by "broadening and elevating followers' goals and providing them with confidence to perform beyond the expectations specified in the implicit or exchange agreement".

Transactional leadership, unlike transformational leadership, is a leadership style based on a contractual exchange of rewards and corrective actions between leaders and followers (Avolio et al., 1999). It has been proposed that transformational leaders enhance individual creativity and organizational innovation (Bass & Riggio, 2006), for example, by championing creative ideas and by heightening followers' motivation (Mumford et al., 2002). Similarly, in their study of 46 R&D companies, Shin and Zhou (2003) found that transformational leadership positively relates to employee creativity.

Some researchers have studied the link between transformational leadership and creativity by taking experimental approaches. In a two-session, longitudinal experimental study with 36 student work groups (using computer-supported group decision systems), Sosik, Avolio, Kahai, and Jung (1998) trained leaders to exhibit high- and low-transformational leadership behaviors in two conditions, respectively. The high-transformational-leadership session groups produced more unique ideas and exhibited greater imagination, innovation, and value-added capability than did the low-transformational-leadership session groups. Redmond, Mumford, and Teach (1993) obtained similar results in their study of leader behavior and subordinate creativity.

However, the evidence on the effect of transformational leadership on creativity is not conclusive. For example, Jaussi and Dionne (2003) reported a negative relationship between transformational leadership and team creativity. They concluded that transformational leadership, if it overemphasizes the positive attributes of groups, may cause groups to ignore other options and strategies in problem-solving processes. P. Wang and Rode (2010) were unable to establish a relationship between transformational leadership and follower creativity in their multilevel modeling analysis. They identified a three-way interaction among transformational leadership, follower identification with the leader, and innovative climate. The innovative climate strengthened the relationship between transformational leadership and follower creativity only when follower identification was high (i.e., when followers were more committed and sensitive to leaders' expectations). The findings of these two studies point to the importance of examining the factors that mediate and moderate transformational leadership and creativity.

LEADER RELATIONSHIP–BASED APPROACHES

Leader–Member Exchange

Leader–membership exchange (LMX) theory is a relationship-based approach to leadership that focuses on the dyadic relationship between leaders and followers (Gerstner & Day, 1997; Graen & Uhl-Bien, 1995). These relationships are differentiated in a team or unit because the leader develops a unique relationship with each follower. In low-quality LMX relationships, based primarily on employment contracts, the leader–follower interaction is formal and impersonal. In high-quality LMX relationships, leaders and followers exchange mutual trust, respect, liking, and influence beyond the terms and conditions of the employment contracts. It has been suggested that the quality of the LMX relationship between a leader and a follower relates to creativity. Leaders in high-quality LMX relationships may provide more opportunities for follower autonomy (Liden & Maslyn, 1998), leading to creativity (Amabile et al., 1996). Mumford and Gustafson (1988) proposed that high-quality LMX relationships influence follower creativity because of the heightened sense of advocacy and trust (i.e., in such relationships, leaders may provide followers with increased support and autonomy).

However, there are only a few empirical studies on LMX and creativity. One of them is Tierney et al.'s (1999) study of 191 employees and 40 supervisors in the R&D areas of large chemical corporations. They found that the quality of the LMX relationship relates positively to leader-rated follower creativity, in particular as measured by the number of invention disclosure forms. Similarly, in a longitudinal study, Atwater and Carmeli (2009) found that a positive relationship exists between employees' perceptions of high-quality LMX relationships and their involvement in creative tasks. Based on these findings high-quality LMX relationships may be positively associated with creative performance.

FACTORS THAT CONNECT LEADERSHIP WITH FOLLOWER CREATIVITY

Several researchers have emphasized the need for an increased understanding of the factors that either moderate or mediate the relationship between leaders and various outcomes (Hunt & Conger, 1999; Yukl, 1999). By moderating factors, we mean the factors that either strengthen or weaken the relationship between two other factors. By mediating factors, we mean the factors that act as an intermediary step between two other factors. In the examination of mediating factors, our expectation is that we will gain a better understanding of how leadership behavior influences creativity. In addition to our recognition of the importance of moderating and mediating factors, we acknowledge the importance of multiple levels of analysis where organizational processes are nested in different levels (Drazin, Glynn, & Kazanijan, 1999; Ford, 1996; Hemlin et al., 2008). For example, the effect of leaders on followers may be contingent on the contextual climate of the team and the culture of the organization (Hunter, Bedell, & Mumford, 2007).

MEDIATING FACTORS

Self-Efficacy

Self-efficacy is the perception of effectiveness in a particular domain. In an experimental study, Redmond et al. (1993) manipulated participants' feelings of self-efficacy by asking leaders to tell participants whether their scores in a pretest were 'well above average' or 'average'. Those participants who were told they performed 'well above average' exhibited greater confidence and greater creativity in solving the problems that were subsequently posed. In Tierney and Farmer's (2004) study, leaders' expectations also influenced employees' self-efficacy perceptions that in turn were related to employees' creativity. Ford (1996) suggests that high self-efficacy may contribute to increased motivation. Thomas and Velthouse (1990) link high self-efficacy to a greater eagerness to pursue unique ideas and to more effective use of cognitive resources. Thus, individual self-efficacy can be considered a mediating factor between leadership and creativity.

At the team level, Sosik, Avolio, and Karhar (1997) studied leadership influence on team perceptions of group potency, a concept that refers to team members' shared beliefs that the team can be effective. In a longitudinal experimental study, they trained two leaders to exhibit transformational leadership and transactional leadership. The leaders then interacted with teams that were tasked with generating ideas in separate conditions. As the authors hypothesized, the results showed that transformational leadership influences potency perceptions and hence team creativity. The effect on potency perceptions and creativity was lower in the transactional leadership condition.

Intrinsic Motivation

Amabile (1996) proposed that individual intrinsic motivation is a pre-condition for creativity since creative people are typically more stimulated by the challenge of solving new or difficult problems than by external rewards associated with their solution. Shin and Zhou (2003) found that intrinsic motivation is significantly related to supervisor-rated creativity and, in addition, acts as a mediating variable between transformational leadership and creativity. In research related to intrinsic motivation, Atwater and Carmeli (2009, p. 265) studied the effect of high-quality LMX relationships on employee energy levels (i.e., the extent to which a person feels 'capable and eager to engage in a particular behavior or undertake a task'). They found that energy levels mediated the relationship between LMX and creative work. They suggested that positive and high-quality LMX relationships may increase employees' energy by strengthening their sense of belonging, their competence, and their autonomy. Therefore, we propose that individual intrinsic motivation is a mediating factor between leadership and creativity.

Psychological Empowerment

Two recent studies show that leaders may influence employee creativity by enabling psychological empowerment. When individuals are empowered, they experience a heightened sense of meaning, competence, and determination (Spreitzer, 1995). In their measurement of psychological empowerment in relation to transformational leadership, Gumusluoglu and Ilsev (2009) found that psychological empowerment mediates between transformational leadership and creativity. Zhang and Bartol (2010) measured the concept of 'empowering leadership' (e.g., sharing influence, providing autonomy, expressing confidence). They found such leadership was positively related to creativity and was mediated by psychological empowerment. Interestingly, Jung et al. (2003) found that transformational leadership was positively related to psychological empowerment; however, they also found that psychological empowerment was negatively related to organizational innovation as measured by R&D expenditures and the number of patents obtained. Because we distinguish creativity (which is only a part of the innovation process) from innovation, these mixed findings from the research call for further investigation of psychological empowerment as it relates to creativity.

MODERATING FACTORS

Conservation

Shin and Zhou (2003) confirmed their hypothesis that an individual's degree of conservation (i.e., the individual's propensity to conform to tradition and to maintain the status quo) moderated the relationship between transformational leadership and individual creativity. A high degree of conservation strengthens the relationship between transformational leadership and creativity, and vice versa. Shin and Zhou explained this result by the fact that individuals with higher degrees of conservation may be more sensitive to stimulation from their leaders; such individuals are more likely to respect superordinate–subordinate hierarchical relationships and to act in accordance with their subordinate role. Another explanation may be that individuals who have a lower degree of conservation may already be the creative drivers in an organization. Thus, the conservation of individuals is a moderating factor in the relationship between leadership and follower creativity such that the relationship is stronger when followers exhibit higher degrees of conservation.

Cognitive Ability and Style

It is proposed that certain styles of cognitive problem solving may influence creativity (Woodman et al., 1993). For example, divergent thinking—the combination of information from diverse sources—generally facilitates creativity (Barron & Harrington, 1981). The way in which an individual approaches a problem may also moderate the effects of leadership on innovation. Tierney et al. (1999) examined followers' cognitive problem-solving styles using Kirton's Adaption-Innovation Instrument (KAI).[4] Their study showed that, according to supervisor ratings and certain objective criteria, LMX quality only affects follower creativity if the follower has an adaptive cognitive problem-solving style. Thus, cognitive style may be a moderating factor between leadership and creativity.

Team Heterogeneity

Researchers have proposed that heterogeneity in team members' skill sets, knowledge, and cognitive problem-solving styles affect team innovation because of the divergent thinking that is stimulated by the team members (Hülsheger, Anderson, & Salgado, 2009; Keller, 2001). Shin and Zhou (2007) found that heterogeneity in education specialization (e.g., different perspectives, knowledge, and skills) positively relates to team creativity if leaders adapt their styles to match team members' requirements. The fact that team heterogeneity is positively related to creativity when transformational leadership is high suggests that leaders with a transformational leadership style may encourage a team to use its members' diverse skill sets. Shin and Zhou's study proposes that when team heterogeneity is high, the prerequisites for divergent thinking are more or less present; however, transformational or participative leadership may be critical to facilitating team creativity. When team heterogeneity is low, the leader may need to take a more direct approach in promoting team reflection, for example, by encouraging disagreement, questions, and discussion. It is clear that team heterogeneity is a moderating factor in leadership for creativity such that more directive leadership is required when heterogeneity among team members is low. Too much heterogeneity in teams may create unwanted tension that stifles the creative process (Hemlin et al., 2008; Reiter-Palmon, de Vreede, & de Vreede, this book).

Creative Climate

P. Wang and Rode (2010) used a multilevel approach (hierarchical linear modeling [HLM]) in their study in which they hypothesized that transformational leadership is positively related to employee creativity. They found that this relationship holds only when employees identified strongly with their leaders and when the level of creative climate was high. They described the creative climate as "the extent to which the organization supports innovative change and the pursuit of new ideas" (p. 1109). Although earlier studies (e.g., Ekvall, 1996; Ekvall & Ryhammar, 1999) have positively related different aspects of team climate to employee creativity, P. Wang and Rode found that climate has a moderating effect on the relationship between leadership and creativity. However, some research in the field of innovation has found that team climate mediates between leadership and innovation (e.g., West et al. 2003). The question then is whether the leader influences team climate or is influenced by it.

CONCLUSION

The Leadership and Creativity Link

The major conclusion from this chapter reviewing team leadership and creativity is that leaders who want to stimulate creativity in their team members should adapt their behavior to their teams' creative potential. The empirical studies that we list in Table 3.1 provide evidence that leaders who are role models for creative performance provide cognitive and emotional support to their followers. In addition, leaders may stimulate creativity among their followers by showing that they expect creativity from followers; such expectations increase their followers' sense of self-efficacy (i.e., their belief in their own creative abilities).

The application of the theory of transformational leadership creativity studies yields mixed results as far as the direct relationship between leader and follower creativity. However, transformational leaders may stimulate creativity by the two main mediating mechanisms: (a) intrinsic motivation and (b) individual and team self-efficacy. The findings are, however, mixed regarding the mediating role of psychological empowerment.

This review also suggests that while leaders may stimulate creativity among followers who are not naturally creative, leaders have limited influence on followers who have demonstrated creativity. For instance, Shin and Zhou (2003) show that the relationship between transformational leadership and creativity was low for individuals with a low degree of conservation, (i.e., individuals who challenged the status quo and resisted conformity with existing norms and regulations). In addition, Tierney el al. (1999) argued that there is no relationship between LMX and follower creativity among followers who have innovative cognitive styles.

Flexible leaders who support followers' creativity act as role models for creativity, set expectations for creativity, and recognize the importance of followers' creative abilities, self-efficacy, and intrinsic motivation. These leaders may be most likely to stimulate followers' creative performance (see also Verbree, van der Weijden, & van den Besselar, ch. 5, this volume). We believe that both transformational and LMX leadership theories address such flexible leadership. However, flexible leadership alone is insufficient for the promotion of the more creative work place. Therefore, the second major conclusion of this review is that the promotion of creative performance depends on a number of factors. These factors, which are influential both at the individual level and at the team level, may either strengthen or weaken the connection between leadership and creativity. We have seen that the relationship between transformational leadership and creativity appears strongest when followers have a high degree of conservation, when the team is heterogeneous, and when there is a supportive team climate.

Table 3.2 Leader Behaviors That Stimulate Creativity With Followers Directly or by Context Factors on the Individual and Team Levels

Level	Leader behavior	Direct influence	Mediating factors	Moderating factors
Individual	Support	yes		
	Role modeling	yes		
	Expectations of creativity	yes	Creative requirements Individual self-efficacy	
	Transformational		Individual self-efficacy	Conservation
	leadership		Intrinsic motivation, identification with leader, psychological empowerment	Creative climate
	LMX	yes		Autonomy
				Advocacy
				Trust
				Cognitive style
Team	Support	yes		
	Role modeling	yes		
	Transformational leadership	yes	Team self-efficacy	Heterogeneity

Studies using LMX theory provide evidence that leaders may stimulate individual creativity through a high-quality leader–follower work relationship. In such a relationship, leaders may allow followers greater discretion, autonomy, and influence (Liden & Maslyn, 1998; Scott & Bruce, 1994). It should be noted also that there is an association between transformational leadership and LMX (Bass, 2008). The union of the two theories on leadership behavior theories (consistent with our "flexible leadership" idea) may be the best way to promote creativity among followers taking into consideration factors in the work context. See Table 3.2 for an overview of the ideas presented in this literature review.

Future Implications for Creative Leadership

Hemlin et al. (2008) argue that a creative knowledge environment (CKE) should be created and supported in organizations that wish to develop their employees' creativity in environments where individual and team creative activities are conducted. In such environments, it is clear that the work design, as well as the social and organizational characteristics at the micro- and meso-levels, can have a crucial influence on the creative processes. To a great extent, because the creation of a CKE is the leader's responsibility, it is essential that leaders identify the specific environmental factors that encourage creativity (see also Puccio, Mance, & Zacko-Smith, this volume, on creative leadership).

We propose six steps that leaders may take to create and support a CKE. First, top management, including its teams, should establish and promote an organization policy for creative performance. Leaders at different levels in the organization should communicate this policy to followers, making clear that creative performance will be rewarded (Mumford & Gustafson, 1988). Second, careful consideration should be given to the team composition in CKEs. Both the team members' creativity and their heterogeneity (Reiter-Palmon & Illies, 2004) should be selection factors in composing the team. However, care should be taken that the team is neither too heterogeneous nor too homogeneous because dissension among team members may arise in either instance (see also Reiter-Palmon et al., this volume). Third, leaders should promote a team climate of psychological safety, respect and even joy. In such climates, followers may express dissenting opinions even in an atmosphere of some tension that do not lead to personal conflicts (Hemlin et al., 2008). Such team climates have been found to be crucial in establishing a fruitful, psychosocial atmosphere that enhances creative processes. (Amabile, 1996; Anderson & West, 1998; Ekvall, 1996). Fourth, followers should be given the necessary autonomy that leads to idea generation and creative problem solving (Pelz & Andrews, 1966; Verbree et al., ch. 5, this volume). Fifth, reasonable time limits should be set for solving problems. In work phases when results are expected, setting deadlines can lead to the selection of the best solutions (Basadur, 2004; Mumford et al., 2002; Puccio et al., 2007). Last and sixth, team leaders should closely engage in the evaluation of creative processes by providing feedback to followers on their ideas and by suggesting better solutions to problems (Hemlin & Olsson, 2011; Mumford et al., 2002).

Limitations of the Study

We note four limitations of our literature review. First, our review is confined to creative processes and products and does not directly address innovation, which is a desired goal in many organizations. We deliberately excluded innovation from our review because we believe there are additional and different processes that drive innovations in organizations. For example, intrusive commercial considerations may suppress creativity in situations where imaginative and divergent thinking is required. To the contrary, such commercial considerations may encourage innovations. However, without creativity, there would be few, if any, innovations because the first steps in innovative development require creativity.

Second, we selected only high-quality journal articles for our review of the leader–creativity relationship. We recognize that by choosing only journal articles we may have ignored some high-quality dissertations and books on the same subject. However, we assume that these studies will eventually be published in quality journals and will be covered in subsequent literature reviews.

Third, we limited our review of the literature to journal articles published after 1980 in order to focus only on studies with methodologies subject to mediator–moderator analysis.

Fourth, the conclusions drawn from studies of this review must be evaluated in terms of the methodologies used by their authors. Most of these studies used followers' and leaders' self-reports to measure leadership, creativity and the associated moderating and mediating factors. Furthermore, many of the studies used cross-sectional designs, which cannot be used to establish causal relationships, thus limiting their conclusions.

Suggestions for Future Research on Leadership and Creativity

Stages of the Creative Process

Many studies in this review examine leadership in relation to the creativity of the individual (e.g., Tierney et al., 1999) or to the creative product (e.g., Redmond et al., 1993). Further research is, however, needed on the effect of leadership on the creative process—that is, on the different stages in creative problem solving. Specific problem definition and construction of these stages (Reiter-Palmon & Ilies, 2004) as well as the generation, evaluation, and promotion of ideas and solutions (Basadur, 2004; Lubart, 2001; Puccio et al., 2007) are all areas for additional research. Although leadership plays a vital role in providing a structure for the advancement of creative performance, little is known about how leaders facilitate the cognitive and emotional processes involved in such work (Amabile et al., 2004; Hemlin & Olsson, 2011; Mumford et al, 2002; Reiter-Palmon & Ilies, 2004). In the earlier stages of creative problem solving, leaders may need to adopt an unconventional leadership style that encourages broad and explorative integration of knowledge and ideas. In the later stages, when a solution is within reach, a more traditional leadership style that has a stronger, more forward-moving focus may be more useful. Future research should therefore examine the different aspects of leadership, its context, and the different stages of the creative process.

Hard Measures of Creativity

Researchers have used both 'soft' and 'hard' measures to measure creativity. We define soft measures as those that measure psychological phenomena (e.g., attitudes and judgments about creativity) and hard measures as those that measure objective phenomena (e.g., patents and new products, or scientific publications on creativity). Academic prizes exemplify the difference between the two measures. The number of academic prizes is a hard measure of creativity whereas the subjective evaluations of their quality, novelty, and usefulness are soft measures. We argue that researchers need to use more hard measures of creativity for three reasons. First, hard measures of creativity, which reflect large areas of the creative process, can be used as proxies to evaluate creative problem-solving, creative performance, and technological success (Archibugi & Pianta, 1992, 1996; Fai, 2007; Organisation for Economic Co-operation and Development, 1994; Pelz & Andrews, 1966). Second, hard measures are objectively determined unlike soft measures that are more subject to bias. Third, as both soft and hard measures refer to the same constructs of creativity, hard measures can be used to validate soft measures.

Negative Leadership Influence

In the articles of this review, studies by Amabile et al. (2004), Choi et al. (2009), and Hemlin et al. (2008) address the possible negative effects of leadership on creativity. These authors propose that certain actions by leaders may stifle creativity, for example, by failing to show support, poor explanations and non-constructive feedback (Amabile et al., 2004; Hemlin et al., 2008). In addition, it is likely that many of the creativity-promoting behaviors or traits that research has shown to be effective may, if taken to a higher degree, become obstacles to creativity. For example, we expect that individuals with extremely high self-efficacy may be less creative because they are not easily influenced and may indeed consider themselves exceptional, despite others' opinions of them. Consequently, there is a need for more research into the circumstances in which leadership is detrimental to creativity. For example, to what extent are leaders limited in their influence on creative processes? And which types of leadership inhibit followers' creativity? Further, research should be directed to identify boundary conditions where certain leadership behaviors, taken too far, are detrimental to creativity. An example of this is when leaders monitor their followers too closely (Choi et al., 2009).

Mechanisms Involved in the Leader–Follower Relationship

There is a gap in our knowledge as to how other moderating or mediating factors (in addition to those discussed in this review) affect leadership and creativity. Researchers have proposed various requirements for creativity: the need for autonomy (Feist & Gorman, 1998), the need for a sense of achievement (Barron & Harrington, 1981), proactivity (Seibert, Kraimer, & Crant, 2001), locus of control (Woodman et al., 1993), and individual goal setting (Hemlin, 2006b; Shalley, 1995). In addition, organizational reward systems have been proposed as a linkage between creativity and leadership (Eisenberger & Armeli, 1997; Shalley & Gilson, 2004). Therefore, leaders at different levels in the organization should understand that creating and supporting a CKE means taking a number of these context factors into consideration if they wish to encourage creative performance. However, some of the factors that depend on individual and group traits must be considered in assigning people to teams.

There is much to be learned about leader and follower creativity in teams from the recent research. Researchers engaged in this area can benefit from a careful reading of the 22 articles we have reviewed as well as the other references we have cited.

NOTES

1. In another paper we review the mechanisms influential leaders use to promote innovation (Denti & Hemlin, 2012).
2. Journal impact factor (JIF) is calculated as the number of citations to articles published in the previous 2 years divided by the number of articles published in the previous 2 years.
3. We were unable to locate two articles we identified as of interest.
4. KAI (Kirton, 1976) distinguishes between innovators (individuals who seek and integrate information from multiple domains, reframe problems, and devise unconventional solutions) and adaptors (individuals who accept a problem and devise conventional solutions).

REFERENCES

The 22 articles on leadership and creativity in this review are marked with an asterisk.

Amabile, T. M. (1996). *Creativity in context.* Boulder, CO: Westview Press.

Amabile, T. M., Conti, R., Coon, H., Lazenby, J., & Herron, M. (1996). Assessing the work environment for creativity. *Academy of Management Journal, 39*, 1154–1184.

*Amabile, T. M., Schatzel, E. A., Moneta, G. B., & Kramer, S. J. (2004). Leader behaviors and the work environment for creativity: Perceived leader support. *The Leadership Quarterly*, *15*, 5–32.

Anderson, N., De Dreu, C. K. W., & Nijstad, B. A. (2004). The routinization of innovation research: A constructively critical review of the state-of-the-science. *Journal of Organizational Behavior*, *25*, 147–173.

Anderson, N. R., & West, M. A. (1998). Measuring climate for work group innovation: Development and validation of the team climate inventory. *Journal of Organizational Behavior*, *19*, 235–258.

Archibugi, D., & Pianta, M. (1992). *The technological specialization of advanced countries: A report to the ECC on international science and technology activities*. Dordrecht, the Netherlands: Kluwer Academic Publishers.

Archibugi, D., & Pianta, M. (1996). Measuring technological change through patents and innovation surveys. *Technovation*, *16*, 451–468.

*Atwater, L., & Carmeli, A. (2009). Leader-member exchange, feelings of energy, and involvement in creative work. *The Leadership Quarterly*, *20*, 264–275.

Avolio, B. J., Bass, B. M., & Jung, D. I. (1999). Re-examining the components of transformational and transactional leadership using the multifactor leadership questionnaire. *Journal of Occupational and Organizational Psychology*, *72*, 441–462.

Barron, F., & Harrington, D. M. (1981). Creativity, intelligence and personality. *Annual Review of Psychology*, *32*, 439–476.

Basadur, M. (2004). Leading others to think innovatively together: Creative leadership. *The Leadership Quarterly*, *15*, 103–121.

Bass, B. M. (1985). *Leadership and performance beyond expectations*. New York, NY: Free Press.

——— (2008). *The Bass handbook of leadership. Theory, research & managerial implications*. New York, NY: Free Press.

Bass, B. M., & Riggio, R. E. (2006). *Transformational leadership* (2nd ed.). Mahwah, NJ: Erlbaum.

Bohmer, R.M.J., Edmondson, A. C., & Roberto, M. A. (2004). *Columbia's final mission* (Business case No. 9-304-090). Cambridge, MA: Harvard Business School.

*Carmeli, A., & Schaubroeck, J. (2007). The influence of leaders' and other referents' normative expectations on individual involvement in creative work. *The Leadership Quarterly*, *18*, 35–48.

*Choi, J. N., Anderson, T. A., & Veilette, A. (2009). Contextual inhibitors of employee creativity in organizations. The insulating role of creative ability. *Group & Organization Management*, *32*, 330–357.

Csikszentmihalyi, M. (1999). Implications of a system perspective for the study of creativity. . In R.J. Sternberg (Ed.), *Handbook of creativity* (pp. 313–334). New York, NY: Cambridge University Press.

Denti, L., & Hemlin, S. (2012). Leadership and innovation in organizations: A systematic review of factors that mediate or moderate the relationship. *Journal of Innovation Management, 16*, 1–20. doi:10.1142/S1363919612400075

Drazin, R., Glynn, M. A., & Kazanjian, R. K. (1999). Multilevel theorizing about creativity in organizations: A sensemaking perspective. *Academy of Management Review*, *24*, 286–307.

Dvir, T., Eden, D., Avolio, B. J., & Shamir, B. (2002). Impact of transformational leadership on follower development and performance: A field experiment. *Academy of Management Journal*, *45*, 735–744.

Eden, D. (1984). Self-fulfilling prophecy as a management tool: Harnessing Pygmalion. *Academy of Management Review*, *9*, 64–73.

Eisenberger, R., & Armeli, S. (1997). Can salient reward increase creative performance without reducing intrinsic creative interest? *Journal of Personality and Social Psychology*, *72*, 652–663.

Ekvall, G. (1996). Organizational climate for creativity and innovation. *European Journal of Work and Organizational Psychology*, *5*, 105–123.
Ekvall, G., & Ryhammar, L. (1999). The creative climate: Its determinants and effects at a Swedish university. *Creativity Research Journal*, *12*, 303–310.
Elkins, T., & Keller, R.T. (2003). Leadership in research and development organizations: A literature review and conceptual framework. *The Leadership Quarterly*, *14*, 587–606.
Fai, F. (2007). A structural decomposition analysis of technological opportunity, corporate survival, and leadership. *Industrial and Corporate Change*, *16*, 1069–1103.
Feist, G. (2006). *The psychology of science and the origin of the scientific mind*. New Haven, CT: Yale University Press.
Feist, G. J., & Gorman, M. E., (1998). The psychology of science: Review and integration of a nascent discipline. *Review of General Psychology*, *2*, 3–47.
*Fodor, E. M., & Roffe-Steinrotter, D. (1998). Rogerian leadership style and creativity. *Journal of Research in Personality*, *32*, 236–242.
Ford, C. M. (1996). A theory of individual creative action in multiple social domains. *Academy of Management Review*, *21*, 1112–1142.
Gerstner, C. R., & Day, D. V. (1997). Meta-analytic review of leader-member exchange theory: Correlates and construct issues. *Journal of Applied Psychology*, *82*, 827–844.
George, J. M, & Zhou, J. (2001). When openness to experience and conscientiousness are related to creative behaviour: An interactional approach. *Journal of Applied Psychology*, *86*, 513–524.
*Gong, Y., Huang, J-C., & Farh, J-L. (2009). Employee learning orientation, transformational leadership, and employee creativity: The mediating role of employee creative self-efficacy. *Academy of Management Journal*, *52*, 765–778.
Graen, G. B., & Uhl-Bien, M. (1995). Relationship-based approach to leadership: Development of leader-member exchange (LMX) theory of leadership over 25 years: Applying a multi-level multi-domain perspective. *The Leadership Quarterly*, *6*, 219–247.
*Gumusluoglu, L. & Ilsev, A. (2009). Transformational leadership, creativity, and organizational innovation. *Journal of Business Research*, *62*, 461–473.
Hemlin, S. (2006a). Creative knowledge environments for research groups in biotechnology: The influence of leadership and organizational support in universities and business companies. *Scientometrics*, *67*, 121–142.
——— (2006b). Managing creativity in academic research. How could creative action and management be reconciled in research? *Science Studies*, *19*, 3–12.
Hemlin, S., Allwood, C. M., & Martin, B. R. (2008). Creative knowledge environments. *Creativity Research Journal*, *20*, 196–210.
Hemlin, S., & Olsson, L. (2011). Creativity-stimulating leadership: A critical incident study of leaders' influence on creativity in research groups. *Creativity and Innovation Management*, *20*, 49–58.
*Hirst, G., Van Dick, R., & Van Knippenberg, D. (2009). A social identity perspective on leadership and employee creativity. *Journal of Organizational Behavior*, *30*, 963–982.
Hülsheger, U. R., Anderson, N., & Salgado, J. F. (2009). Team-level predictors of innovation at work: A comprehensive meta-analysis spanning three decades of research. *Journal of Applied Psychology*, *94*, 1128–1145.
Hunt, J. G., & Conger, J. A. (1999). From where we sit: An assessment of transformational and charismatic leadership research. *The Leadership Quarterly*, *10*, 335–343.
Hunter, S. T., Bedell, K. E., & Mumford, M. D. (2007). Climate for creativity: A quantitative review. *Creativity Research Journal*, *19*, 69–90.

*Jaussi, K. S., & Dionne, S. D. (2003). Leading for creativity: The role of unconventional leader behavior. *The Leadership Quarterly*, *14*, 475–498.

*Jung, D. I., & Avolio, B. J. (1999). Effects of leadership style and followers' cultural orientation on performance in group and individual task conditions. *Academy of Management Journal*, *42*, 208–218.

Jung, D. I., Chow, C., & Wu, A. (2003). The role of transformational leadership in enhancing organizational innovation: Hypotheses and some preliminary findings. *The Leadership Quarterly*, *14*, 525–544.

Keller, R. T. (2001). Cross-functional project groups in research and new product development: Diversity, communications, job stress, and outcomes. *Academy of Management Journal*, *44*, 547–555.

Kirton, M. (1976). Adaptors and innovators: A description and measure. *Journal of Applied Psychology*, *61*, 622–629.

*Kratzer, J., Leenders, R. T. A. J., & Van Engelen, J. M. L. (2008). The social structure of leadership and creativity in engineering design teams: An empirical analysis. *Journal of Engineering and Technology Management*, *25*, 269–286.

Krause, D. E. (2004). Influence-based leadership as a determinant of the inclination to innovate and of innovation-related behaviors: An empirical investigation. *The Leadership Quarterly*, *15*, 79–102.

Liden, R. C., & Maslyn, J. M. (1998). Multidimensionality of leader–member exchange: An empirical assessment through scale development. *Journal of Management*, *24*, 43–72.

Lubart, T. I. (2001). Models of the creative process: Past, present and future. *Creativity Research Journal*, *13*, 295–308.

Mumford, M. D., & Gustafson, S. B. (1988). Creativity syndrome: Integration, application, and innovation. *Psychological Bulletin*, *103*, 27–43.

Mumford, M. D., Scott, G. M., Gaddis, B., & Strange, J. M. (2002). Leading creative people: Orchestrating expertise and relationships. *The Leadership Quarterly*, *13*, 705–730.

Oke, A., Munshi, N., & Walumba, F. O. (2009). The influence of leadership on innovation processes and activities. *Organizational Dynamics*, *38*, 64–72.

*Oldham, G. R., & Cummings, A. (1996). Employee creativity: Personal and contextual factors at work. *Academy of Management Journal*, *39*, 607–634.

Organisation for Economic Co-operation and Development. (1994). *The measurement of scientific and technological activities. Using patent data as science and technology indicators*. Paris: Author.

Pelz, D., & Andrews, F. (1966). *Scientists in organizations: Productive climates for research and development*. New York, NY: Wiley.

Puccio, G. J., Murdock, M. C., & Mance, M. (2007). *Creative Leadership: Skills that drive change*. Thousand Oaks, CA: Sage.

*Redmond, M. R., Mumford, M. D., & Teach, R. (1993). Putting creativity to work: Effects of leader behavior on subordinate creativity. *Organizational Behavior and Human Decision Processes*, *55*, 120–151.

Reiter-Palmon, R., & Illies, J. J. (2004). Leadership and creativity: Understanding leadership from a creative problem-solving perspective. *The Leadership Quarterly*, *15*, 55–77.

Scott, S. G., & Bruce, R. A. (1994). Determinants of innovative behavior: A path model of individual innovation in the workplace. *Academy of Management Journal*, *37*, 580–607.

Seibert, S. E., Kraimer, M. K., & Crant, J. M. (2001). What do proactive people do? A longitudinal model linking proactive personality and career success. *Personnel Psychology*, *54*, 845–874.

Shalley, C. E. (1995). Effects of coaction, expected evaluation, and goal setting on creativity and productivity. *Academy of Management Journal*, *38*, 483–503.

Shalley, C. E., & Gilson, L. L. (2004). What leaders need to know: A review of social and contextual factors that can foster or hinder creativity. *The Leadership Quarterly*, *15*, 33–53.

*Shin, S. J., & Zhou, J. (2003). Transformational leadership, conservation, and creativity: Evidence from Korea. *Academy of Management Journal*, *46*, 703–714.

*——— (2007). When is educational specialization heterogeneity related to creativity in research and development teams? Transformational leadership as a moderator. *Journal of Applied Psychology*, *92*, 1709–1721.

*Sosik, J. J., Avolio, B. J., & Kahai, S. S. (1997). Effects of leadership style and anonymity on group potency and effectiveness in a group decision support system environment. *Journal of Applied Psychology*, *82*, 89–103.

*Sosik, J. J., Avolio, B. J., Kahai, S. S., & Jung, D. I. (1998). Computer-supported work group potency and effectiveness: The role of transformational leadership, anonymity, and task interdependence. *Computers in Human Behavior*, *14*, 491–511.

Spreitzer, G. M. (1995). Psychological empowerment in the workplace: Dimensions, measurement and validation. *Academy of Management Journal*, *38*, 1442–1465.

Thomas, K. W., & Velthouse, B. A. (1990). Cognitive elements of empowerment: An "interpretive" model of intrinsic task motivation. *Academy of Management Review*, *15*, 666–681.

*Tierney, P., & Farmer, S. M. (2004). The Pygmalion process and employee creativity. *Journal of Management*, *30*, 413–432.

*Tierney, P., Farmer, S. M., & Graen, G. B. (1999). An examination of leadership and employee creativity: The relevance of traits and relationships. *Personnel Psychology*, *52*, 591–620.

*Unsworth, K. L., Wall, T. D., & Carter, A. (2005). Creative requirement: A neglected construct in the study of employee creativity? *Group & Organization Management*, *30*, 541–560.

*Wang, P., & Rode, J. C. (2010). Transformational leadership and follower creativity: The moderating effects of identification with leader and organizational climate. *Human Relations*, *63*, 1105–1128.

West, M. A., Borril, C. S., Dawson, J. F., Brodbeck, F., Shapiro, D. A., & Haward, B. (2003). Leadership clarity and team innovation in health care. *The Leadership Quarterly*, *14*, 393–410.

Woodman, R. W., Sawyer, J. E., & Griffin, R. W. (1993). Toward a theory of organizational creativity. *Academy of Management Review*, *18*, 293–321.

Yukl, G. (1999). An evaluation of conceptual weaknesses in transformational and charismatic leadership theories. *The Leadership Quarterly*, *10*, 285–305.

*Zhang, X., & Bartol, K. M. (2010). Linking empowering leadership and employee creativity: The influence of psychological empowerment, intrinsic motivation, and creative process engagement. *Academy of Management Journal*, *53*, 107–128.

4 Leadership, Innovation, and Technology

The Evolution of the Creative Process

Samuel T. Hunter, Lily Cushenbery, Nicole Ginther, and Joshua Fairchild

A recent survey given to over 12,000 leaders across 2,600 organizations and 74 countries asked upper level executives to describe and outline their anticipated needs for the future (Boatman & Wellins, 2011). Emerging quite clearly from this cross-industry survey was that managing change via the facilitation of creativity and innovation will prove critical to organizational success (Boatman & Wellins, 2011). This point is echoed by several other recent surveys, such as those conducted by the Boston Consulting Group (Andrew, Manget, Michael, Taylor, & Zablit, 2010) who found that across executives, innovation facilitation was listed as a top requirement of organizational leaders. Similarly, a survey conducted by IBM in its *Capitalizing on Complexity* (2010) report also found that the ability to drive innovation was listed as the single most desired leadership quality among the range of management characteristics considered. Perhaps even more significant is that when asked what skills leaders lacked, fostering creativity and innovation was rated as the weakest and in need of the greatest improvement (Boatman & Wellins, 2011). The aggregate of these efforts illustrates a growing trend in the applied world: the increasing need to be more innovative and an inability to meet this need successfully. Being innovative, it seems, may be more challenging than it might appear on the surface.

One of the reasons that it has proved so difficult for organizations to successfully innovate is that creative idea generation and implementation are complex endeavors (Mumford & Hunter, 2005). In fact, innovation is most accurately represented as a range of processes rather than a singular event or entity (Runco, 2008). Thus, to effectively innovate, organizations must perform well on a number of activities ranging from identifying high-potential opportunities to successfully implementing and monitoring creative ideas (Baughman & Mumford, 1995). Taken a step further, being innovative requires success across a host of *interconnected*, and often highly *challenging*, processes. As such, gaining insight into what is required for successful innovation will fundamentally require a discrete consideration of these activities and, more centrally, of what might have an impact on each of them uniquely. Although a sizable body of literature exists on the social, psychological, and cognitive influences of creativity, there is a substantial knowledge gap with regard to more external influences. In an attempt to fill this growing knowledge gap, the aim of the present chapter is to consider the role of an ever-increasing influence on our lives: technology.

Just as the cotton gin changed the world of agriculture (Ramussen, 1962) and the steam engine shaped the production as well as the shipment of goods (Wren, 1994), technology has and will continue to shape the landscape of work. In the current digital age, technology and associated software have a fundamental impact on the way we interact with clients, customers, and coworkers. Smartphones connect us as easily to the Web as to each other via text, chat, or phone. Databases are compiled tracking consumer habits that range from the movies we watch to the products we buy (Liu & Lee, 2010). Few of us know a day that is not touched by digital technology, and even fewer of us conduct our business day free of technological influence.

Central to the present chapter, we contend that technology currently has an impact on much of the creative process and, more important, will do so to an even greater degree in the future. Consider, as an illustration, how rapid prototype machines have given designers a tool to quickly test and evaluate their ideas (Chua, Leong, & Lim, 2003). This increase in testing capability, in turn, allows designers to test multiple versions and iterations of a design in the same time that it might have taken to test a single design ten years ago. Rapid prototyping increases the pool of possible ideas to draw from and significantly increases product development cycles—both of which shape successful innovation. As another illustration, consider the growth of popularity in using a distributed network of individuals as creative problem solvers (i.e., 'crowdsourcing') as a mechanism to harness the creative power of an organization's consumer base. Crowdsourcing can potentially increase the pool of novel ideas a hundredfold (Brabham, 2008; Howe, 2006). These influences are discussed in greater detail later in the chapter, but serve now to simply illustrate that technology is an understudied, yet critical, influence of the creative process. As leaders continue to ask themselves what might be done to improve their organizations' innovative successes, it will be essential to ask what role technology may play in this process.

The framework of the chapter is as follows: First, we outline the creative process as a means to establish a baseline set of activities that organizations must successfully engage in to innovate. With the creative process in place, we can begin to explore emerging technologies that are shaping and having an impact on each of the creative processes. Finally, we turn to the role that leaders play in shaping appropriate use of technology in the innovation process. We conclude with a discussion on the five primary themes emerging from the chapter.

CREATIVE PROCESS MODELS

Early Models

Before turning to our discussion on the early creative process models, it would appear prudent to operationalize what is meant by the terms creativity and innovation, respectively. Creativity is defined in this chapter as the generation of ideas that are both (a) novel and (b) useful or serve a purpose (Amabile, Conti, Coon, Lazenby, & Herron, 1996; Mumford & Gustafson, 1988). Innovation, in contrast, represents the instantiation of these creative ideas (Amabile, Schatzel, Moneta, & Kramer, 2004). In this sense creativity serves as a necessary but not sufficient precondition for innovation. Moreover, within the core definitions of creativity and innovation we get the sense that a *process* underlies their conceptualization. That is, creativity involves coming up with original, useful ideas and innovation takes those ideas to the next step of development. The two concepts are fundamentally linked procedurally.

The Eight-Stage Model

In what might be argued as the most comprehensive creative process model, Mumford and colleagues (Mumford, Baughman, Supinski, & Maher, 1996; Mumford, Supinski, Baughman, Costanza, & Threlfall, 1997; Mumford, Mobley, Uhlman, Reiter-Palmon, & Doares, 1991) developed the eight-stage model of creative problem solving. In his review of creative process models, Lubart (2001, p. 299) refers to the eight-stage framework as a 'subprocess' approach. That is, whereas the Wallas (1926) four-stage approach and others like it (e.g., Amabile, 1996) describe the overarching activities involved in creative idea production, the eight-stage model depicts the full range of processes characterizing creativity and innovation. The activities are composed of (1) problem identification and construction, (2) information gathering, (3) concept selection, (4) conceptual combination, (5) idea generation, (6) idea evaluation, (7) implementation planning, and (8) monitoring. The model has seen considerable support in the literature, demonstrating that skill in these processes predict creative performance over and above general cognitive ability, scholastic achievement, and divergent thinking ability (Mumford, Supinski, et al., 1997). Moreover, each stage of the creative process has been examined extensively across a range of studies (e.g., Mumford & Baughman, 1996; Mumford, Baughman, Maher, Costanza, & Supinski, 1997; Mumford et al., 1997).

Integrating Process Models and Multilevel Frameworks

As originally conceptualized, the eight-stage creative process model focused largely on within-person cognition and behavior. Put another way, the early model was primarily an individual-level conceptualization of the innovation process. Recently, however, scholars have begun integrating the eight-stage model with more multilevel frameworks. Driven by the work of Mumford and Hunter (2005), for example, Hunter and Cushenbery (2011) proposed an integrated model that depicts multilevel influences along the stages of the innovation process. Although Mumford et al.'s (Mumford, Supinski, Baughman, Costanza, & Threlfall, 1997; Mumford, Supinski, Threlfall, &

Baughman, 1996; Mumford et al., 1991) eight-stage model describes the innovation process for an individual, Hunter and Cushenbery's (2011) model incorporates influence from leaders, teams, and organizations. For both models, early stages of creativity consist of the cognitive process of conceiving novel ideas and are largely individually driven activities. In contrast, as ideas become more developed and formalized, they are typically vetted and discussed in project teams. Accordingly, middle-stage activities such as idea generation and evaluation become more group-oriented processes. Finally, as ideas progress to implementation stages later in the innovation process, organization influences become more pronounced.

Before moving any further in the chapter, two points must be made about the proposed model. First, across the range of activities, individual, team, and organization influences will shape the innovative process. Thus, although we suggest that multilevel influences play greater or lesser roles across the entire innovation process, (e.g., organization influences are stronger drivers later in the process) we must fully acknowledge the complex role played by multilevel influences. Second, congruent with the two-stage genoplore (derived from the terms *generate* and *explore*) model of creativity put forth by Finke, Ward, and Smith (1992), the innovation process should not be taken to be a lockstep series of activities. Rather, as ideas are evaluated they may be taken back to the conception stage and reshaped in a continually evolving fashion. This perspective is in contrast to other models of problem solving, such as the garbage can model of decision making, which suggests that solutions can sometimes be created before a problem is identified to solve them (Cohen, March, & Olsen, 1972). A primary tenet of the garbage can model is that the problem-solving process is not sequential when there is a high level of uncertainty. We concur that the problem-solving process is unlikely to be perfectly sequential; however, we believe that there is in fact a basic progression, or set of processes, that is often demonstrated by creative problem solvers. The eight-stage model does not suggest that innovative thinking always occurs in a lockstep model, but instead acts as a general guide. The point must be made that creativity is hardly a linear endeavor and is more accurately depicted as an iterative set of activities.

An iterative, multilevel perspective of the innovation process is important to our discussion here for at least three related reasons. First, as we move forward with our discussion on technology's influence on creativity and innovation, bearing in mind that some forms of technology will shape activities at varying levels of analysis will allow for a more accurate picture of technology and innovation to be presented. That is, it is useful to recognize that some forms of technology will have an impact on team activities such as idea exchange whereas others might influence individual activities such as cognition. Second, and along related lines, failing to acknowledge multilevel influences of creativity can lead to, at best, incomplete depictions of the creative process and, at worst, incorrect characterizations of how the innovation process operates (see Mumford & Hunter, 2005). Third and finally, many forms of technology are *intended* to have an impact on differing levels of analysis and, more precisely, have been expressly designed to do so. For example, although 'cloud computing' programs such as Google Docs are intended to allow documents to be shared and edited by multiple team members, haptic devices such as those used by sculptors and designers are intended to influence the individual designer primarily (Vashisth & Mudur, 2008). As such, it would appear useful to acknowledge the original intent of the technology as well as the adaptations made by individuals, allowing for a more complete picture as to how it might have an impact on the innovation process.

In summary, to understand the role technology might play in shaping innovative outcomes, it is important to acknowledge the process perspectives underlying creativity and innovation. In doing so, we can more clearly see the discrete activities that are ripe for influence by technology. Moreover, to further elucidate the processes driving innovation, it is useful to consider the multilevel nature of the creative process noting, in particular, that creative products are the result of individuals, teams, and organizations. Through this lens we can begin to more fully understand how technology shapes the entire creative process, rather than one or two activities. We turn now, then, to our discussion on technology and the processes composing innovation.

THE INNOVATION PROCESS AND TECHNOLOGY

Although multiple definitions exist, we define technology in accordance with *Merriam-Webster*, who suggests that a technology represents "a manner of accomplishing a task especially using technical processes, methods, or knowledge" (Technology, n.d.). In essence, we contend that a technology is best characterized as a tool or mechanism that shapes or has an impact on how we work and interact with each other. In particular, we focus our discussion in this chapter on how varying tools and mechanisms might have an impact on the creative process, specifically. One point to bear in mind as we move forward in the chapter is that we do not draw a distinction between hardware and software as fundamentally differing forms of technology. Rather, we see that both hardware and software have the equal potential to have an impact on the creative process. In the sections that follow, we define each phase of the innovation process before turning to the technologies that are currently, or have the potential to, impact the activities driving innovation. We begin with a discussion on problem construction.

PROBLEM CONSTRUCTION AND IDENTIFICATION

The innovative process must begin at a very basic level, with identifying a problem to be solved (Mumford, Reiter-Palmon, & Redmond, 1994; Reiter-Palmon, Mumford, Boes, & Runco, 1997). Innovative solutions, however, are only possible when a problem is sufficiently ambiguous, complex, and ill defined to permit the *application* of an original idea that might solve it. The actual process of problem construction and identification centers on appropriately framing and conceiving schematic knowledge structures that provide structure to these ill-defined problem spaces. This framing and structure can enhance creativity by narrowing the perceived scope of the problem and provide a conceptual space on which to focus (Boden, 1994). Thus, it is necessary to have elements of both ambiguity and structure for effective problem construction. In particular, having (a) appropriate information about goals or outcomes, (b) understanding key information needed to define the problem, (c) being aware of the procedures needed to solve the problem, and (d) having knowledge of possible challenges to solutions presented have been shown to help guide successful problem construction (Mumford et al., 1994).

Given the complexity characterizing these types of problems, it is often necessary to pare down the problem into these key components using heuristic or schematic knowledge. As an illustration, Redmond, Mumford, and Teach (1993) asked participants to develop a marketing campaign for a three-dimensional (3D) television. Half of the participants were asked to immediately generate solutions, and half were instructed to focus on problem construction. The researchers found that creativity was significantly higher when time was spent on actively constructing the problem. Thus, forms of technology that shape this framing process are likely to be of the greatest influence in this innovative activity. We discuss one of these via the Web-driven availability of user feedback and input.

User Feedback and Input

There is certainly no lack of ambiguous, ill-defined problems in need of creative solutions in today's world. The challenge, rather, lies in pairing expertise and talent with these types of problems—allowing the appropriate individuals to define and generate solutions. One emerging means of doing so is through the analysis of user feedback and input. Today, there is growing support for systems specifically designed to generate creative ideas via end-user feedback (e.g., Bragge, Merisalo-Rantanen, & Hallikainen, 2005). However, other, less formally structured systems are already in place that can provide a wealth of novel problems and potential creative solutions. Large online retailers such as Amazon.com actively retain user input and feedback that are made available, publically, to users of their site (Sugiyama, Hatana, & Yoshikawa, 2004). These databases contain sizable amounts of information about problems and challenges users are facing with products. Key here is that these products and their related challenges are stored using categories established by the retailer. In essence, the retailer is applying a number of schemata to these data, much like individuals do in traditional forms of problem construction (Mumford et al., 1994). Potential problem solvers must simply search and analyze user feedback within these predetermined heuristic structures. In essence, the data contained in and organized by these databases can supplement designers' problem construction efforts by allowing them to observe and address needs and problems they would not have otherwise seen. Simply put, a wealth of complex, ill-defined problems in need of creative solutions lie in wait and are available with only a few keystrokes—a method not available even 10 years ago.

INFORMATION GATHERING

Continuing with a sequential perspective on the creative problem solving process, once problems have been defined, work can begin on generating novel solutions. At times, expertise can serve as a rich source of information on the problem, helping to guide later idea generation (Sweller, 1989). There is also evidence, moreover, that access to a *wider* range of information—even irrelevant information—can lead to novel solutions (Alissa, 1972; Hunter, Friedrich, Bedell-Avers, & Mumford, 2007). It makes intuitive sense, then, that technologies which shape the information gathering process can have a sizable impact on the creativity of final solutions. We discuss two forms of technology that are shaping this creative process: the Web (and Web 2.0) and electronic databases.

Web and Web 2.0

The Internet has simply transformed the way information is shared, and it continually evolves as a means and mechanism to do so. This is well illustrated by the significant trend away from print media to electronic consumption of news and data (e.g., Alterman, 2008). With regard to information gathering, there is perhaps no greater influence on the creative process than the introduction to the Internet. The Web in its next iteration—often called Web 2.0—will bring with it an even greater level of access to data. Although the current Web and Web 2.0 overlap significantly, one distinguishing feature of Web 2.0 is a stream of continually updated data tailored for the user. Thus, rather than searching for a current news story, news stories are updated on our favorite websites or RSS feeds. These updates are continual and seamless. A brave new world of data access, for better or worse, simply cannot be denied.

Electronic Databases and Content

While the internet provides us with access to a range of content, at times overwhelming content, there has also been a quiet revolution underway regarding access to scientific and academic information. Using the termed the 'Invisible Web', Devine and Egger-Sider (2004, p. 265) discuss the limitations of search engines such as Google and Yahoo!, noting that there is a sizable collection of information that cannot be found simply by typing keywords into a search queue. This Invisible Web houses much of the academic literature developed by the scientific and literary community. These databases, however, are not only accessible via a local computer within a given library. Rather, through specific Web portals, many of us have access to these databases electronically. Due to the digital nature of these databases, they are searchable and explorable in ways unimaginable 50 years ago. In our youth, many of us can recall the joy of having a full set of encyclopedias at our disposal. The contrast to what is available to us now, on one Web-connected machine, is simply staggering.

Along the lines noted earlier is another quiet revolution driven by the search engine powerhouse Google. The company has spent recent years attempting to electronically scan, and make available, every book ever published (Conway, 2010). Although it did not succeed in its plan as originally conceived (Sarno, 2011), the attempt highlights the possibilities awaiting us in the future, namely, that we can, at a minimum, be pointed in the appropriate direction to find a piece of information we might need or—perhaps one day—have free and open access to the wide array of thoughts and ideas that compose the fabrics of our societies. The notion of every printed book at our fingertips is not far off.

CONCEPT SELECTION

With a sizable pool of information in place (i.e., from gathering information), the creative task changes from gathering data to choosing which aspects of the data pool to focus on. This narrowing process is known as concept selection (Baughman & Mumford, 1995; Mumford, Supinski, et al., 1997). The key question, however, becomes, What aspects of the data pool are used to narrow the information down? That is, what helps to define the categories or schemas we use in this winnowing process? There is research evidence to suggest that this process may be driven by linkages among data points (Chi, Glaser, & Farr, 1988) or possibly by higher order goals underlying the categories chosen (Carlson & Gorman, 1992). What is clear is that part of the answer to how concepts are successfully selected for creativity is tied to expertise, or having a rich and well-organized understanding of a content domain. It stands to reason, then, that forms of technology that aid—or even augment—this categorization process are most likely to shape the creative process. We discuss an emerging technology that is having a significant impact on conceptual combination and illustrates the potential it has on the creative process: heuristic algorithms.

Heuristic Algorithms

At the conclusion of watching a movie through Internet-based mechanisms such as Netflix, we are typically asked to provide a rating indicating the extent to which we enjoyed the movie. When we provide enough seed ratings, Netflix's heuristic algorithm provides us with recommendations on additional movies or television shows we might also like (Liu & Lee, 2009). Central to the success of this business model is the identification of key concepts that are useful for identifying movies we might like. In the case of Netflix such categories might be, for example, genre (e.g., Westerns or horror), actors in the film, or directors guiding the project. More than likely, it is a combination of a series of categories weighted to maximize the likelihood that we will enjoy the recommended suggestions. Another product example is Pandora radio, which also takes a 'seed'—in this case a favorite song—and begins to generate recommendations from our initial input. Pandora has registered more than 22 million users, and more than 2 million users listen to the radio station each day. It seems that their heuristic algorithm is working rather well. Interestingly, Pandora differs somewhat from Netflix in that the organization employs experts to help provide ratings along with the users.

Central to both of these examples is that software is helping to conduct concept selection. That is, a series of algorithms helps determine which aspects of movies or music are most useful in making recommendations to users. Other programs are being utilized to recommend products we might like or books we might enjoy (Herlocker, Konstan, Terveen, & Reidl, 2004). Another illustration is Stumbleupon.com, which asks for seed categories from users themselves and then matches users based on categories. Users make ratings regarding their preference for a website and then software uses this information, paired with rater similarity, to propose additional websites that a given user might enjoy. The process is continually refined and adjusted based both on category and on similarity among users. Although other examples exist, the preceding should suffice to make the basic point: Technology is providing its own unique form of concept selection and time will tell how such approaches may be integrated into the creative process.

CONCEPTUAL COMBINATION

Conceptual combination is the melding of two or more independent entities to form a novel category or product. In the creative problem-solving process, once concepts have been pared down and only the key elements remain, work can begin on combining concepts in new and novel ways (Mumford, Baughman, et al., 1997). A designer may use conceptual combination when searching for a novel solution to a problem, such as by creating a new product from two existing, seemingly unrelated, products. This process differs from more formal idea generation in that the work here is still largely abstract and conceptual. More precisely, Mumford and Baughman (1996) suggest that the conceptual combination process is composed of (a) identifying central features of key concepts, (b) mapping features from one category onto a differing category, (c) using shared features to develop a new category, and (d) elaborating on the new concept through refinement. It seems reasonable to suggest, then, that technologies that shape and contribute to the activities involved in concept combination are likely to have the largest impact on the creative and innovative process. We discuss one such form of technology-based idea generation: the concept generator.

Concept Generator

Emerging work from the University of Missouri–Rolla in collaboration with other universities such as Bucknell, Virginia Tech, and Penn State University is a new type of software referred to as a concept generator (Bryant, McAdams, Stone, Kurtoglu, & Campbell, 2006; Vucovich et al., 2006). The concept generator is a virtual repository of elements of engineering design (e.g., springs, hinges, etc.) that helps users connect products from one design to another. Products elements are categorized, labeled, and ultimately stored electronically in the design repository and may be culled out and combined with other components or concepts once users has identified which components they would like integrated into their project. A user may, for example, tell the software that he or she needs a mechanism to link two pieces of metal together. The software will then bring up appropriate linkages from the design repository using stored categories. The designer, then, is able to draw from these conceptual combinations driven by the software. Work continues on this project but this example serves a useful point here: to illustrate the means by which software and complex algorithms may be applied to aid directly in the creative design process.

IDEA GENERATION

With concepts selected and combined via implicit or explicit means, work can begin on formal idea generation (Baughman & Mumford, 1995). During this stage of the innovation process, ideas are formally developed and gain a level of concreteness not seen in the conceptual combination stage (Hunter, Bedell, & Mumford, 2006). Idea generation may be an individual level activity, although most organizations have moved to team project–based approaches to creative problem solving. As such, early ideas are often generated by an individual but revised, refined, and shaped by the group (Paulus & Yang, 2000). In moving along in our discussion, it will be useful to keep in mind that technological influences may shape both individual-level and team-level idea generation. We discuss three forms of technology that have the potential to significantly shape idea generation, namely, crowdsourcing, cloud computing, and computer-aided design tools or CAD.

Crowdsourcing

The term *crowdsourcing* was coined by Jeff Howe and Mark Robinson in an issue of *Wired* magazine in 2006 to describe an emerging business model for creativity (Brabham, 2008; Howe, 2006). In essence, an organization presents a problem to a distributed network of individuals, often providing some reward for successfully solving that problem. Crowdsourcing is effectively creative bounty hunting in that organizations simply propose that individuals or teams that solve the problem (often within specific parameters) win a prize. There is relatively little cost to the organization, save the bounty itself, and the approach is quickly emerging as a means to generate creative ideas with very little risk and a high potential for reward.

In perhaps the most well known crowdsourcing event, Netflix offered a $1 million bounty to anyone who could improve their Cinematch software for recommending movies to users. Specifically, a potential bounty hunter had to improve the software by at least 10%, and the winner would be determined by who improved the recommending capability the furthest beyond the minimum 10%. The race came down to who was able to submit their solution first—a million-dollar prize determined by only a few minutes (Bell, Bennett, Koren, & Volinsky, 2009). Also of note is that although many teams worked separately, initially, some teams began to develop alliances and share ideas. Once a subset of teams formed group alliances, the remaining teams also adopted this strategy. In the end, the race was between a grouping of combined teams. The real winner, however, was Netflix, which—with a modest relative investment—was able to significantly improve its Cinematch software.

Other, albeit less dramatic, attempts at crowdsourcing have also proved valuable to organizations. Threadless.com, for example, asks users to design their own T-shirts on its website (Brabham, 2008). These T-shirts are rated by users, and the winning T-shirts are made and sold on the website. The shirts are priced affordably given the low cost of design, and winners receive cash prizes of $1,500.00 and $500.00 in store credit. As an organization, Threadless has done very well, on track in 2006 to gross nearly $18 million with a profit margin near 35% (Howe, 2006). Although other examples exist (e.g., iStockphoto), the preceding examples should suffice to illustrate the basic point. The Internet has allowed for a form of collective idea generation not possible even 20 years ago. By tapping the public as a source of idea generation and providing, relatively speaking, low-cost rewards for novel thinking, organizations can increase their innovative output without investing in the creative infrastructure many organizations rely on.

Cloud Computing

Cloud software refers to the general form of computing that allows users to access data or software in an on-demand fashion using any machine connected to the web. Both software and data are stored on an external server rather than a specific, local machine. E-mail providers such as Gmail or Yahoo! Mail are classic examples of cloud computing—users of these e-mail providers have access to their e-mail accounts from any Internet-enabled computer. What is useful in the creative process, specifically, is the ability to place documents or design projects in the 'cloud' so that team members can access and work on the project in real time (i.e., simultaneously). Two designers might, for example, rapidly exchange sketch pen strokes when creating a new chair—one refining and shaping the other's work in iterative fashion. These types of virtual exchanges also have implications for who is accessible for projects. Because travel is minimized and ease of idea exchange is maximized, experts that might not have been available for collaboration might join a project, thereby increasing the talent pool available.

CAD Tools

Unfortunately, space does not permit a detailed discussion of all the digital tools currently available to designers, but their general use certainly warrants some discussion. Under the broad rubric of CAD, these tools offer a wide range of options to the user with regard to support in the idea generation process (Schneiderman, 2007). An architect might, for example, sketch a building in a software package and then develop a virtual 3D model of it. This 3D perspective allows the architect to examine the building from multiple angles and even change the landscape by season, examining how the building fits into its surroundings across four dimensions. For example, the building information modeling (BIM) technique can create elaborate, multiple-dimension digital models of a building project (Azhar, Hein, & Sketo, 2008). New CAD tools incorporate virtual reality components that can enhance the accuracy and the intuitiveness with which designers can view and manipulate digital models. These virtual reality tools have visualization components such as LCD glasses as well as 3D-interaction components such as data gloves that track hand gestures (Seth, Smith, Shelley, & Jiang, 2005). CAD tools are particularly useful because of their ability to allow for multiple, low-cost iterations of the creative process.

One CAD tool that was designed to enhance the collaborative process, specifically, is the Caretta system (Sugimoto et al., 2004). The system was developed to help urban planners make broad-level design decisions. In urban planning, each stakeholder often has wants, desires, and needs for his or her respective contingencies. There is a need in such a design task to allow for both individual and social creativity. Caretta involves a multiple-input sensing board used as a shared space among designers. The sensing board is similar in appearance to a large game board and allows designers to manipulate physical objects using radio frequency identification (RFID) technology (Warr & O'Neill, 2007). In conjunction with the 'group' sensing board are a series of individual personal digital assistants (PDAs; e.g., a mobile electronic device that operates as a handheld personal information manager, often with Internet connection capability) held by each individual user. Individuals can share their personal input with the group via transferring ideas from their PDAs to the shared sensing board, can interact with the sensing board directly, or can work on their own ideas privately on their

PDAs. The aim of the system is to provide seamless transitions between individual- and team-level creative idea generation (Fisher, Giaccardi, Eden, Sugimoto, & Ye, 2005).

IDEA EVALUATION

Relative to the other stages of the creative process, idea evaluation has tended to receive less research attention (Hunter et al., 2006). It is proving to be, however, one of the more critical aspects of the innovation process (Lonergan, Scott, & Mumford, 2004). Ideas that are sent out for production without being properly vetted and tested are doomed to fail, thereby limiting the likelihood of success from the project team in the future. Similarly, groundbreaking ideas that are squelched by decision makers before they have found their way to market also stand as significant losses to the organization. It seems reasonable to suggest, then, that technology permitting appropriate idea evaluation and testing will have a sizable impact on creativity and innovation. We discuss three forms of technology that have the potential to shape this creative process: rapid prototyping, user feedback, and digital simulation.

Rapid Prototyping

Founder of innovation powerhouse IDEO, David Kelley (2001) coined the phrase 'fail faster to succeed sooner'. This statement nicely illustrates the advantage of rapid prototyping (von Hippel, 1994). Rapid prototyping involves using technology to quickly construct pared-down versions of a design, using low-cost materials (Chua et al., 2003). For instance, designers can use 3D printers to construct mock-ups of a design out of inexpensive polymers. Such technology can take ideas and test them quickly, permitting a rapid evaluation of ideas. Key issues, problems, and challenges can be identified and the ideas refined in quick succession. In essence, the iterative cycle of the innovation process is substantially improved. Pragmatically, this means that products can be delivered more quickly and that organizations can capitalize on momentum and excitement among team members. Moreover, rapid prototyping allows designers to test a range of possible solutions rather than focusing on one or two ideas because of the resource constraints of traditional prototyping.

User Feedback

There is perhaps no better way of evaluating ideas than to give a product or idea to consumers and let them provide their input on what works, and what does not. In today's digital age, there is no shortage of feedback from users and consumers about the quality of a given product. Nearly every organization that sells products via the web employs a user-feedback option (e.g., Amazon, Walmart, Lowe's, Target). From the organization's standpoint, this allows consumers to have greater confidence in the products they are buying. Rather than relying on what our neighbor recommended, we could see what others, sometimes hundreds of others, have to say about a given product before making a choice to purchase it. From a designer's perspective, these quantitative ratings provide insight into what went right and what went wrong, allowing the designer to 'test out' new product ideas. Moreover, qualitative input via consumer feedback also provides detailed indications about specific issues that might have arisen. This feedback can be used to modify or improve new products before excessive time and money is spent on widespread deployment.

Digital Simulation

Physical prototypes are not the only way ideas can be rapidly tested. By utilizing differing types of CAD programs (e.g., BIM), designs can be stored—and represented—digitally. The growing use of physics (or akin) simulators allows designers to test their ideas in these virtual environments (Otto & Wood, 2004). Architects, engineers, and contractors can use CAD techniques such as BIM at all stages of the building process (Azhar et al., 2008). These simulators can be particularly useful when projects are expensive (e.g., designing an airplane) or real-life testing is dangerous (e.g., designing building-wide fire-control systems). Moreover, simulations are not only useful in the design process, but also in the manufacturing process, to determine the most efficient and effective means of producing a given product.

IDEA IMPLEMENTATION

Once individuals and teams have evaluated an idea, they must turn to the organization to determine how to build, make, or instantiate the idea. Planning is particularly important during this stage of the innovation process as it provides a necessary framework for implementing novel ideas (Hunter, Cassidy, & Ligon, in press). Specifically, plans (a) help maximize limited resources, (b) help creative ideas align with broader business strategies, and (c) are important for implementation of ideas over a prescribed period (Hunter et al., 2006). As ideas move from conception to actual implementation, the stakes get higher as greater resources are required for these later stages of the innovation process. Sketching an idea on a napkin is cheap—building 10,000 in a factory can be very costly. Thus, resources, and how they are acquired, will play a significant role in successful implementation. It stands to reason that forms of technology, which might impact resource acquisition and allocation, would play a nontrivial role in shaping this creative process. We discuss one emerging mechanism that may significantly shape the funding process, in particular: crowd-sourced funding platforms.

Crowd-Sourced Funding Platforms

One of the key challenges to the creative design process is obtaining the funds to actually build or create the product (Helfat, 1999). Larger organizations, of course, often have the capital to fund sizable R&D and must simply turn to their coffers for support. Upstart entrepreneurs, on the other hand, have traditionally been forced to seek out a venture capitalist (or two) that would be willing to invest in their idea. Recently, however, there has been the introduction of a rather unique source of support for novel ideas. Broadly framed as crowd-sourced funding platforms, these websites allow users to upload an idea along with a target amount needed for funding their product development (Taylor, 2011). At a popular crowd-sourced funding platform, kickstarter.com, potential funders can visit the site and decide if they would like to provide a much smaller donation than a venture capitalist might. Contrasting the venture capitalist approach, the site succeeds through the *volume* of visitors. Funders are rewarded with the product itself and, at times, an extra incentive such as a personal phone call of thanks from the developer. By design, individuals who fund projects are generally motivated by wanting to see a given product made, rather than seeing a return on their investment (this is actually not permitted on the site, explicitly). On sites such as kickstarter.com, ideas are ranked, and more popular projects find their way to the more prominent spaces on the site. As sites like these grow, it seems that a greater number of budding designers will have access to the capital to see their ideas ultimately implemented (Petrecca, 2011).

MONITORING

With creative ideas implemented, attention can turn to ensuring the success of these ideas (Baughman & Mumford, 1995). Through continual monitoring, organizations can assess whether ideas need further refinement or are leading to new and emerging opportunities. Carson and Carson (1993), for example, suggest that creative performance in one round of product development can help improve creative idea generation in later successions. With regard to technology, those elements that aid the monitoring process—particularly the ease, richness, and speed of monitoring—will likely be most influential on creative achievement. We discuss two forms of such monitoring: feedback from products themselves and feedback from consumers.

Feedback From Products

Advances in technology are allowing many products to evolve in such a way that they are capable of directly monitoring and providing feedback on their own performance. For example, Hughes and Stapleton (2005) describe software titled Mixed Simulation Demo Dome, which can be used to monitor data about a newly built roller-coaster ride. The software captures real-time data such as safety, crowd control, staffing assignments, materials management, merchandise stocking, and technical maintenance. At a more basic level, many Web-enabled devices can be accessed by organizations as a means to assess what an emerging problem might be. A consumer might, for example, call the cable company to have them access the cable box to determine why the channels are missing. Another illustration may be seen when our computers freeze up or when errors emerge while working with specific types of software. Most of us have been given the option to send the data back to software vendors to help them understand how and why their product might have failed. Such information is vital to the continual development and refinement of current and future products. Cataloguing patterns of error reports over time allows designers to identify and act on trends that would not be evident in single-case, cross-sectional data. Such assistance in problem identification feeds directly into the creative process, and may enable designers to detect and propose creative solutions to product-related problems that would not otherwise have been identified.

Feedback From Consumers

As has been noted earlier in our discussion, consumer feedback is readily available to organizations and designers. As discussed previously, such feedback is beneficial during the idea evaluation phase of the creative process. However, data collected from consumers and end-users can also be extremely useful during the monitoring phase. More specifically, many websites allow users to leave comments about the products or services they are using. Such sites provide consumers with the option of anonymously sending feedback via virtual drop boxes or less anonymously on the websites themselves. No longer is it necessary to put pen to paper and stamp to envelope to pass along negative information to an organization. Streamlining this process of soliciting consumer feedback has permitted a richer data pool of continually updated information about new products. In essence, monitoring is easier and faster than ever before.

TECHNOLOGY AND LEADER INFLUENCE

The impact of leadership on innovation is well established (e.g., Amabile et al., 2004; Byrne, Mumford, Barrett, & Vessey, 2009; Mumford, Hunter, Eubanks, Bedell, & Murphy, 2007; Tierney, Farmer, & Graen, 1999) and the reasons for this impact are readily apparent. Throughout the eight-stage process, employees are under the direction of the leaders who shape problems to be solved and the ideas that emerge, as well as the nature, scope, and direction of the work being done. Given the focus of this chapter on technology and creativity, it is important to consider the influence leaders have on innovation, specifically via technology use (for a detailed examination of the characteristics of leaders for innovative endeavors, see Mumford, Peterson, & Robledo, this volume). As such, we propose that the impact of leaders on innovation through technology is threefold. First, leaders can directly encourage and role model followers' adoption of acceptance of technology (King & He, 2006). Second, leaders can shape innovation via technology by developing missions and visions that indirectly require follower utilization and adoption of technology. Third, leaders can shepherd employees away from forms of technology that may harm creativity. We elaborate on each of these propositions next.

LEADER INFLUENCE ON TECHNOLOGY ADOPTION

As we propose in this chapter, technology can potentially shape creativity across all stages of the innovation process. However, individuals in creative fields are often reluctant to adopt new technologies when their familiar methods of working seem to be producing sufficient results—a point that is particularly true for older employees (Morris & Venkatesh, 2000). Moreover, employees may be unaware of the technology, perceive the cost of training as too high, or simply do not perceive the technology as useful (King & He, 2006; Straub, 2009). Given the potentially positive role, technology use can have on creative performance, such nonadoption may be a missed opportunity for greater creative discovery and therefore a potentially fruitful avenue for leadership intervention.

It is clear that technology can have many benefits for innovation, such as improving the accessibility and organization of information (Torkzadeh & Doll, 1999). However, one of the greatest barriers for successfully integrating technology into the innovation process is the adoption of the technology by an organization (King & He, 2006; Straub, 2009). Leaders can help persuade followers to consider new technology through three primary avenues of influence: (a) promoting the technology's fit with the organization's goals, (b) adopting the technology themselves and thus making the use of technology normative, and/or (c) creating incentives for technology adoption. We begin by discussing leader promotion of technology use.

Promoting Technology Adoption

A leader's job is to provide direction for followers to work more efficiently. The initial presentation of a technology provides a significant opportunity for leaders to have an impact on employee technology adoption. When a new technology is introduced, employees look for cues to interpret or 'construct reality' about how the technology will have an impact on them (Edmondson, 2003). Leaders are in the best position to provide these cues to encourage technology adoption and frame the new technology as a way to advance innovation.

It should be noted, however, that technology adoption does not occur in a unilateral direction from leader to follower. Often, leaders must understand followers' perceptions of the pros and cons of technology use in order to adjust their promotion message. The adoption of technology is often the result of tensions between managers and groups, emerging organically over time (Nielsen, 2008). For example, Heracleous and Barrett (2001) conducted a qualitative study of brokers and underwriter adoption of an electronic trading technology in the London Insurance Market. To improve technology adoption, leaders held meetings, road shows, trainings, and peer advice sessions in an effort to increase the adoption of the new technology. Although leaders made an effort to promote the electronic trading system, brokers perceived it as a disruption to the face-to-face interactions during negotiation with clients. This understanding was achieved only after managements engaged in communication with employees about why the technology was not perceived as an effective tool, and managers adjusted procedures to incorporate both technology and face-to-face interactions. This type of adoption conflict between managers and employees occurs at the interorganizational, interdepartmental, and interpersonal levels (Nielsen, 2008). The conflict often stems from the tension of replacing previous processes and differences in opinion for how best to proceed (Nielsen, 2008). Thus, leaders should initiate conversations with employees to monitor any problems with technology adoption.

Modeling Technology Adoption

As role models for appropriate behavior, leaders often set the standard for acceptable actions in the workplace (Jaussi & Dionne, 2003). Edmondson (2003) suggests that leaders are the most visible spokespersons for a new technology. Along these lines, leader adoption of technology can produce a notable influence for follower adoption of technology. For example, studies of adoption on sales automation software (Homburg, Wieseke, & Kuehnl, 2009) and pharmaceutical companies (Mathieu, Ahearne, & Taylor, 2007) found that leaders' commitment to a new sales technology had an impact on whether their followers used it. A recent case also illustrates the impact leaders can have on technology adoption and perception. During the early phases of his presidential campaign, President Obama levered social media to raise funds, send messages, and drum up support. His embracing of technology as a fundamental component of his campaign led to unique and fruitful avenues for success (Talbot, 2008).

Unfortunately, as the United States president Barack Obama has attested to after having to relinquish his personal BlackBerry smartphone when taking office, serving as a role model for technology adoption can be somewhat problematic for leaders. Leaders are often limited in their time as a resource and the nature of their job focuses more on managing people than on completing everyday tasks (Mintzberg, 1978; Tengblad, 2006). In addition, the nature of innovation is somewhat ambiguous, and leader use of technology should present only one example to followers for how the technology can be used to solve problems. Followers may become over-reliant on their leader's method of implementing technology because leaders tend to have a greater influence on innovation initiatives than do other project team members (Cooper, 1999). As a result, leaders should encourage followers to find new ways to use the technology to accomplish creative work. Leaders must strike a careful balance when acting as role models for adopting technology to solve creative problems; they can encourage followers to adopt a technology, but they should not be too specific regarding how the technology should be used. Thus, leaders face a number of challenges with regard to serving as role models for technology use. Those that are able to commit to such endeavors, however, are likely to produce nontrivial influences in their followers' adoption of novel technologies.

In line with the preceding, it would appear useful to discuss the antecedents that may help managers and supervisors adopt new technology. For example, managers are more likely to accept a new technology when they believe that it will be useful and fairly easy to use (Mathieu et al., 2007). In other words, the perceived benefits must be greater than the perceived cost. Notably, one study found that manager perceptions of technology were a stronger predictor of follower adoption than follower's own perceptions of technology (Homburg, Wieseke, & Kuehnl, 2009). This implies that follower perceptions may be malleable and receptive to influence, perhaps by other environmental variables such as coworker adoption and perceived adoption pressure. As such, a worthy strategy to improve technology adoption throughout an organization may be to increase buy-in at the upper echelons and allow it to trickle down to the lower levels. Accordingly, it should be stressed that in order for leaders to model such adoption, they have a personal responsibility to remain up-to-date on new technologies that may be relevant to the organization.

Leader Developed Incentives for Technology Adoption

Given the time constraints faced by leaders, role modeling alone may not be enough to persuade followers to adopt a new technology, and a more direct method may, at times, prove to be more useful (Hunter & Cushenbery, 2011; Klotz, Wheeler, Halbesleben, Brock, & Buckley, in press). That is, leaders may offer direct incentives for appropriate technology use (Selker, 2005). Such incentives may be adequate training and access to developmental opportunities aimed at improving necessary skill sets or direct bonuses for success in using specific tools deemed important by the leader. One thing to bear in mind when developing incentives is that followers have individual differences in their proclivity for technology adoption, and it is important to remember that individuals each have unique personalities and experiences with technology (Venkatesh, 2000). Thus, leaders should attempt to create an environment that rewards risk taking as well as more personalized attempts designed to motivate followers to adopt technology for their own benefit. For followers who are slower to adopt new technology, leaders should make an effort to explain the specific benefits for work flow as early as possible (Agarwal & Prasad, 1998) and support those efforts as skills and comfort with technology are built.

LEADERS CREATE TECHNOLOGY USE THROUGH VISION

Modeling and incentives will likely drive technology adoption, but leaders can also shape this use via more subtle avenues (Hunter & Cushenbery, 2011). Specifically, leaders can create a vision for innovation that indirectly requires the use of technology (Byrne et al., 2009). This vision should be broad enough to allow for solutions that are novel, but still provide enough guidance to enhance problem construction and concept selection. A recent qualitative study of innovative architect Frank Gehry's company suggests an interesting example of the influence of vision on technology adoption and innovation (Boland, Lyytinen, & Yoo, 2007). According to the study, Frank Gehry himself "does not use computers and dislikes their screen images" (Bo et al., 2007, p. 644). Despite this bias, Gehry's vision was so complex that the proposed design could not be completed unless his team of engineers, contractors, builders, and fabricators worked together and utilized novel technologies and advanced 3D-modeling software. As such, although the designer was not directly championing the use of novel technologies, the design he envisioned could not have been completed without such tools.

In another example of leader vision encouraging innovation through technology, Nielsen (2008) described the interchange between information and technology (IT) employees and bankers in promoting the adoption of internet banking in Norway. Originally, banks wanted to improve the retention and attraction of customers. The solution was to implement online banking, which required ways to make banking technology more user-friendly and convenient. Thus, IT employees, marketing managers, and other employees within the banking organizations had to work together utilizing emerging forms of technology to establish a novel online banking system for consumers.

LEADERS STEER AWAY FROM TECHNOLOGY WHEN IT MAY HINDER INNOVATION

As with any new product or process, simply because a technology is available does not mean it is suitable for every task in an organization—a point particularly true for the idiosyncratic and complex phenomena of creativity and innovation. More precisely, for some projects the adoption of technology may actually delay a project or force followers into locked modes of thinking (Edwards, 2001; Kroll, 1995). It is also wrong to assume that forcing technology on an engineer or designer will always make them more efficient. Hinds and Kiesler (1995) argue that 'rich' channels of communication are best for communication when data are missing or meaning is difficult to interpret, an environment that is common in creative work. The 'richest' form of communication is face-to-face communication because it includes many more cues from which to infer meaning. Thus, technology may sometimes get in the way of communication depending on the task that the team is trying to accomplish. Given that leaders are more privy to company schedules, information about resources, and knowledge about the norms in their organizations, we propose that they are in an appropriate position to decide whether a certain type of technology is a good fit for a project. Moreover, it is the responsibility of the leader to understand the needs of his or her followers and how technology can best serve those needs.

In some cases, followers may have their own unique systems that already serve the same function as the technology under consideration (Kao, 2007). An elaborate system of files and notes, for example, may be superior to a complex computer system for someone who is not a fast typist. Leaders should understand when technological adoption is not functional and accommodate when necessary. For example, previous research suggests that leaders with a more directive style elicit higher performance from salespeople who do not regularly depend on technology (Mathieu et al., 2007). This directive style can act as a substitute for technology, but requires highly individualized effort from the leader that is difficult to replicate for every follower.

In addition to accommodating the needs of individual followers, leaders should also recognize when technology may be hindering the creative process more broadly, as a system (Amy, 2008). Discussed earlier, Internet searches tend to replicate similar results as search engines continue to personalize their output to individual users based on previous search history. Such a trend might result in reduced access to diverse knowledge structures that are needed for novel idea production (Hunter, Bedell, Ligon, Hunsicker, & Mumford, 2008). As such, leaders should also encourage followers to seek inspiration from a variety of sources with emphasis on nontraditional information gathering. For example, engineers have been known to look to the muscle and bone groupings of animals when designing new movement mechanisms (Benyus, 1997). The broader point to bear in mind is that although technology can be a useful tool for innovation, it is not a panacea for creative solutions. Leaders should continue to ask followers to seek solutions in a variety of ways and only rely on technology as a tool that aids their thinking, not replaces it.

CHALLENGES TO LEADING INNOVATION THROUGH TECHNOLOGY

The preceding sections argued that leaders influence technology use and innovation through three primary means: direct mechanisms (e.g., rewards), indirect mechanisms (e.g., through a vision or mission) and vis-à-vis directed guidance away from harmful or unbeneficial forms of technology. It is important to bear in mind, however, that these influences may vary in impact by a number of factors. We discuss three such influences that appear worthy of attention, below.

To begin, leaders should recognize that different technologies will require varying levels of their input for adoption and application for innovation. Leaders may be especially necessary when technology is complex, requires more time or resources to use, or does not have a clear fit for employee tasks. In addition, leadership may be particularly important in creating structure via technology when managing many projects concurrently. For example, many companies are recognizing that working on a family of innovative products, some of which are only slight improvements to proven products, can be an effective method of increasing successful product output (Cormican & O'Sullivan, 2004). Moreover, leadership may be more useful for a radical change to procedures than an incremental change in product. A greater change creates more ambiguity for employees, creating a greater need for leaders to provide direction. Thus, leaders must make crucial decisions not only about which technologies to promote, but also how involved they will be in promoting them.

Second, it is important to keep in mind that leaders in different levels of the organization can have varying impacts impact on technology adoption. CEOs and other top-level leaders, for example, are more likely to scan systems, consult with outside constituencies, and gathering information from within the company (Dutton & Ashford, 1993). In other words, top-level leaders may have a more macro view of how technology can aid innovation in the long run and, as such, may largely impact technology adoption via vision and mission formation. Mid-level managers, in contrast, are more involved in day-to-day operations and may play be in a better position to influence technology adoption through role modeling or direct incentives such as support and encouragement. Thus, the role of leadership in facilitating innovation through technology use may vary by level and position of leadership in organizations.

Third and finally, by embracing appropriate technology adoption, leaders must be prepared to relinquish some level of control and influence that may be replaced by the technology itself (Kerr & Jermier, 1978). As varying forms of technology provide feedback to the user, for example, employees may require less guidance and structure-oriented behaviors from leaders. On the more extreme end, advanced artificial intelligence (AI) programs may even offer supportive and consideration-oriented behaviors to employees (Nilsson, 1998), limiting the amount of requisite behaviors from leaders themselves. Thus, although differing forms of technology will have an impact on leadership differentially, the key point here is that technology itself may augment the role and shape of leadership in the future (Luftman, Bullen, Liao, & Nash, 2003). More precisely, as software, hardware, and varying forms of gadgetry continue to shape the workplace, so too will the face of leadership be altered. Given the increasing role of technology in innovation, this point rings particularly true in the enhancement of creative endeavors.

CONCLUSION

Limitations

Before turning to the conclusions of the chapter a number of limitations should be borne in mind. First, our discussion on the various forms of technology is necessarily limited. The number of tools available that may shape the innovation process is substantial and space constraints did not permit discussion of all of them. Moreover, such a discussion would prove nearly impossible given the continual, rapid emergence of new technologies. We believe our discussion does illustrate *trends* in how technology might shape the innovation process, but our list was not comprehensive and should be noted as such.

Second, although the eight-stage creative process model represents a reasonably detailed list of the innovation process, it is important to note that other conceptualizations do exist and have merit in their own right. For example, Finke, Ward, and Smith (1992) proposed the 'geneplore' model. In this model, creativity is composed of both information-generating and exploratory processes. Additionally, Treffinger (1995) proposes a three-stage model, consisting of identifying a problem, generating ideas, and planning for implementation, with each of these stages consisting of several related substages (see Lubart, 2001, for a review). We do believe that the eight-stage process model serves as a solid foundation for discussing the role of technology on the innovation process, but it is appropriate to acknowledge it was one choice among many.

Third, our discussion focused largely on product development rather than on process or even artistic creative endeavors. Although much of the discussion is certainly applicable to these creative activities and outcomes, we did not focus on them expressly and this stands as a limitation of our chapter. Fourth and finally, we have painted a rather positive picture with regard to technology having an impact on the creative process. There certainly exists the possibility that some forms of technology may have a negative influence on novel idea generation. If a company has an automated process for solving all problems in the same manner, then there may not be room for creative problem solving. It is important to note that this potential exists and that technological influences can produce complex relationships with innovative outcomes. With these limitations in mind, we can turn our attention to the general conclusions emerging from the chapter.

Emergent Themes

Even with these limitations in mind, we believe the present chapter makes a number of contributions to our understanding of technology, creativity, and leadership. Specifically, our discussion reveals five core themes. First, technology's primary role in influencing creativity appears to be improving speed and at times, efficiency. That is, much of the creative process would appear to fundamentally remain the same and is simply faster as a result of emerging technology. By enabling designers to navigate the creative process more quickly, they should be able to propose, test, and implement a larger number of innovative solutions than they would be able to without technological assistance. This increase in speed, however, does have possible implications for how organizations *approach* creativity. Namely, when possible, organizations should embrace technologies such as CAD, digital simulation, and rapid prototyping in an effort to permit iterative testing on products. These processes can allow organizations to bring more refined products to market thereby improving the reception of novel products. Moreover, organizations can—and should—pursue *numerous* ideas in early stages, particularly when preproduction costs are low. By accelerating the design process by utilizing new, work-saving technologies, designers may be able to reduce the cognitive load they experience while navigating the creative process. By reducing this burden, they may be able to generate a larger number of ideas than would otherwise be possible, further enhancing the potential for the organization to develop innovative solutions. By building a larger pool of possible products and solutions, chances are improved that they will lead to radically different and successful ideas. Organizational leaders, moreover, would appear well served to encourage the pursuit of radically different ideas at these early stages if innovation.

The second theme emerging from our discussion is that there is some indication that technology may be shaping the creative process in ways beyond simply improving efficiency and speed. More precisely, technology—at least to some degree—may be altering the nature of the creative process. For example, the computational power behind the heuristic algorithms at Netflix and Pandora may produce idea combinations that are not possible via traditional forms of creative processing. Netflix's Cinematch software, for example, processes billions of bits of data in a given session. It is unlikely that even the foremost experts could provide that level of conceptual combination. Along related lines, as tools such as the concept generator continue to improve we may see design options that would not have emerged from humans, alone. It is unlikely that technology, computers in particular, will become the primary generators of our creative output—but the combination of person and machine in the creative process may become inextricably linked in the future.

The third emergent theme is that, in the future, organizations will be leveraging input and idea generation from the public or 'crowd'. Netflix's approach to crowdsourcing was so successful that they immediately put out a call for a second bounty to improve suggestions for viewers who did not provide input on their movie preferences, instead relying on demographic data alone. Moreover, other organizations have certainly taken notice and have begun their own crowdsourcing events activities (see discussion about Threadless.com, earlier). In addition to crowdsourcing, which is essentially a form of idea generation, input from customers via feedback is also proving invaluable to the innovation process. Organizations can monitor, change, and adapt their designs rapidly to meet consumer demands. When Apple releases its next iteration of phone or tablet, we can be sure that design changes (e.g., including a camera, improving battery time) are based, in no small part, on the sizable amount of feedback it has been able to receive from customers via emerging technology.

Fourth, the aggregate of the discussion illustrates a trend toward a blending of the activities comprising the creative process. A prototype that is developed very quickly, for example, can redefine the problem itself, beginning the innovation process anew (Fischer, Giaccardi, Eden, Sugimoto, & Ye, 2005). Similarly, as products are monitored for their success (or failure) these data can be combined into useful categories that help shape the next set of ideas to be generated. Finally, the use of the customer as possible idea developer (e.g., crowdsourcing) has also begun to blur the lines between consumer and developer. The point emerging here is that the creative process is a continually evolving and morphing entity. Time will tell if our current ways of characterizing innovative activities are as applicable tomorrow as they are today given the growth of technological advances.

Fifth and finally, it is apparent that leaders can shape innovation both directly (Byrne et al., 2009) as well as through the use and application of technology. The complex nature of creativity, innovation, and technology necessitates guidance and structure if useful innovations are to be produced (Simon, 2006). The ever-increasing use of technology means that leaders, in particular, will be needed to guide appropriate application of emerging tools and software. At times, leaders must push their employees to adopt new technologies; at others, their role is to pull back the reigns and rely on more traditional methods. Strong decision making, it seems, is not a leadership quality that is likely to go out of fashion any time soon.

To summarize, the aim of the present chapter was to outline how leadership and the use of novel technologies can shape the various stages of the creative process. In doing so it is clear that creativity and technology are inextricably linked. This is not to deny the importance of 'old-school' approaches. For example, the technology-savvy organization Pixar pioneered computer animation and set the standard for animated film. Its artists, however, continue to develop much of their work using paper-and-pencil-driven storyboards. Rather, it is important to bear in mind that at some point in the creative process technology will shape the activities composing that process. It is unlikely that this will change in the future and the linkages will become even more intertwined.

In closing, our final note pertains to future research on the study of creativity and technology. As is illustrated by the range of sources utilized in this chapter, the study of creativity and technology cannot succeed via a single disciplinary effort. Focusing only on psychology literature, for example, will result in a narrowed view of cognitive mechanisms underlying creativity, whereas focusing only on design literature will ignore the significant historical work on the creative process. We must continue to cross disciplinary lines and enter our various silos to begin to understand the issues outlined in this chapter.

REFERENCES

Agarwal, R., & Prasad, J. (1998). The antecedents and consequents of user perceptions in information technology adoption. *Decision Support Systems*, *22*, 15–29.

Alissa, I. (1972). Stimulus generalization and over-inclusion in normal and schizophrenic subjects. *Journal of Clinical Psychology*, *34*, 182–186.

Alterman, E. (2008, March 31). Out of print: The death and life of the American newspaper. *The New Yorker*, http://www.newyorker.com/reporting/2008/03/31/080331fa_fact_alterman?currentPage=1.

Amabile, T., Conti, R., Coon, H., Lazenby, J., & Herron, M. (1996). Assessing the work environment for creativity. *Academy of Management Journal*, *39*, 1154–1184.

Amabile, T. M., Schatzel, E. A., Moneta, G. B., & Kramer, S. J. (2004). Leader behaviors and the work environment for creativity: Perceived leader support. *The Leadership Quarterly*, *15*, 5–32.

Amy, H. (2008). Leaders as facilitators of individual and organizational learning. *Leadership and Organizational Development Journal*, *29*, 212–234.

Andrew, J. P., Manget, J., Michael, D. C., Taylor, A., & Zablit, H. (2010, April). *Innovation 2010: A return to prominence—and the emergence of a new world order.* Boston, MA: The Boston Consulting Group.

Azhar, S., Hein, M., & Sketo, B. (2008, April). Building information modeling (BIM): Benefits, risks and challenges. Paper presented at the 44th ASC Annual Conference, Auburn, AL.

Baughman, W. A., & Mumford, M. D. (1995) Process analytic models of creative capacities: Operations involves in the combination and reorganization process. *Creativity Research Journal*, *8*, 37–62.

Bell, R., Bennett, J., Koren, Y., & Volinsky, C. (2009, May). The million dollar programming prize. *IEEE Spectrum*, 28–33.

Benyus, J. (1997). *Biomimicry: Innovation inspired by nature.* New York, NY: William Morrow.

Boatman, J. E., & Wellins, R. S. (2011). *Global leadership forecast 2011.* Pittsburgh, PA: Development Dimensions International.

Boden, M. A. (1994). *Dimensions of creativity.* Cambridge, MA: MIT Press.

Boland, R. J., Lyytinen, K., & Yoo, Y. (2007). Wakes of innovation in project networks: The case of digital 3-D representations in architecture, engineering, and construction. *Organizational Science*, *18(*4), 631–647.

Brabham, D. C. (2008). Crowdsourcing as a model for problem solving. *International Journal of Research Into new Media Technologies*, *14*, 75–90.

Bragge, J., Merisalo-Rantanen, H., & Hallikainen, P. (2005). Gathering end-user feedback for continuous development of information systems: A repeatable and translatable e-collaboration process. *IEEE Transactions on Professional Communication*, *48*, 55–67.

Bryant, C. R., McAdams, D. A., Stone, R. B., Kurtoglu, T., & Campbell, M. I. (2006, September). A validation study of an automated concept generator design tool. Paper presented at the proceedings of the 2006 ASME IDETC/CIE. Philadelphia, PA.

Byrne, C. L., Mumford, M. D., Barrett, J. D., & Vessey, W. B. (2009). Examining the leaders of creative efforts: What do they do, and what do they think about? *Creativity and Innovation Management*, *18*, 256–268.

Carlson, W. B., & Gorman, M. E. (1992). Cognitive framework to understand technical creativity: Bell, Edison, and the telephone. In R. J. Weber & D. N. Perkins (Eds.), *Inventive minds: Creativity in technology* (pp. 48–79). New York, NY: Oxford University Press.

Carson, P. O., & Carson, E. D. (1993). Managing creativity enhancement through goal-setting and feedback. *The Journal of Applied Psychology*, *27*, 36–45.

Chi, T. H., Glaser, R., & Farr, M. J. (1988). *The nature of expertise*. Hillsdale, NJ: Erlbaum.

Chua, C. K., Leong, K. F., & Lim, C. S. (2003). *Rapid prototyping: Principals and applications* (2nd ed.). Singapore: World Scientific Publishing.

Cohen, M. D., March, J. G., & Olsen, J. P. (1972). A garbage can model of organizational choice. *Administrative Science Quarterly*, *17*, 1–25.

Conway, P. (2010). Preservation in the age of Google: Digitization, digital preservation, and dilemmas. *Library Quarterly*, *80*, 61–79.

Cooper, R .G. (1999). From experience: The invisible success factors in product innovation. *Journal of Product Innovation Management*, *16*, 115–133.

Cormican, K., & O'Sullivan, D. (2004). Auditing best practice for effective product innovation management. *Technovation*, *24*, 819–829.

Devine, J. & Egger-Sider, F. (2004). Beyond Google: The invisible web and the academic library. *Journal of Academic Librarianship*, *30*(4), 265–269.

Dutton, J. E., & Ashford, S. J. (1993). Selling issues to top management. *Academy of Management Review*, *18*, 397–428.

Edmondson, A. (2003). Framing for learning: Lessons in successful technology implementation. *California Management Review*, *45*, 34–54.

Edwards, S. M. (2001). The technology paradox: Efficiency versus creativity. *Creativity Research Journal*, *13*, 221–228.

Finke, R. A., Ward, T. B., & Smith, S. M. (1992). *Creative cognition: Theory, research, and applications*. Cambridge, MA: MIT Press.

Fischer, G., Giaccardi, E., Eden, H., Sugimoto, M., & Ye, Y. (2005). Beyond binary choices: Integrating individual and social creativity. *International Journal of Computer Studies*, *63*, 482–512.

Helfat C. E. (1997). Know-how and asset complementarity and dynamic capability accumulation: the case of R&D. *Strategic Management Journal*, 18(5), 399–360.

Heracleous, L., & Barett, M. (2001). Organizational change as discourse: Communicative actions and deep structures in the context of information technology implementation. *Academy of Management Journal*, *44*, 755–778.

Herlocker, J. L., Konstan, J. A., Terveen, L. G., & Riedl, J. T. (2004). Evaluating collaborative filtering recommender systems. *ACM Transactions on Information Systems*, *22*, 5–53.

Hinds, P., & Kiesler, S. (1995). Communication across boundaries: Work, structure, and use of communication technologies in a large organization. *Organization Science*, *6*, 373–393.

Homburg, C., Wieseke, J., & Kuehnl, C. (2009). Social influence on salespeople's adoption of sales technology: A multilevel analysis. *Journal of the Academy of Marketing Science*, *38*(2), 159–168.

Howe, J. (2006). The rise of crowdsourcing. *Wired*, *14*(6), 5–12.

Hughes, C. E., & Stapleton, C. B. (2005, July). *The shared imagination: Creative collaboration in augmented virtuality*. Paper presented at HCI International, Las Vegas, NV.

Hunter, S. T., Bedell, K. E., Ligon, G. S., Hunsicker, C. M., & Mumford, M. D. (2008). Applying multiple knowledge structures in creative thought: Effects on idea generation and problem-solving. *Creativity Research Journal*, *20*, 137–154.

Hunter, S. T., Bedell, K. E., & Mumford, M. D. (2006). Dimension of creative climate: A general taxonomy. *International Journal of Creativity and Problem Solving, 15*, 97–116.

Hunter, S. T., Cassidy, S. E., & Ligon, G. S. (in press). Planning for innovation: A process-oriented perspective. In M. D. Mumford (Ed.), *Handbook for organizational creativity*. Oxford, England: Elsevier.

Hunter, S. T., & Cushenbery, L. (2011). Leading for innovation: Direct and indirect influences. *Advances in Developing Human Resources, 13*(3), 248–265.

Hunter, S. T., Friedrich, T. L., Bedell-Avers, K. E., & Mumford, M. D. (2007) Creative cognition in the workplace: An applied perspective. In M. J. Epstein, T. Davila, & R. Shelton (Eds.), *The creative enterprise – managing innovative organizations and people* (pp. 171–193). New York, NY: Praeger.

IBM Global Business Services. (2010). *Capitalizing on complexity: Insights from the global chief executive officer study.* Somers, NY: IBM Corporation.

Jaussi, K. S., & Dionne, S. D. (2003). Leading for creativity: The role of unconventional behavior. *The Leadership Quarterly, 14*, 351–368.

Kao, J. (2007). *Innovation nation: How America is losing its innovative edge, why it matters and what we can do to get it back*. New York, NY: Free Press.

Kelley, T. (2001). *The art of innovation: lessons in creativity from IDEO, America's leading design firm*. New York, NY: Currency/Doubleday.

Kerr, S., & Jermier, J. M. (1978). Substitutes for leadership: Their meaning and measurement. *Organizational Behavior and Human Performance, 22*, 375–403.

King, W. R., & He, J. (2006). A meta-analysis of the technology acceptance model. *Information & Management, 43*, 740–755.

Klotz, A. C., Wheeler, A. R., Halbesleben, J. R. B., Brock, M. E., & Buckley, M. R. (in press). Can reward systems influence the creative individual? In M. D. Mumford (Ed.), *Handbook for organizational creativity*. Oxford, England: Elsevier.

Kroll, J. (1995, February 27). Computers and creativity: The arts. *Newsweek*, p. 68.

Liu, F., & Lee, H. J. (2009). Use of social network information to enhance collaborative filtering performance. *Expert Systems with Applications, 37*, 4772–4778.

Lonergan, D.C., Scott, G. M., & Mumford, M. D. (2004). Evaluative aspects of creative thought: Effects of idea appraisal and revision standards. *Creativity Research Journal, 16*, 231–246.

Lubart, T. I. (2001). Models of the creative process: Past, present and future. *Creativity Research Journal, 13*(3), 295–308.

Luftman, J., Bullen, C. V., Liao, D., & Nash, E. (2004). *Managing the information technology resource leadership in the information age.* Upper Saddle River, NJ: Pearson Education.

Mathieu, J., Ahearne, M., & Taylor, S. R. (2007). A longitudinal cross-level model of leader and salesperson influences on sales force technology use and performance. *Journal of Applied Psychology, 92*(2), 528–537.

Mintzberg, H. (1975). The manager's job: Folklore and fact. *Harvard Business Review, 53*, 49–61.

Morris, M. G., & Venkatesh, V. (2000). Age differences in technology adoption decisions: Implications for a changing workforce: Implications for a changing workforce. *Personnel Psychology, 53*, 375–403.

Mumford, M. D., Baughman, W. A., Threlfall, V., Supinski, E. P., & Costanza, D. P. (1996). Process-based measures of creative problem-solving skills: I: Problem construction. *Creativity Research Journal, 9*, 63–76.

Mumford, M. D., Baughman, W. A., Maher, M. A., Costanza, D. P., & Supinski, E. P. (1997). Process-based measures of creative problem-solving skills: IV: Category combination. *Creativity Research Journal*, 10, 59–71.

Mumford, M. D., Baughman, W. A., Supinski, E. P., & Maher, M. A. (1996). Process-based measures of creative problem-solving skills: II. Information encoding. *Creativity Research Journal, 9*(1), 77–88.

Mumford, M. D., & Gustafson, S. B. (1988). Creativity syndrome: Integration, application, and innovation. *Psychological Bulletin*, *103*, 27–43.

Mumford, M. D., & Hunter, S. T. (2005). Innovation in organizations: A multi-level perspective on creativity. In F. J. Yammarino & F. Dansereau (Eds.), *Research in multi-level issues: Vol. IV. Multi-level issues in strategy and methods* (pp. 11–74). Oxford, England: Elsevier.

Mumford, M. D., Hunter, S. T., Eubanks, D. L., Bedell, K. E., & Murphy, S. T. (2007). Developing leaders for creative efforts: A domain-based approach to leadership development. *Human Resource Management Review*, *17*, 402–417.

Mumford, M. D., Mobley, M. I., Uhlman, C. E., Reiter-Palmon, R., & Doares, L. M. (1991). Process analytic models of creative capacities. *Creativity Research Journal*, *4*, 91–122.

Mumford, M. D., Reiter-Palmon, R., & Redmond, M. R. (1994). Problem construction and cognition: Applying problem representations in ill-defined domains. In M. A. Runco (Ed.), *Problem finding, problem solving, and creativity* (pp. 3–39). Norwoord, NJ: Ablex.

Mumford, M. D., Supinski, E. P., Baughman, W. A., Costanza, D. P., & Threlfall, V. (1997). Process-based measures of creative problem solving skills: V. Overall prediction. *Creativity Research Journal*, *10*, 73–85.

Nielsen, J. (2008) Models of change and the adoption of web technologies: Encapsulating participation. *Journal of Applied Behavioral Science*, *44*, 263–286.

Nilsson, N. (1998). *Artificial intelligence: A new synthesis*. San Francisco, CA: Morgan Kaufmann.

Otto, K., & Wood, K. (2001). *Product design: Techniques in reverse engineering and new product development*, Upper Saddle River, NJ: Prentice Hall.

Paulus, P. B., & Yang, H. (2000). Idea generation in groups: A basis for creativity in organizations. *Organizational Behavior and Human Decision Processes*, *82*, 76–87.

Petrecca, L. (2011, May). Crowdfunding and peer-to-peer lending help small businesses. *USA Today*. Retrieved from *http://usatoday30.usatoday.com/money/smallbusiness/2011-05-06-creative-financing-for-small-businesses_n.htm*

Rasmussen, W. D. (1962). The impact of technological change on American agriculture. *Journal of Economic History*, *22*, 578–591.

Redmond, M., Mumford, M.D., & Teach, R. (1993). Putting creativity to work: Effects of leader behavior on employee creativity. *Organizational Behavior and Human Decision Processes*, *55*, 120–151.

Reiter-Palmon, R., Mumford, M. D., O'Connor Boes, J., & Runco, M. A. (1997). Problem construction and creativity: The role of ability, cue consistency, and active processing. *Creativity Research Journal*, *10*, 9–23.

Runco, M. A. (2008). Commentary: Divergent thinking is not synonymous with creativity. *Psychology of Aesthetics, Creativity, and the Arts*, *2*, 93–96.

Sarno, D. (2011, March 22). Federal judge rejects Google digital book settlement. *Los Angeles Times*. Retrieved from http://articles.latimes.com/2011/mar/22/business/la-fi-google-books-20110323

Schneiderman, B. (2007). Creativity support tools. *Communications of the ACM*, *45*, 116–120.

Selker, T. (2005). Fostering motivation and creativity for computer users. *International Journal of Human-Computer Studies*, *63*, 410–421.

Seth, A., Smith, S.S., Shelley, M., & Jiang, Q. (2005). A low cost virtual reality human computer interface for CAD model manipulation. *Engineering Design Graphics Journal*, *69*, 31–38.

Simon, L. (2006). Managing creative projects: An empirical synthesis of activities. *International Journal of Product Management*, *24*, 116–126.

Straub, E. T. (2009). Understanding technology adoption: Theory and future directions for informal learning. *Review of Educational Research*, *79*, 625–649.

Sugimoto, M., Hosoi, K., & Hashizume, H. (2004). Caretta: A system for supporting face-to-face collaboration by integrating personal and shared spaces. In *Proceedings of CHI 2004, Vienna, Austria* (pp. 41–48). New York: ACM Press.

Sugiyama, K., Hatano, K., & Yoshikawa, M. (2004). Adaptive web search based on user profile constructed without any effort from users. In *Proceedings of the WWW Conference* (pp. 675–684). New York: ACM Press.

Sweller, J. (1989). Cognitive technology: Some procedures for facilitating learning and problem solving in mathematics and science. *Journal of Educational Psychology, 81*(4), 457–466.

Talbot, D. (2008). How Obama really did it: The social-networking strategy that took an obscure senator to the doors of the White House. *MIT Technology Review, 111*(5), 78–83.

Taylor, K. (2011, February 4). Funding site lets organizations "curate." *The New York Times*. Retrieved from http://artsbeat.blogs.nytimes.com/2011/02/04/funding-site-lets-nonprofits-curate/

Technology. (n.d.). In *Merriam-Webster's online dictionary* (11th ed.). Retrieved from http://www.m-w.com/dictionary/technology

Tengblad, S. (2006). Is there a "new managerial work"? A comparison with Henry Mintzberg's classic study 30 years later. *Journal of Managerial Studies, 43*, 405–435.

Tierney, P., Farmer, S. M., & Graen, G. B. (1999). An examination of leadership and employee creativity: The relative of traits and relationships. *Personnel Psychology, 52*, 591–620.

Torkzadeh, G., & Doll, W. J. (1999). The development of a tool for measuring the perceived impact of information technology on work. *Omega International Journal of Management, 27*, 327–339.

Treffinger, D. J. (1995). Creative problemsolving: Overview and educational implications. *Educational Psychology Review, 7*, 301–312.

Vashisth, A., & Mudur, S. (2008). Deforming point-based models using an electronic glove. In *Proceedings of the 2008 C3S2E conference* (pp. 193–197). New York: ACM Press.

Venkatesh, V. (2000). Determinents of perceived ease of use: Integrating control, intrinsic motivation, and emotion into the technology acceptance model. *Information Systems Research, 11*, 343–351.

von Hippel, E. (1994). "Sticky information" and the locus of problem solving. *Management Science, 40*(4), 429–439.

Vucovich, J., Bhardwaj, N., Ho, H., Ramakrishna, M., Thakur, M., &Stone, R., (2006, September). *Concept generation algorithms for repository-based early design*. Paper presented at the 26th Computers and Information in Engineering Conference, Philadelphia, PA.

Warr, A., & O'Neill, E. (2007, June). *Tools to support collaborative creativity*. Paper presented at the 4th ACM Conference on Creativity and Cognition, Washington, D.C.

Wren, D. A. (1994). *The evolution of managerial thought*. New York, NY: John Wiley & Sons.

Part II

Empirical Section

5 Academic Leadership of High-Performing Research Groups

Maaike Verbree, Inge van der Weijden, and Peter van den Besselaar

Academic excellence depends not only on the availability of creative, innovative, and enthusiastic researchers, but also on the opportunities provided by the environment they operate in (e.g., Allison & Long, 1990; Andrews, 1979b; Heinze, Shapira, Rogers, & Senker, 2009; Hemlin, Allwood, & Martin, 2008; Pelz & Andrews, 1966). The work environment of most researchers is the research group. We focus on research groups as organizational units embedded in a research organization with researchers and support staff as group members, a research agenda, a research budget, and headed by an academic group leader (e.g., Andrews, 1979b; Beaver, 2001; B. P. Cohen, Kruse, & Anbar, 1982; De Haan, 1994; Laredo, 2001; Laredo & Mustar, 2000; Rey-Rocha, Martin-Sempere, & Garzon, 2002; Stankiewicz, 1976). The job of an academic group leader is to manage his or her group by motivating the researchers, not only by creating the necessary facilities, by defining and implementing the mission and strategy, and by positioning the group internally and externally (Sousa & Hendriks, 2008), but also by acquiring resources. Management of research groups is particularly important for achieving high performance because group leaders affect the conditions for a productive research environment (Bland & Ruffin, 1992). Although a number of scholars underscore the positive influence of leadership and management on research performance (e.g., Babu & Sing, 1998; Bland & Ruffin, 1992; Harvey, Pettigrew, & Ferlie, 2002; Knorr, Mittermeir, Aichholzer, & Waller, 1979; Mumford, Peterson, & Robledo, 2013, this volume; Mumford, Scott, Gaddis, & Strange, 2002; Stankiewicz, 1976), studies in which specific management activities and leadership practices have a positive relation with performance are scarce. In a previous study, one of the current authors investigated the relation between specific management activities and performance, and showed that a diverse constellation of management activities positively relates to the performance of research groups (Van der Weijden, 2007; Van der Weijden, de Gilder, Groenewegen, & Klasen, 2008).

Research groups are increasingly expected to perform excellent research. This emphasis on excellence by science policy is reflected, for instance, in specific funding programs for individual top researchers and in the concentration of the best researchers in 'centers of excellence'. In this chapter, we address the question whether research groups with outstanding academic performance—termed high-performing research groups—have group leaders with particular academic leadership characteristics. Our approach is innovative in that we do not conduct correlation analyzes between variables; instead, we compare groups with different performance levels in order to identify the leadership and management characteristics of high-performing research groups. If high-performing research groups differ from other research groups, we may be able to improve the leadership, management, and organization conditions that facilitate an excellent research performance.

ACADEMIC LEADERSHIP

We define academic leadership as organizing, managing, and leading researchers and the research group; it refers to a variety of opinions, tasks, and practices of academic group leaders. We distinguish four components of overall academic leadership. The first component, *resource strategy*, is the task of acquiring and combining resources for the research group. According to Zaleznik (1977), the leadership role that requires an open and spontaneous attitude for developing creative and innovative ideas is rather different from the management role required for a well-structured organization. *Leadership* in a narrow sense, as second component of academic leadership, refers to how researchers are directed and stimulated by the inspiration and vision of the leader. The third component, *group management*, concerns the tools that are used to manage and coordinate the research process. In addition to their internal role, leaders have external tasks. More specifically, the activities undertaken to position the research group in the academic and societal environment to obtain legitimacy, reputation, and visibility is what we call *network management.*

A research group is a heterogeneous set of people with varying capabilities and preferences, interests and priorities, which have to be reconciled by the group leader. Hackett (2005) identified six areas of tension that academic leaders have to deal with in managing and leading their groups. First, leaders have to find an appropriate balance between autonomy and coordination. Second, leaders have to deal with the paradox of risk: working on a research line that seems low risk can be very risky when it is not innovative. A third area of tension can be found in role-induced conflict. For instance, leaders educate the new generation of researchers and want to give them the freedom to choose what to explore. However, this freedom may have to be limited as leaders are responsible for attaining both group goals and the strategic goals of the organization. Fourth, leaders have to determine when to share information both within and between groups, and when further time is needed for more in-depth research. The fifth area of tension can take place when leaders are personally active as a *researcher* in the lab, doing experiments and analyzing data. Simultaneously, however, they need to *manage* their groups, build external networks for collaboration, and position their group in the environment. Finally, there is the accumulation of advantage within groups. Past decisions about research technologies and research directions partly determine later performance and group recognition. This is the tension between continuity and innovativeness.

Leaders have to cope with these six tensions to stimulate creativity within their group. In the following sections we review several aspects of the academic leadership, addressing the abovementioned areas of tension. We provide an overview of previous empirical studies about the relationship between academic leadership and research performance.

Resource Strategy

From the perspective of resource-based theory, one of the main goals of a research group is to achieve a productive combination of resources as inputs for research (Grant, 1996; Pfeffer & Salancik, 1978). In the eyes of research leaders, it is also one of the most challenging goals (Mets & Galford, 2009). Examples of resources required for scientific research are human capital, physical capital, technology, accumulated knowledge and experience, and social networks (Horlings & Versleijen, 2009). Research groups must have access to tacit knowledge, skills, social relations, information (access to journals, databases and libraries), equipment, (large-scale) facilities, and technology (e.g., Babu & Sing, 1998; Bozeman & Corley, 2004; Crewe, 1988; Dundar & Lewis, 1998; Horlings, Gurney, Somers, & Van den Besselaar, 2009; Melin, 2000). How research leaders deal with these resources is a matter of administrative control (Omta & De Leeuw, 1997). Administrative control includes, for instance, administrative procedures for the appointment of staff, for the procurement of equipment, and for reallocating personnel and material resources to a new research line. Resource strategy is the group leader's management task undertaken to acquire and combine resources for his or her group. We address two key resources for conducting research, namely, funding and human capital.

Research groups need *funding* to perform in the first place, and lack of funding is clearly harmful (e.g., Babu & Sing, 1998; Creswell, 1985; Culpepper & Franks, 1983; Meltzer, 1956; Pruthi, Jain, Wahid, & Nabi, 1993). The more intriguing question is whether the available funding sources, rather than the amount of the funding, influence the performance of research groups. In addition to traditional institutional funding, external funding sources have become increasingly important. External funding and its availability can be seen as an opportunity for new initiatives (Auranen & Nieminen, 2010). This is especially the case for international funding (Geuna, 2001; Van der Weijden, Verbree, Braam, & Van den Besselaar, 2009), in which external funding sources seem to increase research output (Gulbrandsen & Smeby, 2005; Johnes, 1988). A comparison of various external funding sources provides evidence that funding by National Research Councils is positively related to research performance, whereas third-party funding (contract research) has no or even a negative relationship with performance (Carayol & Matt, 2006; Cherchye & Abeele, 2005; Groot & García-Valderrama, 2006). This difference could be explained by the fact that applied research for third parties is not always translatable into academic output (Groot & García-Valderrama, 2006). Conversely, Gulbrandsen and Smeby (2005) showed that professors who receive funding from the industry publish more than do their colleagues who receive other types of external funding. Yet here the causal relation may run in the opposite direction. As Carayol (2003) argued, researchers with a high reputation can pick and choose industry collaborations that suit their own research aims. In other studies, no relation was found between the productivity of graduate students and postdocs and financial support from industry (Louis, Holdsworth, Anderson, & Campbell, 2007). Despite the possible positive effect of competitive research council funding on performance, an increase in this funding source can hamper creative science when stable institutional funding is not guaranteed (Heinze et al., 2009). It should be noted that, although funds acquired in competitions may be important for stimulating the academic careers of talented young researchers, they are not necessarily related to their research performance (Bornmann, Leydesdorff, & Van den Besselaar, 2010; Hornbostel, Böhmer, Klingsporn, Neufeld, & Von Ins, 2009; Van den Besselaar & Leydesdorff, 2009). In addition, Auranen and Nieminen (2010) showed that that the notion of competition for funding as a driver of productivity in university research is not self-evident. Another intriguing question is whether more funding sources lead to higher research output, or the other way around: whether higher output attracts more funding sources. Sandstrom (2009) could not confirm this either way.

Concerning *human capital*, we introduce two aspects. First, what is the relation between the size and performance of research groups? Second, which combination of researchers and support staff is required for research groups to achieve a good performance? According to Von Tunzelmann, Ranga, Martin, and Geuna (2003), both small and large groups have different benefits for research performance. On one hand, 'big is better', because it leads to 'economies of scale' (i.e., greater output per unit input) and 'economies of scope' (i.e., synergies arising from conducting allied activities). On the other hand, the advantage of smaller groups is 'dynamic scale economies' through greater agility and responsiveness to change. Urwick (1956) argues that smaller groups have a narrower span of control; that is, coordination costs are reduced for group leaders. This leads to better cooperation, greater efficiency, improved morale, and a sense of unity. The two opposing views on large versus small groups are both supported by studies on size and performance. Several studies find support for the positive influence of a small group size (Carayol & Matt, 2004, 2006; Heinze et al., 2009). Other studies demonstrate the positive influence of a large group size (Louis et al., 2007;

Pineau & Levy-Leboyer, 1983; Spangenberg et al., 1990). A third view suggests that size, small or large, does not play a key role in achieving high performance (J. E. Cohen, 1991; Crewe, 1988; Hoare, 1995; Kyvik, 1995; B. Martin, Skea, & Ling, 1992). An alternative perspective on size is that its positive effect diminishes as size increases (Dundar & Lewis, 1998; Jordan, Meador, & Walters, 1988, 1989; Omta, 1995). Blackburn, Behymer, and Hall (1978) argue that a minimum size (between 11 to 15 group members on average) is necessary to facilitate communication with and stimulation from colleagues. However, there seems to be an upper limit for group size, above which having more staff does not lead to a proportional increase in the groups' research output. The critical mass threshold appears to be quite low (Stankiewicz, 1979; Von Tunzelmann et al., 2003). The review by Von Tunzelmann et al. (2003, p. iii) concludes that "productivity seems to rise as the team size increases to about six or eight persons, above which there is usually little or no extra gain per capita". The threshold can vary among scientific fields. For arts and humanities, no threshold was found. In fields where applied subjects dominate (e.g., clinical medicine), the threshold is somewhat higher, whereas the threshold is lower in more theoretical fields (mathematics). Earlier studies also note that the effect of size on performance depends on the discipline (Baird, 1986; Kyvik, 1995).

The second human capital issue we address here is the optimal group size. Research groups should be composed in such a way that the full set of knowledge and capabilities needed to achieve the groups' aims are available (Harvey et al., 2002; Heinze et al., 2009; Jones, 2010; Stevens & Campion, 1994). This requires heterogeneity that can be obtained from many sources (Carayol & Matt, 2004; García-Romero, 2006; Hemlin, 2006; Smith, 1971), including a mix of disciplines, functional areas, and types of research and methods, as well as a mix of researchers in terms of educational specialization, experience, position, and gender (Baird, 1986; Carayol & Matt, 2004, 2006; De Haan, 1994; Dundar & Lewis, 1998; Fox & Mohapatra, 2007; García-Romero, 2006; Groot & García-Valderrama, 2006; Keller, 2001; Pelz, 1956; Pineau & Levy-Leboyer, 1983; Rey-Rocha et al., 2002; Rey-Rocha, Garzon-Garcia, & Martin-Sempere, 2006; S. J. Shin & Zhou, 2007; Smeby & Try, 2005). Heterogeneity in membership provides research groups with divergent perspectives, different expertise, knowledge and skills, and access to external (communication) networks (for an overview, see the meta-analysis of Hülsheger, Anderson, & Salgado, 2009).

An essential goal of research groups is to educate the next generation of researchers. This means that research groups need to attract PhDs, but they also need to have more experienced researchers to supervise the new generation. The best publication of scientists, that is, with the highest impact, tends to appear between their late 30s (in the case of mathematicians) and early 40s (for medical scientists), whereas on average their last major contribution is published in their 50s (Simonton, 2004, p. 69, Fig. 3.5). Recent research shows that the age at which researchers make their most important contributions has risen over time. On average, the rise has been about 8 years during the last century. This may be due to the increasing complexity of science, which therefore requires longer training, and the collaboration with others who have different competences (Jones, 2010). Longer training increases the need for intensive and high-quality supervision of young researchers. Several studies show the importance of sponsorship or mentorship for the performance and development of young researchers (Anderson & Shannon, 1988, p. 40). The sponsor or mentor roles in research groups are provided by the more experienced members (i.e., senior researchers and group leaders), who supply supervision. Supervision can be important for stimulating creativity, (predoctoral or early-career) research productivity, self-efficacy, grants, collaboration and the professional networking of young researchers (Bozeman & Corley, 2004; Cameron & Blackburn, 1981; Cronan-Hillix, Gensheimer, Cronan-Hillix, & Davidson, 1986; Green, 1991; Hemlin & Olson, 2011; Long & McGinnis, 1985; Paglis, Green, & Bauer, 2006; Reskin, 1979; J. C. Shin & Cummings, 2010; Van Balen, 2010). Affiliation to a prestigious department seems to be of even more importance for the research performance of new generation researchers, probably because these departments can provide better access to eminent researchers for sponsorship (Cameron & Blackburn, 1981; Crane, 1965; Reskin, 1979).

Leadership

Leaders try to lead and inspire their researchers. Research involvement, or motivation, is dedication to one's work. Pelz and Andrews (1966) were the first to thoroughly study the relation between motivation and individual research performance. They discovered a mildly positive relationship, which was also confirmed by later studies (Andrews, 1979a; Babu & Sing, 1998; Fox, 1983; García-Romero, 2006; Harris & Kaine, 1994; Ramsden, 1994). Researchers can be motivated by internal as well as external sources (Amabile, 1993). Tien (2000) showed that both internal sources (e.g. curiosity) as well as external rewards (e.g. promotion) tend to increase productivity (measured in terms of published articles). Amabile (1993) found that both intrinsic motivators—autonomy, challenges, and excitement about the work—and extrinsic motivators—rewards, clear project goals, and feedback—are important for research performance and creativity. In addition, Tien also showed that different sources of motivation lead to different types of research performance. The more researchers are motivated by a desire to show their scholarly excellence, the higher the probability that they will publish books. The more researchers see an increase in their personal income as important, the more likely it is that they will apply successfully for research grants (Tien, 2000). Although both internal and external motivations are important for research performance, Pelz and Andrews (1966) found that researchers who are internally stimulated by their own ideas are more productive, while researchers who are externally stimulated by a supervisor are less productive. However, Pelz and Andrews (1966) stipulated that self-motivated researchers are not isolated. Their internal motivation is stimulated by a variety of external sources, such as practical problems, technical literature, colleagues, and previous research. Specific sources of motivation do not seem to relate positively to performance, but to an underlying factor of 'intellectual self-reliance', that is, confidence in one's own ideas.

If internal motivation is generated by confidence in one's own ideas, can it be externally stimulated by an academic group leader? According to Pelz and Andrews (1966), it can. First, it is important to enable researchers to demonstrate their contributions through presentations of their own ongoing work, the publication of reports or papers, registering patents or reviewing progress on (technical) designs. Second, researchers should have the freedom to define and pursue individual scientific interests. Third, commitment to one's own work can be reinforced by enthusiastic feedback from supervisors and colleagues. The academic group leader has a special role and should show strong commitment to his or her members' research by, for example, contributing technical competence, showing an interest in the projects, remaining informed, and participating in the research (Andrews & Farris, 1967). In addition, scientists will be more involved when the leader encourages participation in defining the problems to be pursued and the approaches used in addressing these problems (Mumford et al., 2002). Andrews (1979a) demonstrated that motivation is not only a characteristic of individual researchers, but it can also be considered a characteristic of the research group. He found that researchers' motivation levels can vary between research groups. Research groups with highly motivated group members (group leader, senior researchers, and technical staff) have higher performance; that is, these groups are evaluated by their members as more productive and innovative.

Research activities, besides conducting experiments and analyzes, include attending research meetings, participating in (external) research projects, supervising PhDs, publishing papers, giving presentations, and participating in the national and international scientific community. Time for research activities is scarce because it is constrained by other tasks that research groups have to accomplish, particularly teaching in universities (Gottlieb & Keith, 1997; Leisyte, Enders, & de Boer, 2009). A stronger preference for research and a lower preference for teaching is positively associated with performance, as are other research activities such as obtaining grants, supervising master's or PhD students, participating in informal discussions, acting as a reviewer or editor, presenting papers at conferences, maintaining influential gatekeeping roles in national and international scientific communities, and frequent contact with national and international colleagues (Blackburn et al., 1978; Fox, 1992; Gottlieb & Keith, 1997; Harris & Kaine, 1994; Porter & Umbach, 2001; Prpic, 1996; Ramsden, 1994; J. C. Shin & Cummings, 2010).

It is often questioned whether research and teaching are mutually reinforcing or competing tasks. Empirical evidence refers to research and teaching as different dimensions, which implies a trade-off between these activities (Bellas & Toutkoushian, 1999; Fox, 1992; Porter & Umbach, 2001). As Fox says, "Good—or at least more productive—researchers have less classroom contact with students, spend fewer hours preparing for courses, and consider teaching much less important than research" (1992, p. 301). Time spent on research and time spent on teaching are negatively correlated as shown in a meta-analysis by Hattie and Marsh (1996) and by a follow-up study by the same authors (Marsh & Hattie, 2002). However, they found no relationship between teaching and research. Good researchers can either be good or bad teachers and good teachers can either be good or bad researchers. Likewise, Harris and Kaine (1994) found that high-performing researchers do not regard their teaching commitments as a constraint on their research activities. Too much time spent teaching tends to be negatively related to research performance, but teaching as such does not seem to constrain research performance. In fact, too much time spent on research is not favorable for research performance either. Pelz and Andrews (1966) found that lesser performing researchers tend to have a '9-to-5' mentality, whereas researchers working 1 or 2 hours extra a day performed somewhat better. Nonetheless, more hours does not necessarily mean better performance. Highly involved researchers are not obsessed by their work 24 hours a day. They have periods of relaxation when they spend time on activities unconnected to work. Indeed, the more effective researchers do not spend all their working hours on research. It seems to be more productive to also spend some time on teaching and administration (Pelz & Andrews, 1966). Interestingly, extra research time does not necessarily translate into higher research and user effectiveness, as Omta (1995) found. He compared the research time of researchers from preclinical and paraclinical units with researchers from clinical units. Clinical researchers spent half as much time on research, due to patient-care duties, but on average they published a similar number of scientific papers as preclinical and paraclinical researchers.

Group Management

Leaders can use various tools to manage the research process. The balance between academic freedom and coordination is one of the main issues of internal management (Barnowe, 1975; Hackett, 2005; Pettigrew, 1979). In general, coordination and academic freedom are both necessary for performance. Freedom is needed to enhance creativity in approaching research problems. Coordination is needed since interaction with academic leaders stimulates researchers, especially the young (Pelz, 1956; Pelz & Andrews, 1966). Coordination of research through standard procedures (such as lab protocols) is positively related to the performance of medical PhD students (Van der Weijden, De Gilder, Groenewegen, & Geerling, 2008). However, too much coordination can hamper innovation (Andrews & Farris, 1967). In high-performing biomedical laboratories, Pineau and Levy-Leboyer

(1983) found a management approach in which supervisors gave researchers moderate freedom but still formally controlled their researchers' work. Formal control can be done in assessment procedures such as the annual individual job evaluation or the preliminary evaluation of research proposals before these are submitted for external funding (Mets & Galford, 2009; Van der Weijden, 2007), although little is known about the effects on the performance of academic groups (Van der Weijden, de Gilder, Groenewegen, & Klasen, 2008).

Gaston (1975) argues that

> the generally accepted model of a scientist as a person who goes about his work virtually free to choose the scientifically relevant research problems that he wants to attack with few if any major constraints is, if not erroneous, at least an exaggeration. (p. 228)

He remarks that the autonomy of researchers in British university departments is related to academic rank. Professors, as heads of a department, have influence in many decision areas, such as staff appointments and promotion, financial support of research activities and the purchase of equipment. Scientists of lower ranks have far less influence in decisions, but they are satisfied with this lack of voice because of socialized expectations: Young scientists do not expect to have any influence on departmental affairs as long as their research needs are met, and their work load is reduced by not involving them in decision-making processes (Gaston, 1975). Additionally, Leisyte, Enders, and de Boer (2008) note that individual freedom is restricted by 'collective' freedom of choice. To be more precise, the individual freedom of researchers is bound by the research agenda of the research unit and by the thematic choices of the academic group leader, although there is still room for individual choice and consultation. This 'collective freedom of choice' stimulates creativity (Heinze et al., 2009). An early study by Haraszthy and Szántó (1979) found some significant positive correlations between the perceptions of researchers on the effectiveness of the research group and their perceptions on the adequacy of research planning and the coherence of the research program. Junior researchers mainly follow the professor's preferences (Leisyte et al., 2008). Particularly for junior researchers, supervision is an important way to motivate and improve performance. For instance, Van der Weijden, De Gilder, Groenewegen, and Geerling(2008) showed that medical PhD students who were more satisfied with their supervision received more prizes with a higher total value. In line with this, Katz (1978a, 1978b) found autonomy to be negatively related to job satisfaction for employees just starting a new job or entering the organization (i.e., junior researchers). He argues that new employees first need to discover the relevance of their work within the group before issues like autonomy become more important. It seems that young researchers first need to get socialized in research before their need for supervision decreases.

The *reward system* is an important management tool for motivating researchers. As Merton (1957) pointed out long ago, the institution of science has developed a reward system designed to give recognition and esteem to those scientists who have best fulfilled their roles; to those who have made genuinely original contributions to the common stock of knowledge. Incentive mechanisms can be used to motivate researchers with both material and intangible forms. The presence of a reward system positively influences research performance (Omta & De Leeuw, 1997; Spangenberg et al., 1990). Examples of material rewards are financial bonuses, a salary increase or promotion, and fringe benefits. Examples of intangible rewards are additional opportunities for career planning, facilities for developing skills (courses and conferences), support in achieving recognition through publishing, presentations and fellowships, flexible working hours and locations, possibilities for gaining experience in foreign research groups, and receiving public recognition. Above a certain income level, material incentives are only important in limited circumstances, for example, if researchers have low salaries compared to others (Gustad, 1960). McKeachie (1979) argues that salary has a relative meaning for researchers; it is compared with past raises and those of one's peers. Salary symbolizes the achieved level of competence in the eyes of respected peers. Rewards that motivate a researcher can change over time in the development of a career, and especially in the early years of one's career, the symbolic meaning of a salary raise could be important. A meta-analysis showed that financial incentives can be important in encouraging employee performance, but only for the quantity and not the quality of the output (Jenkins, Mitra, Gupta, & Shaw, 1998). In line with the modest role of material incentives in enhancing performance, academic research leaders seldom use them and concentrate more on intangible incentives. (Van der Weijden et al., 2009). Intangible rewards, especially praise and prizes, positively correlate to several aspects of performance, namely, the number of publications, the proportion of external research funding, and the number of submitted and granted research proposals (Van der Weijden, 2007; van der Weijden, de Gilder, Groenewegen, & Klasen, 2008). Hagstrom (1971) found that honoring scientists for their accomplishments is important; together with memberships of advisory committees, honorific awards account for 44% of the variance in departmental prestige. Researchers seem to value material incentives less than intangible incentives.

Academic group leaders emphasize *communication* with their researchers as a prime task. Indeed, the frequency of interaction within groups has a positive relationship with different aspects of performance (Allen & Sloan, 1970; Harris & Kaine, 1994; Kretschmer, 1985; Ramsden, 1994; Visart, 1979). Internal communication includes group meetings, retreats, research meetings on ongoing projects and feedback on results, meetings to discuss long-term research policy, and an internal website (e.g., Frederiksen, Hemlin, & Husted, 2004; Mets & Galford, 2009; Van der Weijden, 2007). As early as the 1960s, Pelz and Andrews (1966) found that the more communication is sought and received, the better researchers perform, because communication is important for intellectual stimulation, the generation of new ideas, catching errors, coordination, and reflection. Hamming's 'open doors' quotation nicely illustrates the role of interaction with colleagues:

> If you have the door to your office closed, you get more work done today and tomorrow, and you are more productive than most. But 10 years later somehow you don't quite know what problems are worth working on; all the hard work you do is sort of tangential in importance. He who works with the door open gets all kinds of interruptions, but he also occasionally gets clues as to what the world is and what might be important. There is a pretty good correlation between those who work with the doors open and those who ultimately do important things, although people who work with doors closed often work harder. Somehow they seem to work on slightly the wrong thing—not much, but enough that they miss fame. (Erren, 2008, p. 475)

It is easier to establish contact and collaboration with colleagues in a more consolidated research group. A consolidated research group has attained a certain size, composition, duration, autonomy, funding, member involvement, cohesiveness, intra- and interteam collaboration, and competitiveness. Consolidation increases with group age. Being a member of a consolidated research group fosters participation in funded projects and develops the potential to publish in international, mainstream journals (Martin-Sempere, Garzon-Garcia, & Rey-Rocha, 2008; Martin-Sempere, Rey-Rocha, & Garzon-Garcia, 2002; Rey-Rocha et al., 2002, 2006; Rey-Rocha, Garzon-Garcia, & Martin-Sempere, 2007). Consolidation must be balanced by an open attitude toward external sources (e.g., asking an expert), because too much consolidation leads to conservation and insulation ('groupthink'; see Janis, 1972), which lowers creativity. For example, Moorhead and Montanari (1986) showed that groups that felt most insulated from outsiders tended to have lower performance.

Network Management

Group leaders need to manage their groups but also their external relationships. Network management is the dependence on the environment for gaining access to resources and acquiring visibility and a reputation for the research group (Bozeman & Corley, 2004; Oh, Chung, & Labianca, 2004; Oh, Labianca, & Chung, 2006; Reagans, Zuckerman, & McEvily, 2004; Reagans & Zuckerman, 2001). The social capital of group leaders is often crucial to attaining high performance (Mehra, Dixon, Brass, & Robertson, 2006). The smaller the group, the more important it is to build a wide network (Von Tunzelmann et al., 2003, p. 15). Relationships can be built by collaborating with research groups located in the same research institute, in other domestic research institutes, or in research institutes abroad,[1] especially interactions with the scientific community have a positive influence on research performance (e.g., Spangenberg et al., 1990). They offer an intellectual exchange that stimulates creativity (Melin, 2000; Wagner, 2005).

METHODOLOGY

We designed a survey (section survey) that is based on the literature summarized earlier as well as on several open interviews with group leaders. Additionally, we created a performance index to identify the excellent groups (section performance index).

Performance Index

A performance index was constructed to differentiate between high-performing research groups and other, average-performing research groups. For this differentiation it is crucial to define high, or excellent, research performance. However, there is no consensus on an objective definition of research excellence due to its multidimensional character. One way to clarify the concept of excellence is to use it comparatively (Tijssen, 2003). The comparative approach evokes two main issues. The first concerns the set of comparable entities, or, in other words, which frame of reference to use for the comparison. The second issue concerns which indicators to use for comparing academic performance.

The first issue about the frame of reference is important because excellence can only be evaluated in an appropriate context (Bornmann, Mutz, Neuhaus, & Daniel, 2008; Van der Meulen, 1995). We have to deal with two frames of reference that determine the comparison of our set of Dutch health research groups: the international context and the disciplinary context. From an international perspective, overall Dutch health research is judged to be of high quality (Nederlands Observatorium van Wetenscahp en Technologie [NOWT], 2010). Basic and experimental medicine and health-care sciences in the Netherlands score slightly above the global average impact; Dutch biomedical science scores approximately 15% above global average impact, and Dutch clinical medicine about 30% (NOWT, 2010). In other words, high-performing Dutch groups are also excellent from an international perspective.

The disciplinary context was used to distinguish between different types of research in the health research field, since disciplines differ heavily in publication and citation patterns (e.g., Baird, 1986; Martin & Irvine, 1983; Neuhaus & Daniel, 2009; Schubert & Braun, 1996; Van Raan, 2004; Whitley, 2000; Wouters, 1999). In the health sciences, the relation to patient care provides a good framework for discipline classification (Van der Weijden, 2007; Van der Weijden, de Gilder, Groenewegen, & Klasen, 2008). Group leaders were asked to classify their groups' research in terms of one or more of 28 (sub)disciplines in health research (e.g., public health, immunology or oncology; for an overview, see Van der Weijden, 2007, p. 202). On this basis, we classified the research groups in three categories distinguished by the relation to the patients[2]:

- *Preclinical research* is basic (life sciences) research, usually laboratory based, for example, immunology, micro-biology and neurosciences.
- *Clinical research* is application oriented, with direct patient contact, for example, dermatology, nephrology, and psychiatry.
- *Paraclinical research* into health-care systems often has a social sciences perspective: Researchers have an advisory relationship with patients, for example, social medicine, public health, and medical psychology.

The issue of research excellence indicators is more complicated as it involves the multidimensional character of excellence. Academic performance consists of various aspects that can be measured with different indicators (e.g., King, 1987; B. R. Martin, 1996; Schubert & Braun, 1996; Tijssen, 2003). Commonly used indicators that measure academic performance are publication and citation counts (e.g., King, 1987; Martin, 1996; Van Raan, 2004, 2008). We use four types of publication and citation indicators during a 3-year time span (2004 to 2006). Each indicator measures a different aspect of performance. The first indicator is the publication count of the group leader; it reflects the *quantity* of knowledge produced. The second indicator is the citation count of the group leader; it reflects the *visibility* of the output. The third indicator is the publication count divided by the number of researchers in the group; it reflects *productivity* and normalizes output for group size. When a group attracts a lot of citations for each publication, they tend to work at the research front and publish innovative ideas that their colleagues consider interesting (Wouters, 1999). Therefore, we take as a final indicator the number of citations per publication, which reflects the *creativity* of the group.

Data about academic performance of the research groups were collected from three databases. PubMed (US National Library of Medicine's search service) was used to obtain the *publication counts* of research groups over the given 3-year period (2004–2006). Thomson Reuters (formerly ISI) Web of Knowledge was used to obtain publication counts for the same period.[3] Finally, the Nederlandse Onderzoeks Databank (Dutch Research Database) was used to check whether research group leaders were affiliated to more than one organization; if so, we included their publications from both affiliations.[4] We collected publication counts from both databases by searching for the group leaders' surname and first two initials. In addition, we collected *citation counts* and the number of *citations per publication* from the Web of Knowledge[5] over the publications published in the specific 3-year period with a citation span of January 2004 through May 2008.[6] The last performance indicator was calculated by normalizing the number of *publications by group size.*

Our definition of high performance was based on the assumption that the best research groups would have outstanding performance in all four aspects of academic performance. It appeared, however, that some groups had many publications but not an outstanding number of citations, while other groups had many citations but not an outstanding number of publications. This can be explained by a difference in publication and citation strategies (Hemlin & Gustafsson, 1996; Moed, 2000; Whitley, 2000; Wouters, 1999). The four indicators correlate moderately strongly (see Table 5.1).

The first two unweighted indicators measure the absolute performance of research groups: total output and impact. Of course, the larger the group, the more that can be published and the more citations that can be gained; hence, these indicators favor larger groups. The third and fourth indicators measure relative performance: publications per FTE (full-time equivalent) researcher and citations per article. These indicators favor the more efficiently operating groups. As coordination costs increase with size, a key management goal is to combine the benefits of size (high absolute output, visibility, and creativity) with managing the coordination costs in order to remain efficient (low costs per unit of output and impact). Thus, excellence is multidimensional, and we attempt to measure it by combining the four indicators. This opens up the possibility for smaller (but not too small) groups that are excellent, as well as for large groups that are not too inefficient. Indeed, in our selection of high-performing research groups we found both large and small groups. Similar findings are reported in a study about innovation published by West and Anderson (1996), who reported that group size in hospitals is not important for predicting team innovation.

The four indicators were standardized on the maximum score for each indicator and then weighted into an aggregate performance index, separately

Table 5.1 Spearman's rho for Performance Indicators

		Citations	Publications/ group size	Citations/ publication
Publications	Spearmans rho	.848***	782***	.357**
	Sig. (2-tailed)	.000	.000	.000
	N	184	184	184
Citations	Spearmans rho	1.000	.688***	.722***
	Sig. (2-tailed)		.000	.000
	N	184	183	184
Publications/ group size	Spearmans rho		1.000	294***
	Sig. (2-tailed)			.000
	N		184	183

for each discipline (paraclinical, preclinical, and clinical). The performance index formula is

$$\text{Performance index} = \frac{\left[\dfrac{P}{P_{max}} + \dfrac{C}{C_{max}} + \dfrac{X}{X_{max}} + \dfrac{Y}{Y_{max}}\right]}{4}$$

where *P* is the number of publications; *C* is the number of citations; *X* is *C/P* and *Y* is *P*/group size; *P*max is the maximum value of *P* within the sample, which varies by discipline; *C*max is the maximum value of *C* within the sample, which varies by discipline; *X*max is the maximum value of *C/P* within the sample, which varies by discipline, and *Y*max is the maximum value of *P*/group size, which again varies by discipline.[7] The resulting performance index is the skewed distribution represented in Figure 5.1.[8]

The figure shows that the natural cutoff point that separates excellent performing groups from other groups is 0.4. This point corresponds to the 'elbow' visible in the graph. Below the threshold of 0.4, the data points form a more continuous line, and the curves become flatter. Research groups

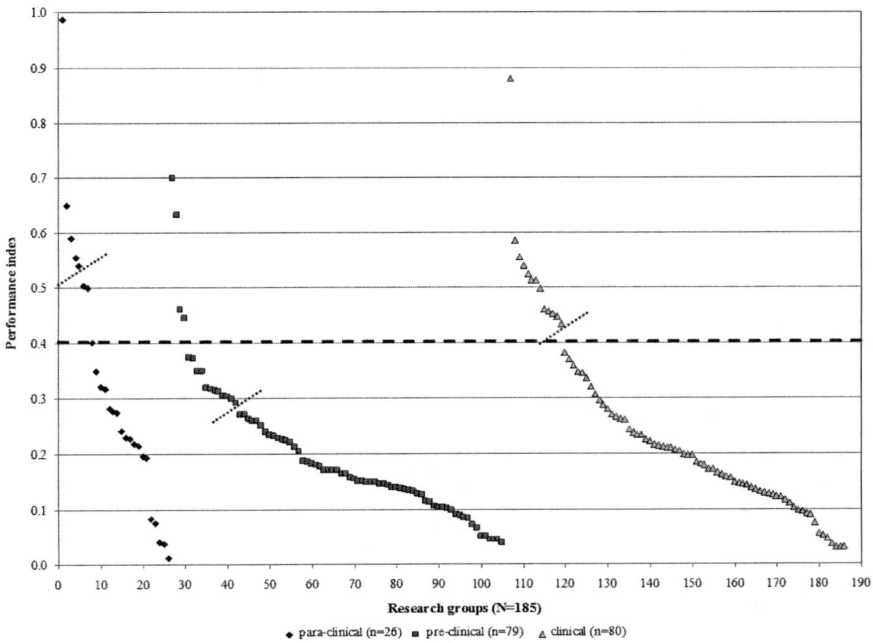

Figure 5.1 Sample distribution

below the threshold are more similar in their academic performance than are the top-performing groups with relatively higher scores. The top of the skewed distribution—above the 0.4 threshold—results in a very small selection of top-performing research groups and is unevenly distributed among the domains: There are 7 paraclinical, 4 preclinical, and 13 clinical groups. The sum total of the top selection—24 research groups—is too small for detecting significant differences between top-performing and other research groups. For statistical reasons, we therefore decided to select a larger proportion of groups. This ensures that the high-performing groups account for a similar proportion of all groups in each domain.[9] The selection of high-performing research groups includes 5 paraclinical (19.2%), 16 preclinical (20.3%), and 13 clinical research groups (16.3%). The difference between high-performing and average groups is indicated by the dashed lines in Figure 5.1, which show a gap between high-performing and other research groups.[10]

To test that we had selected an appropriate set of high-performing research groups, we conducted four analyzes. First, we examined whether groups that were not among the selected high-performing groups still had outstanding scores on two or more indicators. This was not the case. In addition, we asked experts in the Dutch biomedical and health-care fields to evaluate some high-performing research groups (not in our sample) and to compare the publication and citation scores of their groups with our selected high-performing groups. Analysis shows similar publication and citation scores. Thirdly, we compared the publication and citation scores of group leaders who had won the Spinoza Prize (the most prestigious Dutch award) in all three disciplines with our selection of high-performing groups, and these totals were comparable in both cases. Finally, we conducted a sensitivity analysis: a comparative analysis between all high-performing and other research groups, again excluding the weak-performing groups (the 18 lowest-scoring groups on the performance index). This had no consequences for the results. It seems that weak-performing groups were headed by less experienced group leaders; we return to this issue in the Results section.

Survey

The data were collected in 2007 through a survey of health research group leaders employed by university medical centers or by biomedical public research institutes in the Netherlands (Van der Weijden, 2007; van der Weijden, de Gilder, Groenewegen, & Klasen, 2008; Van der Weijden et al., 2009). Names and addresses of group leaders (mainly chaired professors) were obtained from administrative records. We approached 832 group leaders, of whom 131 were ineligible because they had only been involved with their groups for a short time (less than 6 months), or they had not been able to complete the questionnaire because the group did hardly any research (a focus on patient care), or they had filled in the questionnaire as a department

Table 5.2 Description and Descriptive Statistics of Academic Leadership Practices

Variables	Description	Descriptive statistics				
		Mean	Median	Standarddeviation	Minimum	Maximum
Resource strategy						
Funding sources	Percentage of total funding from:					
	–Institutional funding	32.18	30.00	22.49	0	100
	–Competitive funding	20.72	20.00	18.55	0	80
	–Contract funding	17.08	10.00	19.93	0	89
	–Charity funding	17.67	10.00	19.82	0	100
	–European funding	7.78	0	14.33	0	85
	–Other funding sources	4.57	0	10.99	0	90
Number of funding sources	Measure the total number of funding sources	3.57	4.00	1.9	1	6
Group size	Number of scientific and support staff (FTE)	16.59	16.00	8.39	3	40
Scientific staff	Number of scientific staff (FTE)	13.23	12.0	6.76	3	34
Staff constellation	Percentage of total scientific staff:					
	–Professors	11.03	10.00	6.64	0	33
	–Senior staff	30.96	30.77	13.61	0	75
	–PhDs	51.31	50.00	15.11	14	88
	–Other scientific staff	6.70	0	10.66	0	55
	–Support staff (analysts and technicians)	18.32	18.18	14.79	0	60
	Presence of coleader				yes: 73.9%	no: 26.1%
Supervisory capacity	Measure the number of PhDs per professor	5.75	5.0	3.83	0.5	23
Leadership						
Coresearcher versus manager	Measure degree of research involvement of the leader where 1 is *totally disagree* and 5 is *totally agree*: I feel more like a researcher than like a manager	3.32	3.0	1.0	1	5

(*Continued*)

Table 5.2 (Continued)

Variables	Description	Descriptive statistics				
		Mean	Median	Standarddeviation	Minimum	Maximum
Highly skilled scientist	My staff think of me as a highly skilled scientist	3.69	4.0	0.82	1	5
Importance of visibility	Measure degree of importance to obtain visibility in top journals, like *Nature*, *Science*, and the *Lancet*, where 1 is *not important* at all and 5 is *very important*	3.31	3.0	1.0	1	5
Importance of preliminary evaluation of research proposals	Measure importance of quality control for acquiring funding where 1 is *totally disagree* and 5 is *totally agree*: Internal pre-assessments of research proposals generally result in major changes to receiving external funding	4.03	4.0	0.86	1	5
Time allocation	Percentages of time spent on:					
	–research	44.52	42.95	16.19	11.00	87.40
	–education	13.62	10.00	10.61	0	55.50
	–group management	16.53	16.70	8.39	0	48.60
	–patient care	16.74	11.05	18.87	0	78.00
	–percentage of groups				yes: 56.4%	no: 43.6%
	–external management tasks (network management)	8.15	6.20	8.15	0	40.60
Group management						
Preliminary evaluation of research proposals	Dummy variable measuring whether or not leaders evaluate research proposals before they are submitted to funding agencies				yes: 80.2%	no: 19.8%

head instead of as a group leader—we verified this by checking the reported group size (>40 group members) and, if available, their organization's website—or they had just left the research group (emeritus professors). The participating group leaders ($N = 188$) returned the completed questionnaires with an overall response rate of 27%. To maximize the response rate we used the tailored design method (Dillman, 2000). Because of incomplete data, only 185 research groups were used in the analysis.

Nonresponse analysis shows that the respondents can be regarded as a representative sample of the Dutch biomedical and health research groups. First, we checked whether the respondents were evenly distributed among the various research institutions and the disciplines, and this proved to be the case. Second, we compared the scholarly performance of respondents and non-respondents. Since publication counts are highly skewed, a Mann–Whitney U test was used. Respondents did not significantly differ (Mdn = 25.0) from nonrespondents (Mdn = 23.0), $U = 62315.0$, $p = 0.365$, $r = -0.03$. The mean difference between the two groups was also small (33 publications for respondents vs. 31 for nonrespondents).

The questionnaire was designed on the basis of unstructured interviews with experienced research leaders and a literature review. Interview partners were asked to reflect on a list of management activities that might contribute to a group's research performance. We used both interview and literature data to develop and design survey questions and possible answer categories. The questionnaire was composed with the assistance of survey experts, and it was pretested. The final version of the questionnaire was constructed using the comments and suggestions of the pretesters on both individual questions and the questionnaire as a whole. The survey questions dealt with academic leadership practices (see Table 5.2).

Analysis

As the data were not distributed normally (Kolmogorov-Smirnov test), we conducted Mann–Whitney U tests and chi-square tests to compare differences in academic leadership between high-performing and other research groups. See the Appendix for descriptive statistics of academic leadership by performance. When we find disciplinary differences, we report separate results for the three domains.

RESULTS

Characteristics of the Group Leader

The performance of the research groups generally did not seem dependent on their leaders' years of experience, age, or gender. No differences were found in years of experience as group leader in high-performing research groups (*Mdn* = 12.0) compared to other groups (*Mdn* = 10.0), $U = 2{,}150.0$, $p = 0.35$, $r = -0.07$, or in age (*Mdn* = 49.5 and *Mdn* = 54.0, respectively), $U = 2{,}140.0$, $p = 0.13$, $r = -0.11$. Concerning gender, the percentage of groups headed by women did not differ by performance: $\chi 2(1, N = 185) = 1.040$, $p = .31$. It should be noted that most groups are headed by men (87.6%).

Resource Strategy

Funding Sources

Research groups obtain funding from various sources among others institutions, research councils and industry or companies. High-performing research groups from the preclinical and clinical domain obtain their funding from more sources (*Mdn* = 4.0) compared to the other groups (*Mdn* = 3.0), $U = 1{,}178.5$, $p = 0.007$, $r = -0.22$. Also, leaders of high-performing groups—including paraclinical groups—acquire a higher percentage of competitive funding from research councils (*Mdn* = 25.0 and *Mdn* = 20.0, respectively), $U = 1{,}913.0$, $p = 0.067$, $r = -0.14$.

Human Capital

All the groups showed a variance in group size; there are big and small groups in both high-performing and other research groups. The size of most groups varies between 15 and 18 FTE.[11] Approximately three quarters of the research groups have a second group leader.[12] We corrected size for the presence of a second group leader. Also, no significant differences were found between high-performing (*Mdn* = 10.0 FTE) and other research groups for size per group leader (*Mdn* = 8.5 FTE), $U = 2{,}133.0$, $p = 0.137$, $r = -0.11$. Research groups have, on average, 9.8 FTE per group leader and 7.9 FTE scientific staff per group leader.

The group composition—regarding the proportions of scientific and support staff—varies between research groups but not between high-performing (*Mdn* = 0.83) and other groups (*Mdn* = 0.82), $U = 2470.0$, $p = 0.774$, $r = -0.02$. All groups have approximately 82% scientific staff (professors, senior researchers, PhD students, and other scientific staff) and 18% support staff (analysts and technical staff). No differences were found between the three disciplines. For most of the groups (76%), the composition had been stable over a number of years. Other groups have had a yearly change in composition (18%), and some groups have changed even more often (6%). The predominance of composition stability as well as the variety in composition between groups may relate to the type of research conducted in particular groups. For example, some groups need a high proportion of support staff to carry out lab experiments or analyses.

Finally, concerning supervisory capacity, we did discover significant differences between high-performing (*Mdn* = 6.0) and other research groups

(*Mdn* = 5.0), $U = 1{,}838$, $p = 0.084$, $r = -0.13$. Professors in high-performing research groups supervise more junior researchers (an average of 6.7 PhDs) per professor compared to leaders of other groups (5.5 PhDs per professor). Leaders of high-performing research groups tend to attract more young researchers.

Leadership

Research Involvement

Research involvement was measured by the degree in which group leaders felt committed to the research on the shop floor. Leaders of high-performing groups perceived themselves to be more like researchers than did managers (*Mdn* = 4.0) compared to leaders of other research groups (*Mdn* = 3.0), $U = 1{,}826.5$, $p = 0.012$, $r = -0.19$. Moreover, leaders of high-performing groups were more convinced that they are seen as high-skilled scientists (*Mdn* = 4.0) than did leaders of other research groups (*Mdn* = 4.0),[13] $U = 1{,}261.0$, $p < 0.001$, $r = -0.32$.

Leaders of high-performing groups, except for paraclinical groups, also attach more importance to quality. They strive for visibility in top journals (*Mdn* = 4.0) more so than did leaders of other groups (*Mdn* = 3.0), $U = 1{,}229.0$, $p = 0.003$, $r = -0.24$. In addition, it appears that leaders of high-performing research groups attached more value to the preliminary evaluations of research proposals for acquiring external funding (*Mdn* = 4.0) compared to leaders of other research groups (*Mdn* = 4.0),[14] $U = 1166.5$, $p = 0.014$, $r = -0.20$.

Time Allocation

Intuitively, you would expect leaders of high-performing research groups to spend more time on research, given that they also feel more involved with the research. Indeed, leaders of high-performing clinical groups do spend more time on research (*Mdn* = 43.9%) than do leaders of other clinical groups (*Mdn* = 39.3%), $U = 295.5$, $p = 0.068$, $r = -0.20$. However, this does not apply for preclinical and paraclinical groups. Concerning leaders of high-performing preclinical groups, they spend less time on education (*Mdn* = 5.8%) than do leaders of other groups (*Mdn* = 9.0%), $U = 307.0$, $p = 0.016$, $r = -0.27$. For paraclinical groups, no significant differences were found between leaders of high-performing groups and those of other groups. Apparently, the optimal time allocation among tasks varies between disciplines. But it seems to be that leaders of high-performing groups are less distracted by other tasks and focus more on research tasks. In the clinical domain where patient care takes a lot of time, leaders of high-performing groups spend more time on research. In the preclinical domain, leaders of high-performing groups spend less time on education, which thus also allows them to spend more time on research.

Group Management

Coordination Versus Autonomy

Leaders of high-performing research groups highly value quality. But despite the fact that leaders of high-performing groups regarded preliminary evaluations as very important to obtain external funding, they did not conduct more of these evaluations. Preliminary evaluations of research proposals are very common in 80% of the research groups.

To set the group's research agenda, a group leader draws on strategic considerations such as visibility, possibilities for applications in health care, and innovation. When we examine the percentage of strategic considerations that is valued as important, we found that leaders of high-performing research groups—except for the paraclinical groups—attached a statistically significant higher value to the various strategic considerations ($Mdn = 62.5$) than did leaders of other groups ($Mdn = 50.0$), $U = 1{,}183.5$, $p = 0.008$, $r = -0.22$. Leaders of high-performing groups seem to take a broader set of considerations into account when setting their research agendas, which may create more freedom for group members to place their individual research interests on the group's program.

Finally, we analyzed whether researchers participate in meetings to discuss the long-term research policy of the group. The percentage of leaders of high-performing research groups (93.1%) that organize these meetings for their group members is higher than the percentage of leaders of other research groups (80.0%), $\chi^2(1, N = 159) = 2.806$, $p = 0.094$. That is, it is 3.4 times more likely that researchers can express their views on long-term research policy in high-performing research groups than in other research groups. As an overall conclusion, the emphasis is more on autonomy than on coordination in high-performing research groups.

Reward System

No significant differences were found between groups that did or did not award nonfinancial prizes, $\chi^2(1, N = 184) = 0.709$, $p = 0.400$. A little bit more than half of the group leaders (57%) use this management tool to motivate their researchers.

Research Communication

Research communication stands for the organization of internal meetings for giving feedback on research presentations, concept papers, or research proposals. In particular, high-performing research groups hold significantly more feedback meetings ($Mdn = 3.7$) than do other groups ($Mdn = 3.3$), $U = 2{,}004.0$, $p = 0.092$, $r = -0.12$. Also, leaders of high-performing groups are 2.7 times more likely to organize meetings about ongoing projects at least once a week (namely, 70.6% of the groups) than did other research groups (namely, 47.0% of the groups), $\chi^2(1, N = 185) = 6.171$, $p = 0.013$.

Network Management

Network activities refers to the number of days a year that are spent on positioning the research group in the academic and societal environment by giving lectures, attending conferences, participating in editorial boards and participating in assessment committees. The highest network activity is shown in leaders of high-performing research groups—with the exception of paraclinical groups—(Mdn = 50 days) in comparison with leaders of other groups (Mdn = 40 days), U = 1,200.0, p = 0.002, r = –0.24. The amount of time spent on external management tasks in proportion to the total work time is also higher for leaders of high-performing groups (Mdn = 7.1%) than for leaders of other groups (Mdn = 5.9%), U = 663.0, p = 0.078, r = –0.17. This does not apply for leaders of clinical groups, which might be an effect of the higher proportion of time they spend on patient care duties.

Weak-Performing Research Groups

Now we have examined the differences between leaders of high-performing and other research groups, let us come back to the issue of weak-performing groups. Leaders of weak-performing groups did indeed have some different

Table 5.3 Academic Leadership by High-, Other-, and Weak- Performing Research Groups

		No.	Median	Bottomquartile	Topquartile	Mean	Stddeviation
Experience in years	*High*	32	12.0	10.0	15.0	12.5	6.14
	Other	132	10.0	6.0	17.0	12.1	7.59
	Weak	18	6.3	4.9	11.3	8.6	5.95
Percentage of junior researchers	*High*	34	55.8	46.1	66.7	54.6	14.4
	Other	132	50.0	40.2	63.5	51.7	15.0
	Weak	18	46.4	27.7	52.2	41.5	15.3
Network activities in days per year	*High*	34	45.0	30.0	70.0	49.9	24.2
	Other	133	40.0	25.0	50.0	38.8	17.0
	Weak	18	30.0	20.0	40.0	31.4	14.4

	J	z	R
Experience in years	3011.0	–1.801*	-0.13
Percentage of junior researchers	2904.5	2.533* *	-0.19
Network management in days per year	2719.5	–3.176***	-0.23

Note: Jonkeere's test with effects reported at * = $p < 0.10$, ** = $p < 0.05$, and *** = $p < 0.01$. J = test score; z = standardized test score; R = effect size.

characteristics in academic leadership. Most interesting, leaders of weak-performing research groups have less experience as group leaders than do leaders of high-performing and other research groups. The difference in years of experience suggests that these research groups may be in early phases of development, rather than just harboring weak performers. This phase is characterized by a different group composition: They have fewer junior researchers. Also, they spend less time on network activities. In other words, their activity in the (international) scientific community is less well developed (see Table 5.3).

Before summarizing our conclusions in the next section, let us discuss several (potential) limitations of this study. First, we focused on one set of research fields: the biomedical and health-care sciences. However, these fields cover about 40% of all public research in the Netherlands. In addition, the biomedical and health-care sciences are rather heterogeneous and consist of fundamental preclinical research, application-oriented clinical research, and health care research with an approach comparable to the social sciences. Nevertheless, the situation may be different in the technical sciences, for instance, and in parts of the humanities. Second, the study focused on one medium-sized, high-impact science country: the Netherlands. Leadership and management practices may, of course, differ between countries, so comparative studies are needed as a next step. Third, as the research system changes over time, leaders may need to adapt their behavior. In this volume (Verbree, Van der Weijden, & Van den Besselaar, Chapter 6, this volume), we report these changes over time. Finally, the sample was not very large, and this prevents us from conducting more advanced model testing. We hope to be able to do that after a new survey.

CONCLUSION AND DISCUSSION

In this study, we searched for the main organizational, management and leadership factors that are expected to influence performance. Our survey of research leaders included key factors identified by the relevant literature. Our research method adds to the current state of knowledge in that we do not focus on bivariate relations between academic leadership variables and performance, but aim at identifying the productive characteristics of academic leadership by comparing leaders from high-performing and average-performing, that is, good, research groups. From the comparison between high-performing and other research groups, we conclude that leaders of high-performing research groups emphasize both leadership practices and management tools. They are better able to combine diverse activities. Leaders of high-performing research groups are more research oriented; they perceive themselves as highly involved in research as coresearchers rather than as managers, and as highly skilled scientists; they set a high quality standard, and they are less distracted by nonresearch tasks. Furthermore, these leaders are also more devoted to their group management, because they organize internal communication on research more intensively, and provide researchers the freedom to develop their individual research interests. In addition to group management, leaders of high-performing groups spend more time on network management. These activities are focused both on obtaining visibility and a reputation in the scientific community as well as on obtaining resources from various sources, especially from competitive funding agencies. As a general conclusion, leaders of high-performing research groups are versatile: They are *all-rounders*. They set an example by showing strong research commitment, and they are able to manage their group both internally and externally.

Implications

If high-performing and other research groups differ in academic leadership, it might be possible to identify opportunities to improve organizational, management, and leadership conditions for excellent performance. On one hand, it is important for a research leader to have internal and external managerial skills in order to acquire and combine the necessary resources and to direct the research process. On the other hand, successful research leaders need to be committed to research. More specifically, they should be involved in the research of their group by generating ideas, participating in meetings, carrying out analyses, and motivating their researchers. These two tasks may be at conflict. Managing both the *research process* and the *resources* calls for an ability to organize the group in a well-structured way, whereas leading *researchers* requires an ability to handle the open, spontaneous processes needed to develop creative and innovative ideas. Although leaders of high-performing research groups appear to be capable of both, these qualities may be difficult to find in a single person. This may explain why the trend in increasingly larger research groups is to have coleaders; this is a pattern we clearly observed in our population (Van der Weijden et al., 2009). Based on recent interviews with research leaders, we observed that coleaders are young and direct their own research line under the management of the principal group leader. Coleaders provide opportunities for task specialization in order to deal with the possible tension between the leading of *researchers* and the managing of *research inputs and processes*.

Our conclusions about the importance of versatility in academic leadership also apply to the researchers in the group. Previous research by Pelz and Andrews (1966) and Andrews (1979a) showed that diversity in work activities is weakly but positively related to research performance. They distinguished eight types of diversity, including the allocation of time over several tasks such as teaching and administration; interdisciplinary orientation in which one makes use of methods, theories, and other specific elements developed in other fields; and the obtainment of funding from multiple sources. When different aspects of diversity are aggregated, the effect of diversity becomes stronger, and explains roughly ten per cent of the variance in performance (Andrews, 1979a). This implies not only that group leaders should carry out wide-ranging activities; they also need to promote versatility in the work activities for their group members.

This chapter indicates that leaders of top-performing and weak-performing research groups have different characteristics, and this has implications for research evaluation. The findings on weak-performing research groups demonstrate that assessment based solely on bibliometric indicators can disadvantage younger, less experienced group leaders. Our analysis suggests that these groups are generally in an early phase of development rather than simply weak performers. Research evaluation should not overlook the time necessary to develop both a research group and the leadership and management qualities needed for high performance (Braam & van den Besselaar, 2010). Finally, assessing only output and impact is clearly not sufficient in evaluating research. In order to come up with useful recommendations for improving performance, evaluation protocols should include the quality of organization, management, and leadership. This chapter indicates what to focus on in evaluation protocols, for instance, an intensive communication structure, diversity in funding sources, and the research focus of the group leader.

In this chapter, we defined high-performing research groups based on multiple performance indicators, namely, quantity, visibility, productivity, and creativity. These four indicators measure different aspects of performance and correlate moderately strongly (Table 5.1). An open question remains whether leaders of top-creative groups behave differently from leaders of top-productive, top-visible, or top-quantity groups. We investigated this in a follow-up study using multivariate analyses (Verbree, Horlings, Van der Weijden, & Van den Besselaar, 2011).

ACKNOWLEDGMENTS

We are grateful to the 27 participants in a four-hour expert meeting, who provided feedback on draft conclusions and policy recommendations. Participants were research managers at various levels and policy makers representing such organizations as university medical centers, nonuniversity research institutes, research councils, charities and research policy institutes. We wish to thank Robert Braam and Peter Groenewegen for their insightful feedback on earlier versions of the literature section, and Edwin Horlings for his valuable contributions in analyzing the results. Also, we highly appreciate the comments of two anonymous reviewers, which have improved the paper considerably. Finally, this research was not possible without the support of the Rathenau Institute.

NOTES

1. See Katz and Martin (1997) for a conceptualisation of research collaboration and its different forms.

2. Our classification is similar to the NOWT classification: paraclinical research (health sciences), preclinical research (basic and experimental medicine, biomedical science), and clinical research (clinical medicine).
3. PubMed differs from Web of Knowledge in that it also contains Dutch publications. It is important to include Dutch publications because some group leaders focused mainly on a national level and rarely published in international journals. But PubMed is limited by its less advanced search options; it is not possible to select a particular organization, which was a problem with very common names. Web of Knowledge differs from PubMed in that it holds a more varied set of documents, such as conference abstracts, notes, and letters. With Web of Knowledge, it is also possible to connect the names of group leaders to their organization or affiliation. The two databases sometimes gave dissimilar publication counts for the same research group, caused by errors in either database or by a difference in content. After a test for research groups with large differences between the two databases, the lower publication count proved to be the more accurate.
4. Sometimes group leaders were affiliated to two organizations, either because they had moved to another organization in the selected 3-year period or because they actually held two positions.
5. These data were not available in PubMed.
6. This was the biggest time span as possible at the moment of collection (May 12–26, 2008).
7. The specific maximum values are not given in order to protect the anonymity of the research groups. In addition, maximum values of citation counts vary by date of collection—these data were collected within a specific period (May 12–26, 2008)—and are thus not comparable with recent data. Some citation data were corrected after this period, but this had no consequences for the selection of high-performing research groups.
8. None of the high-performing groups scored a 1 on the performance index, which means that none of the high-performing groups scored the maximum in all four performance indicators.
9. By using the larger proportion of high-performing groups, we found the same differences between leaders of high-performing and other groups and a few more than with the smaller top selection, but with the smaller selection the statistical power is not sufficient (the probability of Type II errors increases).
10. By selecting a similar proportion of high-performing groups of each domain, we can be sure that the differences between high-performing and other research groups are not a domain effect.
11. Based on a 95% confidence interval for the mean.
12. The second group leader may have a formal or informal leadership position.
13. Note that the medians for leaders of high-performing and other research groups are similar, but that the distribution is significantly different. Mann–Whitney test differences in (mean) ranks and not in medians.
14. See note 13.

REFERENCES

Allen, T. J., & Sloan, A. P. (1970). Communication networks in R and D laboratories. *R & D Management*, *1*(1), 14–21.

Allison, P. D., & Long, J. S. (1990). Departmental effects on scientific productivity. *American Sociological Review*, *55*(4), 469–478.

Amabile, T. M. (1993). Motivation synergy: Toward new conceptualizations of intrinsic and extrinsic motivation in the workplace. *Human Resource Management Review*, *3*(3), 185–201.

Anderson, E. M., & Shannon, A. L. (1988). Toward a conceptualization of mentoring. *Journal of Teacher Education*, *39*(1), 38–42.
Andrews, F. M. (1979a). Motivation, diversity, and the performance of research units. In F. M. Andrews (Ed.), *Scientific productivity. The effectiveness of research groups in six countries*. Cambridge, England: Cambridge University Press.
——— (1979b). *Scientific productivity. The effectiveness of research groups in six countries*. Cambridge, England: Cambridge University Press.
Andrews, F. M., & Farris, G. F. (1967). Supervisory practices and innovation in scientific teams. *Personnel Psychology*, *20*(4), 497–515.
Auranen, O., & Nieminen, M. (2010). University research funding and publication performance—an international comparison. *Research Policy*, *39*, 822–834.
Babu, A. R., & Sing, Y. P. (1998). Determinants of research productivity. *Scientometrics*, *43*(3), 309–329.
Baird, L. L. (1986). What characterizes a productive research department? *Research in Higher Education*, *25*(3), 211–225.
Barnowe, J. T. (1975). Leadership and performance outcomes in research organizations: The supervisor of scientists as a source of assistance. *Organizational Behavior and Human Performance*, *14*(2), 264–280.
Beaver, D. D. (2001). Reflections on scientific collaboration (and its study): Past, present, and future. *Scientometrics*, *52*(3), 365–377.
Bellas, M. L., & Toutkoushian, R. K. (1999). Faculty time allocations and research productivity: Gender, race and family effects. *Review of Higher Education*, *22*(4), 367–390.
Blackburn, R. T., Behymer, C. E., & Hall, D. E. (1978). Research note—correlates of faculty publications. *Sociology of Education*, *51*(2), 132–141.
Bland, C. J., & Ruffin, M. T. (1992). Characteristics of a productive research environment: literature review. *Academic Medicine*, *67*(6), 385–397.
Bornmann, L., Leydesdorff, L., & Van den Besselaar, P. (2010). A meta-evaluation of scientific research proposals: Different ways of comparing rejected to awarded applications. *Journal of Informetrics*, *4*(3), 211–220.
Bornmann, L., Mutz, R., Neuhaus, C., & Daniel, H. (2008). Citation counts for research evaluation: standards of good practice for analyzing bibliometric data and presenting and interpreting results. *Ethics in Science and Environmental Politics*, *8*, 93–102.
Bozeman, B., & Corley, E. (2004). Scientists' collaboration strategies: implications for scientific and technical human capital. *Research Policy*, *33*(4), 599–616.
Braam, R., & van den Besselaar, P. (2010). Life cycles of research groups: The case of CWTS. *Research Evaluation*, *19*(3), 173–184.
Cameron, S. W., & Blackburn, R. T. (1981). Sponsorship and academic career success. *Journal of Higher Education*, *52*(4), 369–377.
Carayol, N. (2003). Objectives, agreement, and matching in science-industry collaborations: Reassembling pieces of the puzzle. *Research Policy*, *32*(6), 887–908.
Carayol, N., & Matt, M. (2004). Does research organization influence academic production? Laboratory level evidence from a large European university. *Research Policy*, *33*(8), 1081–1102.
Carayol, N., & Matt, M. (2006). Individual and collective determinants of academic scientists' productivity. *Information Economics and Policy*, *18*(1), 55–72.
Cherchye, L., & Abeele, P. V. (2005). On research efficiency: A micro-anaylsis of Dutch university research in economics and business management. *Research Policy*, *34*(4), 495–516.
Cohen, B. P., Kruse, R. J., & Anbar, M. (1982). The social structure of scientific research teams. *Pacific Sociological Review*, *25*(2), 205–232.
Cohen, J. E. (1991). Size, age and productivity of scientific and technical research groups. *Scientometrics*, *20*(3), 395–416.

Crane, D. (1965). Scientists at major and minor universities—a study of productivity and recognition. *American Sociological Review*, *30*(5), 699–714.

Creswell, J. W. (1985). *Faculty research performance: Lessons from the sciences and social sciences*. Washington, DC: Association for the Study of Higher Education.

Crewe, I. (1988). Reputation, research and reality—the publication records of UK Departments of Politics, 1978–1984. *Scientometrics*, *14*(3–4), 235–250.

Cronan-Hillix, T., Gensheimer, L. K., Cronan-Hillix, W. A., & Davidson, W. S. (1986). Students' views on mentors in psychology graduate training. *Teaching of Psychology*, *13*(3), 123–127.

Culpepper, L., & Franks, P. (1983). Family medicine research—status at the end of the 1st decade. *JAMA–Journal of the American Medical Association*, *249*(1), 63–68.

De Haan, J. (1994). *Research groups in Dutch sociology*. Amsterdam, the Netherlands: Thesis Publishers Amsterdam.

Dillman, D. (2000). *Mail and Internet surveys: The tailored design method*. New York, NY: John Wiley & Sons.

Dundar, H., & Lewis, D. R. (1998). Determinants of research productivity in higher education. *Research in Higher Education*, *39*(6), 607–631.

Erren, T. C. (2008). Hamming's "open doors" and group creativity as keys to scientific excellence: The example of Cambridge. *Medical Hypotheses*, *70*(3), 473–477.

Fox, M. F. (1983). Publication productivity among scientists—a critical-review. *Social Studies of Science*, *13*(2), 285–305.

——— (1992). Research, teaching, and publication productivity—mutuality versus competition in academia. *Sociology of Education*, *65*(4), 293–305.

Fox, M. F., & Mohapatra, S. (2007). Social-organizational characteristics of work and publication productivity among academic scientists in doctoral-granting departments. *Journal of Higher Education*, *78*(5), 543–571.

Frederiksen, L. F., Hemlin, S., & Husted, K. (2004). The role of knowledge management in R&D: A survey of Danish R&D leaders' perceptions and beliefs. *International Journal of Technology Management*, *28*(7–8), 820–839.

García-Romero, A. (2006). Assessing the socio-economic returns of biomedical research (I): How can we measure the relationship between research and health care? *Scientometrics*, *66*(2), 249–261.

Gaston, J. (1975). Autonomy in the research role and participation in departmental decision-making. *The British Journal of Sociology*, *26*(2), 227–241.

Geuna, A. (2001). The changing rationale for European university research funding: Are there negative unintended consequences? *Journal of Economic Issues*, *35*(3), 607–632.

Gottlieb, E. E., & Keith, B. (1997). The academic research-teaching nexus in eight advanced-industrialized countries. *Higher Education*, *34*(3), 397–419.

Grant, R. M. (1996). Toward a knowledge-based theory of the firm. *Strategic Management Journal*, *17*, 109–122.

Green, S. G. (1991). Professional entry and the adviser relationship—socialization, commitment, and productivity. *Group & Organization Studies*, *16*(4), 387–407.

Groot, T., & García-Valderrama, T. (2006). Research quality and efficiency. An analysis of assessments and management issues in Dutch economics and business research programs. *Research Policy*, *35*, 1362–1376.

Gulbrandsen, M., & Smeby, J. C. (2005). Industry funding and university professors' research performance. *Research Policy*, *34*(6), 932–950.

Gustad, J. W. (1960). *The career decisions of college teachers*. Washington, DC: Department of Health, Education and Welfare.

Hackett, E. J. (2005). Essential tensions: Identity, control, and risk in research. *Social Studies of Science*, *35*(5), 787–826.

Hagstrom, W. O. (1971). Inputs, outputs and the prestige of university science departments. *Sociology of Education*, *44*(4), 375–397.

Haraszthy, A., & Szántó, L. (1979). Some problems of research planning: Data from Hungary compared to other Round 1 countries. In F. M. Andrews (Ed.), *Scientific productivity. The effectiveness of research groups in six countries*. Cambridge, England: Cambridge University Press.

Harris, G., & Kaine, G. (1994). The determinants of research performance—a study of australian university economists. *Higher Education*, *27*(2), 191–201.

Harvey, J., Pettigrew, A., & Ferlie, E. (2002). The determinants of research group performance: towards mode 2? *Journal of Management Studies*, *39*(6), 746–774.

Hattie, J., & Marsh, H. W. (1996). The relationship between research and teaching: A meta-analysis. *Review of Educational Research*, *66*(4), 507–542.

Heinze, T., Shapira, P., Rogers, J. D., & Senker, J. M. (2009). Organizational and institutional influences on creativity in scientific research. *Research Policy*, *38*, 610–623.

Hemlin, S. (2006). Creative knowledge environments for research groups in biotechnology. The influence of leadership and organizational support in universities and business companies. *Scientometrics*, *67*(1), 121–142.

Hemlin, S., Allwood, C. M., & Martin, B. R. (2008). Creative knowledge environments. *Creativity Research Journal*, *20*(2), 196–210.

Hemlin, S., & Gustafsson, M. (1996). Research production in the arts and humanities—a questionnaire study of factors influencing research performance. *Scientometrics*, *37*(3), 417–432.

Hemlin, S., & Olson, L. (2011). Creativity-stimulating leadership. A critical incident study of leaders' influence on creativity in research groups. *Creativity and Innovation Management*, *20*(1), 49–58.

Hoare, A. G. (1995). Scale economies in academic excellence—an exploratory analysis of the United-Kingdom 1992 Research Selectivity Exercise. *Higher Education*, *29*(3), 241–260.

Horlings, E., Gurney, T., Somers, A., & Van den Besselaar, P. (2009, September). *The economic and societal impacts of large-scale research facilities: A literature review in support of evidence-based decision making*. Paper presented at the Atlanta Conference on Science, Technology and Innovation Policy, Atlanta, GA.

Horlings, E., & Versleijen, A. (2009). *Taking stock of big science: An inventory of large-scale scientific research facilities in the Netherlands*. The Hague, the Netherlands: Rathenau Institute, Science System Assessment.

Hornbostel, S., Böhmer, S., Klingsporn, B., Neufeld, J., & Von Ins, M. (2009). Funding of young scientist and scientific excellence. *Scientometrics*, *79*(1), 171–190.

Hülsheger, U. R., Anderson, N., & Salgado, J. F. (2009). Team-level predictors of innovation at work: A comprehensive meta-analysis spanning three decades of research. *Journal of Applied Psychology*, *94*(5), 1128–1145.

Janis, I. L. (1972). *Victims of groupthink* Boston, MA: Houghton Mifflin.

Jenkins, G. D., Mitra, A., Gupta, N., & Shaw, J. D. (1998). Are financial incentives related to performance? A meta-analytic review of empirical research. *Journal of Applied Psychology*, *83*(5), 777–787.

Johnes, G. (1988). Determinants of research output in economics departments in British universities. *Research Policy*, *17*(3), 171–178.

Jones, B. (2010). *As science evolves, how can science policy?* (NBER Working Paper 16002). Cambridge, MA: National Bureau of Economic Research.

Jordan, J. M., Meador, M., & Walters, S. J. K. (1988). Effects of department size and organization on the research productivity of academic economists. *Economics of Education Review*, *7*(2), 251–255.

Jordan, J. M., Meador, M., & Walters, S. J. K. (1989). Academic research productivity, department size and organization: Further results. *Economics of Education Review*, *8*(4), 345–352.

Katz, R. (1978a). Influence of job longevity on employee reactions to task characteristics. *Human Relations*, *31*(8), 703–725.

Katz, R. (1978b). Job longevity as a situational factor in job satisfaction. *Administrative Science Quarterly*, *23*(2), 204–223.

Keller, R. T. (2001). Cross-functional project groups in research and new product development: diversity, communications, job stress, and outcomes. *The Academy of Management Journal*, *44*(3), 547–555.

King, J. (1987). A review of bibliometric and other science indicators and their role in research evaluation. *Journal of Information Science*, *13*, 261–276.

Knorr, K. D., Mittermeir, R., Aichholzer, G., & Waller, G. (1979). Leadership and group performance: a positive relationship in academic research units. In F. M. Andrews (Ed.), *Scientific productivity. The effectiveness of research groups in six countries*. Cambridge, England: Cambridge University Press.

Kretschmer, H. (1985). Cooperation structure, group-size and productivity in research groups. *Scientometrics*, *7*(1–2), 39–53.

Kyvik, S. (1995). Are big university departments better than small ones. *Higher Education*, *30*(3), 295–304.

Laredo, P. (2001). Benchmarking of RTD policies in Europe: "Research collectives" as an entry point for renewed comparative analyses. *Science and Public Policy*, *28*(4), 285–294.

Laredo, P., & Mustar, P. (2000). Laboratory activity profiles: An exploratory approach. *Scientometrics*, *47*(3), 515–539.

Leisyte, L., Enders, J., & de Boer, H. (2008). The freedom to set research agendas—illusion and reality of the research units in the Dutch universities. *Higher Education Policy*, *21*, 377–391.

Leisyte, L., Enders, J., & de Boer, H. (2009). The balance between teaching and research in Dutch and English universities in the context of university governance reforms. *Higher Education*, *58*(5), 619–635.

Long, J. S., & McGinnis, R. (1985). The effects of the mentor on the academic career. *Scientometrics*, *7*(3–6), 255–280.

Louis, K. S., Holdsworth, J. M., Anderson, M. S., & Campbell, E. G. (2007). Becoming a scientist: The effects of work-group size and organizational climate. *Journal of Higher Education*, *78*(3), 311–336.

Marsh, H. W., & Hattie, J. (2002). The relation between research productivity and teaching effectiveness—complementary, antagonistic, or independent constructs? *Journal of Higher Education*, *73*(5), 603–641.

Martin, B., Skea, J. E. F., & Ling, E. N. (1992). Performance indicators for academic scientific research. Report submitted to the Advisory Board for the Research Councils and the Economic and Social Research Council, SPRU, University of Sussex. In: Hemlin, S. & Gustafsson, M. (1996). Research production in the arts and humanities: A questionnaire study of factors influencing research performance. *Scientometrics*, *37*(3), 417–432.

Martin, B. R. (1996). The use of multiple indicators in the assessment of basic research. *Scientometrics*, *36*(3), 343–362.

Martin, B. R., & Irvine, J. (1983). Assessing basic research. Some partial indicators of scientific progress in radio astronomy. *Research Policy*, *12*, 61–90.

Martin-Sempere, M. J., Garzon-Garcia, B., & Rey-Rocha, J. (2008). Team consolidation, social integration and scientists' research performance: An empirical study in the Biology and Biomedicine field. *Scientometrics*, *76*(3), 457–482.

Martin-Sempere, M. J., Rey-Rocha, J., & Garzon-Garcia, B. (2002). The effect of team consolidation on research collaboration and performance of scientists. Case study of Spanish university researchers in Geology. *Scientometrics*, *55*(3), 377–394.

McKeachie, W. (1979). Perspectives from psychology. In D. Lewis & W. Becker (Eds.), *Academic rewards in higher education* (pp.3–20). Cambridge, MA: Ballinger.

Mehra, A., Dixon, A. L., Brass, D. J., & Robertson, B. (2006). The social network ties of group leaders: Implications for group performance and leader reputation. *Organization Science*, *17*(1), 64–79.

Melin, G. (2000). Pragmatism and self-organization. Research collaboration on the individual level. *Research Policy*, *29*, 31–40.

Meltzer, L. (1956). Scientific productivity in organizational settings. *Journal of Social Issues*, *12*(2), 32–40.

Merton, R. K. (1957). Priorities in scientific discovery: A chapter in the sociology of science. *American Sociological Review*, *22*(6), 635–659.

Mets, B., & Galford, J. A. (2009). Leadership and management of academic anesthesiology departments in the United States. *Journal of Clinical Anesthesia*, *21*(2), 83–93.

Moed, H. F. (2000). Bibliometric indicators reflect publication and management strategies. *Scientometrics*, *47*(2), 323–346.

Moorhead, G., & Montanari, J. R. (1986). An emperical investigation of the groupthink phenomenon. *Human Relations*, *39*(5), 399–410.

Mumford, M. D., Scott, G. M., Gaddis, B., & Strange, J. M. (2002). Leading creative people: Orchestrating expertise and relationships. *Leadership Quarterly*, *13*(6), 705–750.

Neuhaus, C., & Daniel, H. D. (2009). A new reference standard for citation analysis in chemistry and related fields based on the sections of Chemical Abstracts. *Scientometrics*, *78*(2), 219–229.

Nederlands Observatorium van Wetenschap en Technologie. (2010). *Wetenschaps- en Technologie-Indicatoren 2010* [Science and Technology Indicators 2010]. The Hague: Ministry of Education, Culture and Science.

Oh, H. S., Chung, M. H., & Labianca, G. (2004). Group social capital and group effectiveness: The role of informal socializing ties. *Academy of Management Journal*, *47*(6), 860–875.

Oh, H. S., Labianca, G., & Chung, M. H. (2006). A multilevel model of group social capital. *Academy of Management Review*, *31*(3), 569–582.

Omta, S. W. F. (1995). *Critical success factors in biomedical research and pharmaceutical innovation*. Groningen, the Netherlands: Rijks Universiteit Groningen.

Omta, S. W. F., & De Leeuw, A. C. J. (1997). Management control, ucertainty, and performance in biomedical research in universities, institutes and companies. *Journal of Engineering and Technology Management*, *14*, 223–257.

Paglis, L. L., Green, S. G., & Bauer, T. N. (2006). Does adviser mentoring add value? A longitudinal study of mentoring and doctoral student outcomes. *Research in Higher Education*, *47*(4), 451–476.

Pelz, D. C. (1956). Some social factors related to performance in a research organization *Administrative Science Quarterly*, *1*(3), 310–325.

Pelz, D.C., & Andrews, F. M. (1966). *Scientists in organizations. Productive climates for research and development*. New York: John Wiley & Sons.

Pettigrew, A.M. (1979). Studying organizational cultures. *Administrative Science Quarterly*, *24*(4), 570–581.

Pfeffer, J., & Salancik, G. R. (1978). *The external control of organizations: A resource dependence perspective*. New York, NY: Harper & Row.

Pineau, C., & Levy-Leboyer, C. (1983). Managerial and organizational determinants of efficiency in biomedical research teams. In S. R. Epton, R. L. Payne, & A. W. Pearson (Eds.), *Managing interdisciplinary research* (pp. 141–163). New York, NY: John Wiley and Sons.

Porter, S. R., & Umbach, P. D. (2001). Analyzing faculty workload data using multilevel modeling. *Research in Higher Education*, *42*(2), 171–196.

Prpic, K. (1996). Characteristics and determinants of eminent scientists' productivity. *Scientometrics*, *36*(2), 185–206.

Pruthi, S., Jain, A., Wahid, A., & Nabi, S. A. (1993). Problems of management of scientific research—results of a survey. *Journal of Scientific & Industrial Research, 52*(2), 81–94.

Ramsden, P. (1994). Describing and explaining research productivity. *Higher Education, 28*(2), 207–226.

Reagans, R., Zuckerman, E., & McEvily, B. (2004). How to make the team: Social networks vs. demography as criteria for designing effective teams. *Administrative Science Quarterly, 49*(1), 101–133.

Reagans, R., & Zuckerman, E. W. (2001). Networks, diversity, and productivity: The social capital of corporate R&D teams. *Organization Science, 12*(4), 502–517.

Reskin, B. F. (1979). Academic sponsorship and scientists careers. *Sociology of Education, 52*(3), 129–146.

Rey-Rocha, J., Garzon-Garcia, B., & Martin-Sempere, J. (2006). Scientists' performance and consolidation of research teams in Biology and Biomedicine at the Spanish Council for Scientific Research. *Scientometrics, 69*(2), 183–212.

Rey-Rocha, J., Garzon-Garcia, B., & Martin-Sempere, M. J. (2007). Exploring social integration as a determinant of research activity, performance and prestige of scientists. Empirical evidence in the Biology and Biomedicine field. *Scientometrics, 72*(1), 59–80.

Rey-Rocha, J., Martin-Sempere, M. J., & Garzon, B. (2002). Research productivity of scientists in consolidated vs. non-consolidated teams: The case of Spanish university geologists. *Scientometrics, 55*(1), 137–156.

Sandstrom, U. (2009). Research quality and diversity of funding: A model for relating research money to output of research. *Scientometrics, 79*(2), 341–349.

Schubert, A., & Braun, T. (1996). Cross-field normalization of scientometric indicators. *Scientometrics, 36*(3), 311–324.

Shin, J. C., & Cummings, W. K. (2010). Multilevel analysis of academic publishing across disciplines: research preference, collaboration, and time on research. *Scientometrics, 85*(2), 582–594.

Shin, S. J., & Zhou, J. (2007). When is educational specialization heterogeneity related to creativity in research and development teams? Transformational leadership as a moderator. *Journal of Applied Psychology, 92*(6), 1709–1721.

Simonto, D. K. (2004). *Creativity in science. Chance, logic, genius and zeitgeist.* Cambridge, England: Cambridge University Press.

Smeby, J. C., & Try, S. (2005). Departmental contexts and faculty research activity in Norway. *Research in Higher Education, 46*(6), 593–619.

Smith, C. G. (1971). Scientific performance and composition of research teams. *Administrative Science Quarterly, 16*(4), 486–496.

Sousa, C. A. A., & Hendriks, P. H. J. (2008). Connecting knowledge to management: The case of academic research. *Organization, 15*(6), 811–830.

Spangenberg, J. F. A., Starmans, R., Bally, Y. W., Breemhaar, B., Nijhuis, F. J. N., & Vandorp, C. A. F. (1990). Prediction of Scientific Performance in Clinical Medicine. *Research Policy, 19*(3), 239–255.

Stankiewicz, R. (1976). Research groups and the academic research organization. *Sociologisk Forskning, 13*(2), 20–32.

Stankiewicz, R. (1979). The size and age of Swedish academic research groups and their scientific performance. In F. M. Andrews (Ed.), *Scientific productivity. The effectiveness of research groups in six countries* (pp. 191–222). Cambridge, England: Cambridge University Press.

Stevens, M. J., & Campion, M. A. (1994). The knowledge, skill, and ability requirements for teamwork—implications for human-resource management. *Journal of Management, 20*(2), 503–530.

Tien, F. F. (2000). To what degree does the desire for promotion motivate faculty to perform research? Testing the expectancy theory. *Research in Higher Education, 41*(6), 723–752.

Tijssen, R. J. W. (2003). Scoreboards of research excellence. *Research Evaluation, 12*(2), 91–103.

Urwick, L. F. (1956). The manager's span of control. *Harvard Business Review, 34*, 39–47.

Van Balen, B. (2010). *Op het juiste moment op de juiste plaats. Waarom wetenschappelijk talent een wetenschappelijke carrière volgt* [*At the right moment at the right spot: Why scholarly talent decides for an academic career*].The Hague: the Netherlands: Rathenau Instituut.

Van den Besselaar, P., & Leydesdorff, L. (2009). Past performance, peer review, and project selection: A case study in the social and behavioral sciences. *Research Evaluation, 18*(4), 273–288.

Van der Meulen, B. (1995). Understanding evaluation processes in research systems in transition. *Science Studies, 8*(1), 24–35.

Van der Weijden, I. (2007). *In search of performance. Research management within the Dutch public medical and health sector.* Amsterdam, the Netherlands: Vrije Universiteit.

Van der Weijden, I., De Gilder, D., Groenewegen, P., & Geerling, M. (2008). Organizational culture, performance and career choices of Ph.D.s: A case study of Dutch Medical Researchers. *Higher Education Policy, 21*, 305–321.

Van der Weijden, I., de Gilder, D., Groenewegen, P., & Klasen, E. (2008). Implications of managerial control on performance of Dutch academic (bio)medical and health research groups. *Research Policy, 37*, 1616–1629.

Van der Weijden, I., Verbree, M., Braam, R., & Van den Besselaar, P. (2009). *Management en prestaties van onderzoeksgroepen* (SciSA rapport 0913) [*Management and performance of research groups*]. The Hague: the Netherlands: Rathenau Instituut.

Van Raan, A. F. J. (2004). Measuring science. Capita selecta of current main issues. In H. F. Moed, W. Glanzel, & U. Schmoch (Eds.), *Handbook of quantitative science and technology research* (pp. 19–50). Dordrecht, the Netherlands: Kluwer Academic.

——— (2008). Scaling rules in the science system: Influence of field-specific citation characteristics on the impact of research groups. *Journal of the American Society for Information Science and Technology, 59*(4), 565–576.

Verbree, M., Horlings, E., Van der Weijden, I., & Van den Besselaar, P. (2011). *Addressing leadership and management of research groups: A multivariate study* (Science System Assessment Report 1120). The Hague: Rathenau Instituut.

Visart, N. (1979). Communication between and within research units. In F. M. Andrews (Ed.), *Scientific productivity. The effectiveness of research groups in six countries* (pp. 223–252). Cambridge, England: Cambridge University Press/UNESCO.

Von Tunzelmann, N., Ranga, M., Martin, B., & Geuna, A. (2003). *The effects of size on research performance: A SPRU review* (SPRU Science and Technology Policy Research). Brighton, England: SPRU, University of Sussex.

Wagner, C. S. (2005). Six case studies of international collaboration in science. *Scientometrics, 62*(1), 3–26.

West, M. A., & Anderson, N. R. (1996). Innovation in top management teams. *Journal of Applied Psychology, 81*(6), 680–693.

Whitley, R. (2000). *The intellectual and social organization of the sciences* (2nd ed.). New York, NY: Oxford University Press.

Wouters, P. (1999). *The citation culture.* Amsterdam, the Netherlands: University of Amsterdam.

Zaleznik, A. (1977). Managers and leaders. Are they different? *Harvard Business Review, 55*(3), 67–78.

Appendix

Descriptive Statistics of Academic Leadership by Performance

		No.	Median	Bottomquartile	Topquartile	Mean	Stddeviation
Number of funding sources (preclinical and clinical)	*High*	28	4.0	3.3	5.0	4.1	0.9
	Other	123	3.0	3.0	4.0	3.5	1.1
Percentage of competitive funding	*High*	33	25.0	10.0	32.5	24.9	16.9
	Other	143	20.0	0	30.0	19.6	18.9
Supervisory capacity	*High*	33	6.0	4.0	7.5	6.7	4.4
	Other	138	5.0	3.0	7.0	5.5	3.7
Coresearcher versus manager	*High*	33	4.0	3.0	4.0	3.7	0.8
	Other	151	3.0	2.0	4.0	3.2	1.0
High-skilled scientist	*High*	31	4.0	4.0	5.0	4.3	0.6
	Other	151	4.0	3.0	4.0	3.6	0.8
Importance of visibility (preclinical and clinical)	*High*	29	4.0	4.0	4.0	3.9	1.0
	Other	129	3.0	2.0	4.0	3.2	1.1

(*Continued*)

		No.	Median	Bottomquartile	Topquartile	Mean	Stddeviation
Importance of preliminary evaluation of research proposals (preclinical and clinical)	*High*	26	4.0	4.0	5.0	4.4	0.6
	Other	124	4.0	4.0	5.0	4.0	0.8
Time spent on research (clinical)	*High*	13	43.9	36.2	59.0	49.0	16.3
	Other	67	39.3	27.1	46.4	39.7	14.3
Time spent on education (preclinical)	*High*	16	5.8	4.9	7.5	7.9	5.3
	Other	63	9.0	5.9	17.5	13.0	9.8
Importance of considerations for research agenda (in percentage) (preclinical and clinical)	*High*	28	62.5	40.6	75.0	58.9	24.0
	Other	124	50.0	37.5	62.5	47.3	21.7
Participation in researchpolicy meetings (% yes)	*High*	29	93.1				
	Other	130	80.0				
Frequency of feedback meetings	*High*	33	3.7	3.0	4.0	3.6	0.6
	Other	149	3.3	3.0	3.7	3.4	0.6
Project meetings at least once a week (% yes)	*High*	34	70.6				
	Other	151	47.0				
Network activities (preclinical and clinical)	*High*	29	50.0	32.5	70.0	53.3	24.4
	Other	130	40.0	25.0	50.0	38.4	16.5
Time spent on external management tasks (paraclinical and preclinical)	*High*	21	7.1	5.6	18.1	11.5	9.0
	Other	84	5.9	1.0	12.2	8.4	8.6

6 Generation and Life-Cycle Effects on Academic Leadership

Maaike Verbree, Inge van der Weijden, and Peter van den Besselaar

A well-functioning science system requires both creative and high-performing researchers and an environment in which they can flourish. However, a researcher's natural life cycle affects his or her performance. Age and experience have nonlinear relationships with performance, resulting in a performance decline for the oldest and most experienced researchers (Costas, Van Leeuwen, & Bordons, 2010; Gingras, Lariviere, Macaluso, & Robitaille, 2008; Goodwin & Sauer, 1995; Simonton 1988, 2004; Sturman, 2003). Generation cohort membership also influences activities and performance of researchers (Bayer & Dutton, 1977; Kyvik & Olson, 2008; Levin & Stephan, 1991; Rauber & Ursprung, 2008).

The role of the academic group leader is crucial for stimulating the creativity and productivity of researchers (e.g., Babu & Sing, 1998; Bland & Ruffin, 1992; Denti & Hemlin, this volume; Harvey, Pettigrew, & Ferlie, 2002; Knorr, Mittermeir, Aichholzer, & Waller, 1979b; Mumford, Scott, Gaddis, & Strange, 2002; Smith, 1971; Spangenberg et al., 1990; Stankiewicz, 1976; Van der Weijden, 2007; Van der Weijden, De Gilder, Groenewegen, & Klasen, 2008). Special skills are required for leaders of creative professionals, such as researchers. Academic leaders orchestrate people, their relationships, and their expertise in such a way that new ideas are generated (Mumford et al., 2002). For example, leaders provide incentives or rewards for researchers at different professional stages, which can avoid a decline in performance (Gonzalez-Bambrila & Veloso, 2007; Kyvik, 1990; Turner & Mairesse, 2002). Older researchers are more likely to remain productive when they receive professional recognition for their work (Kyvik, 1990).

Knowledge production takes place in increasingly large groups rather than by individual researchers or in small teams, which intensifies the role of the academic group leader. The driving factor behind this increase in scale is the growing complexity of research problems that have to be addressed in multi-, inter- and transdisciplinary research. This development requires larger teams whose members possess a variety of specialties and skills (Börner et al., 2010; Stokols, Hall, Taylor, & Moser, 2008; Stokols, Misra, Moser, Hall, & Taylor, 2008).

Furthermore, a number of changes in the science system have direct consequences for the role of academic group leaders. The traditional tasks of group leaders are extended with entrepreneurial activities (Hansson & Monsted, 2008). In this chapter we highlight four environmental developments of group leaders. First, research is currently shaped by the pressure to achieve excellence. The emphasis on excellence by science policy is reflected in, for example, specific funding programs for individual top researchers and in the concentration of the best researchers in 'centers of excellence' (i.e., Hornbostel, Böhmer, Klingsporn, Neufeld, & Von Ins, 2009; Van Leeuwen, Visser, Moed, Nederhof, & Van Raan, 2003). Second, the societal benefits of scientific research are increasingly important. Knowledge transfer and research activities that extend beyond the academic world are a growing focus of attention in science policy (i.e., De Jong, Van Arensbergen, Daemen, Van der Meulen, & Van de Besselaar, 2011; Etzkowitz & Leydesdorff, 1995; Gibbons et al., 1994; Göktepe-Hultén, 2008; Göransson, Maharajh, & Schmock, 2009). For example, research funding agencies require an explicit indication of societal relevance of the research project within proposals. Third, there is a growing competition between researchers and research organizations for funding. The share of project funding is increasing worldwide (for a variety of countries, see Lepori et al., 2007; for the recent period in the Netherlands, see Van Steen, 2011). Finally, there is a growing emphasis on accountability for investments in science, leading to the implementation of research evaluation procedures and systems.

These developments heavily influence prevailing academic values, norms, and the rules of the science system, which in turn influence, for example, the propensity to collaborate, the quality criteria for output and the strategies of group leaders. The socialization of new generations of researchers occurs in the doctoral phase, when academic values, norms, and rules are learned (Austin, 2002; Bland & Schmitz, 1986; Morrison, Rudd, Picciano, & Nerad, 2011). The period in which researchers are educated, therefore, determines their cohort membership.

This also holds for research leaders. As is true for researchers, life cycle (age) affects leadership; leadership knowledge and skills increase when leaders gain more experience throughout their career (Lord & Hall, 2005; Mumford, Zaccaro, Harding, Jacobs, & Fleishman, 2000). Additionally, cohort membership also influences performance norms and the leadership activities of research leaders (Bayer & Dutton, 1977; Kyvik & Olson, 2008; Levin & Stephan, 1991; Rauber & Ursprung, 2008). For research leaders, the boundaries between generation cohorts should be based on changes in the research environment. We examine generation effects on leadership by identifying the relevant environmental changes in the science system and by establishing relationships between these changes and the behavior of academic group leaders.

To summarize, academic leadership is becoming increasingly complex and demanding, and it requires a broad range of skills and adaptation to changes in the organization of research and of the science system. The influence of academic leadership on research performance is becoming even more crucial. However, life-cycle and generation differences affect academic leadership.

This brings us to our research question: How do life-cycle and generation differences influence academic leadership? This study is the first to address the development of academic leadership over time and discuss generation differences across our specific sample of academic leaders. Generation effects were examined by a natural experiment. We compared two groups of academic leaders who were socialized in periods either before or after large contextual changes took place. We conducted a cross-sectional analysis in which we compared academic group leaders at varying points in their life cycle to investigate differences in leadership behavior. In the following sections we briefly review the literature on generation differences and life cycles and present hypotheses about their relationship with academic leadership.

GENERATION EFFECTS

Within the framework of the generational cohort theory, several generations were distinguished based on the specific time periods people were born in and the time periods they grew up in; for example, the baby boomers (1943–1960) and generation X (1961–1980).This cohort membership influences the behavior and attitudes of leaders (D'Amato & Herzfeldt, 2008; Gursoy, Maier, & Chi, 2008; Sessa, Kabacoff, Deal, & Brown, 2008). In academia, however, 'date of birth as a researcher' is far more important than 'date of birth as a human being', and age is a bad predictor of performance (Bayer & Dutton, 1977; Kyvik & Olson, 2008; Simonton, 2004).

'Date of birth as a researcher' typically begins when one pursues a doctoral degree. During the doctoral phase, one is socialized as a researcher, learning academic values, norms, and rules (Austin, 2002; Bland & Schmitz, 1986; Morrison et al., 2011). When researchers are educated over varied periods, naturally their socialization may differ. As a result, they will develop varying behavior and attitudes, which will also be reflected in their academic leadership practices. As shown by various scholars, differences in performance are also codetermined by generation (Bayer & Dutton, 1977; Kyvik & Olson, 2008; Levin & Stephan, 1991; Rauber & Ursprung, 2008). In addition, cohort effects are discipline specific (Gonzalez-Brambila & Veloso, 2007).

Cohorts should reflect changes in the science system that affect the knowledge base or research practices of the researchers in a given era (Gonzalez-Brambila & Veloso, 2007). To identify such changes and establish the appropriate cohorts, we describe some crucial changes in the Dutch science system as illustrated in Figure 6.1. We also present two hypotheses that specify how cohort membership results in specific academic leadership behavior.

Science policy in the Netherlands (and in most Western countries) during the 1980s was about implementing policy changes that aimed at higher quality, more efficient, relevant, and accountable scholarly research. At the same time, there was a growing desire from the public for transparency in government expenditure and effectiveness in all areas of policy (Westerheijden,

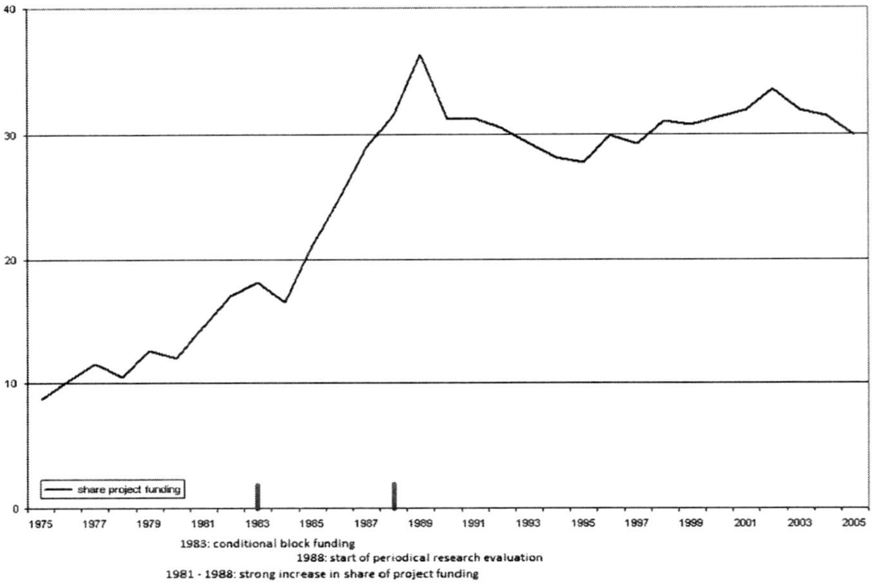

Figure 6.1 Structural changes in the Dutch research environment (based on Versleijn et al., 2007)

1997). Furthermore, at the end of the 1980s, after a period of university budget restrictions, and as a result of direct government interventions, the government and the universities agreed to change their policy relationship to one of 'autonomy and quality': as long as universities were able to show that education and research were of good quality, the government would respect their autonomy. As part of this new implicit contract, the Association of Universities assumed responsibility for the evaluation of research programs and was therefore able to strengthen its position (Van der Meulen, 2009). Such expectations led to changes in criteria for the allocation of research funding, and to the introduction of externally organized research evaluations.

In terms of research evaluations, government, intermediary organizations,[1] and social groups in the Netherlands have increasingly put a strong emphasis on the evaluation of research since the end of the 1980s (Rip & Van der Meulen, 1995; Spaapen, 1995; Van der Weijden, 2007). Consequently, the evaluation of research has become a crucial process throughout the scientific careers of researchers with respect to decisions about appointments, promotion, tenure positions, and the allocation of funding. Research evaluations are also an important factor affecting research groups, departments, faculties, and institutes. As a result, large numbers of research evaluations were conducted at many different levels. In 1988, the Council of Medical Sciences of the Royal Netherlands Academy of Arts and Sciences organized the first national evaluation of medical and health research in cooperation with the Disciplinary Board of the Medical Sciences of the Association of Universities[2] (Koninklijke Nederlandse Akademie van Wetenschappen [KNAW], 1988). The output of research conducted at all faculties of Medical and Health Sciences and at most nonuniversity research institutes in the Netherlands was evaluated by peer review.

Another important change in the Netherlands took place in 1983 (Versleijen et al., 2007). Ex ante evaluations were implemented and linked to university research funding through a new funding scheme, denoted as conditional funding. This funding scheme distinguished between education and research. About half of the block grant was based on teaching components, and the other half on research, although universities retained autonomy in deciding how to distribute the money (Van Arensbergen, 2010). As a condition for funding, each university was required to have a sufficient amount of research incorporated in programs. The assessments were conducted by specially established disciplinary committees. A positive assessment implied a budget protection of the program for 5 years (Versleijen et al., 2007). The purpose of the evaluation was to identify high-quality research within disciplines and to provide such research initiatives with a degree of budgetary security. In 1987, a second round of evaluations to allocate funding was initiated in which the previously established research programs were assessed ex post. This second round of evaluations has evolved over the years into the relatively extensive and formalized system of ex post evaluation of academic research, which is currently used in the Netherlands.[3] As a result of the rise in research evaluation, focus on research became very strong in universities. This understanding then guides the argument for our first hypothesis:

> *Hypothesis 1:* Academic leaders who obtained their PhD after changes in research evaluation had taken effect (cohort young; PhD post-1987) have a higher research focus compared to academic leaders who obtained their PhD before changes in research evaluation had taken effect (cohort old; PhD pre-1988).

Second, another main change that took place in the 1980s was the growth of competitive funding. Block funding traditionally took up around 90% of all university funding, whereas this started to change around 1980. The share of project funding started to rise from 10% to about 30% of total public research funding, as shown in Figure 6.1. This development increased competition and forced researchers to write proposals to secure funding for PhD projects and postdoctoral research from the various 'new' sources. Such changes occurred during the 1980s and were fully in place around 1987. Therefore, we use 1988 as the year that distinguishes between the two different generations. This leads to the second hypothesis:

> *Hypothesis 2:* Academic leaders who obtained their PhD after changes in research funding had taken effect (cohort young; PhD post-1987) focus in

their resource strategy more on external funding compared to academic leaders who obtained their PhD before changes in research funding had taken effect (cohort old; PhD pre-1988).

LIFE-CYCLE EFFECTS

The life cycle of research leaders refers to years of experience. According to Lord and Hall (2005) and Mumford et al. (2000), leadership also changes over one's life cycle. We assume the same holds true for academic leadership. When academic leaders gain experience, their responsibilities change, and this may affect their leadership practices. Furthermore, we expect that changing goals, responsibilities, and interests over their life cycle are related to academic leadership practices. Because little is known about changes in research leadership behavior and the effectiveness over a life cycle, we use the literature on researchers' life cycles to derive our hypotheses.

A considerable number of studies show how *research performance* changes over time (Costas et al., 2010; Gingras et al., 2008; Goodwin & Sauer, 1995; Long, Allison, & McGinnis, 1993; Pezzoni, Sterzi, & Lissoni, 2009; Simonton, 2004; Sturman, 2003). Simonton (2004), for example, developed a model for the productive output of scientists that allows for variation in age at career onset. He proposed a typology of scientific careers: low-creative early bloomers, low-creative late bloomers, high-creative early bloomers, and high-creative late bloomers.

Literature also shows how the *behavior* of researchers is not stable over time. At the start of a research career, the main goal is to obtain a scientific reputation. Each scientist begins with a repository of phenomena, facts, concepts, variables, constants, techniques, theories, laws, questions, goals, and criteria that define his or her domain sample. This initial creative potential is then transformed into published output through a two-step process of ideation and elaboration (Simonton, 2004). Early career success is related to holding a degree in a different field and having freedom to try new things (Eubanks, Palanski, Swart, Hammond, & Oguntebi, this volume). Younger researchers spend more time on research and professional development (Baldwin, Lunceford, & Vanderlinden, 2005). However, more research time does not always directly result in more publications. In the beginning of a researcher's career (under 30 years) the productivity rate is relatively low (Dennis, 1956). The probability of producing prize-winning work is also much lower for young (under 30 years) researchers, because their work might not be immediately recognized as praiseworthy (Stephan & Levin, 1993). On the other hand, high performance in the beginning of one's career is a good predictor of high performance at later stages. Receiving a university and PhD degree at a young age, a postdoctoral position early on in one's career, early publications (before the doctorate), research experience (during university), received citations of early work, rapid publications in top journals (especially in cooperation with coauthors), and getting a higher scientific rank at a young age are all good predictors of research performance later in one's career, and lead to greater integration into the scholarly community (Aggarwal, Schirm, & Zhao, 2007; Blackburn, Behymer, & Hall, 1978; Brancati, Mead, Graves, Levine, & Klag, 1992; Cole & Cole, 1967; Dietz & Bozeman, 2005; Horta, 2009; Prpic, 1996).

When researchers acquire responsibility as group leaders, they again have to work on their reputation, which is based on results obtained earlier as young researchers. So just like up-and-coming researchers, new academic group leaders need to establish their scientific reputation. This leads to the third hypothesis:

> *Hypothesis 3:* Academic leaders in the starting phase perform academic leadership practices aimed at acquiring and sustaining reputation as a creative and productive research leader, such as accomplishing an innovative research agenda.

The role of researchers changes over time, which affects performance (Baldwin et al., 2005; Bayer & Dutton, 1977; Costas et al., 2010; Goodwin & Sauer, 1995; Gingras et al., 2008). During their career, researchers acquire many nonresearch duties and responsibilities such as teaching, administration, project management, funding, and supervision. To have career success in this phase, the challenge is to secure cognitive capacity, that is, by appropriately managing time, people, and resources (Eubanks et al., this volume). As researchers gain experience, their focus on goal executing is substituted by goal setting. A focus on goal setting means that older researchers are responsible for increasing the group's access to scientific funding and other resources for projects. As a result, older and more established researchers are involved in an increasing number of research projects, and therefore invest less time in each individual research project. This change in focus generally results in strong individual publication growth, through coauthoring papers on a variety of research projects (Knorr & Mittermeir, 1980; Knorr, Mittermeir, Aichholzer & Waller, 1979a). This is most likely the role of the academic group leader (Gingras et al., 2008).

This role change might have positive spillover effects on group performance. More experienced group members (i.e., senior researchers and group leaders) who provide supervision, or who act as mentors, positively influence predoctoral or early-career research productivity, self-efficacy, grants, and the collaboration between and professional networks of PhD students and other less experienced researchers (Cameron & Blackburn, 1981; Cronan-Hillix, Gensheimer, Cronan-Hillix & Davidson, 1986; Green, 1991; Long & McGinnis, 1985; Paglis, Green, & Bauer, 2006; Reskin, 1979; Van Balen, 2010). Experienced researchers, such as full professors and leading scholars, may increase the number of successful PhD candidates in the group and may increase access to material resources, such as research funds that allow for the recruitment of competent researchers (Carayol & Matt, 2004; Dundar & Lewis, 1998). The accumulated professional experience of research leaders also enables them to influence the knowledge and values of the group members, to exploit professional contacts and networks, and to help colleagues (Dill, 1982). In short, the role-change from researcher to group leader refers to a more collective perspective, with less supervision from experienced researchers and the securing of research funding as a main task. This leads to the fourth hypothesis:

> *Hypothesis 4:* Academic leaders more mature in experience perform academic leadership practices aimed at preserving and extending group vitality.

Researchers at the end of their careers are the lowest performers (Costas et al. 2010; Goodwin & Sauer, 1995; Sturman, 2003; Gingras et al., 2008). Simonton (2004) shows that supply of new ideas is slowly depleted, resulting in a postpeak decline in individual productivity at the later established career trajectories. He suggests that individual creative potential is consumed faster than it can be replenished because of growing professional and personal responsibilities, a narrower scope of expertise acquisition, and a loss of receptiveness to new ideas (Planck's Principle).

Kyvik and Olson (2008) argue that the drop in performance might be a result of obsolete skills, which hinder older academic leaders in shifting their research focus, leading to less important publications. Indeed, very few researchers above the age of 55 get their work awarded with a Nobel Prize (Stephan & Levin, 1993). The diminished impact of their research can also be caused by less visibility in the scientific community. For example, Van Rijnsoever, Hessels, and Vandeberg (2008) showed that the level of network activities of experienced researchers starts to decline after 20 years of a scientific career. Promoting your research helps to attract citations (Aizenman & Kletzer, 2008). However, the proportion of time spent on research activities declines for late-career researchers, while time spent on teaching increases (Baldwin et al., 2005). Altogether, this suggests that on average, researchers who are near to retirement show a declining productivity and impact with a shift of focus to educational tasks. This leads to the fifth hypothesis:

> *Hypothesis 5:* Academic leaders in the end phase perform academic leadership practices aimed at relinquishing their role as group leader and shifting their focus from research to education.

METHODS

Data

The data for this study were collected in a survey in 2002 among academic leaders of biomedical and health research groups in the Netherlands. These group leaders were working in university medical centers (UMCs) or in biomedical/medical public research institutes in the Netherlands (Van der Weijden, 2007; Van der Weijden et al., 2008; Van der Weijden, Verbree, Braam, & Van den Besselaar, 2009). The participating group leaders (N = 137) returned the completed questionnaires with an overall response rate of 38%. To maximize the response rate, we used the tailored design method (Dillman, 2000). Nonresponse analysis shows that the respondents can be regarded as a representative sample of the Dutch biomedical and health research groups. The mean difference in publication output is very small and nonsignificant (24.35 vs. 24.83 publications, respectively; Van der Weijden, 2007; Van der Weijden et al., 2008). Furthermore, we checked whether the respondents were evenly distributed among the various research institutions and the disciplines, and this proved to be the case (Van der Weijden, 2007).

Independent Variables

To measure generation and life-cycle effects, we distributed the respondents over various groups representing either a specific generation or a specific phase in the life-cycle career of a group leader. Generation membership is defined as the year the PhD was obtained and determines which period socialization in academia has taken place. In the previous section, we noted that the Dutch science system underwent structural changes in funding and research evaluation before 1988. Consequently, we consider 1988 to be the breaking point in time between the two generations. The older generation had obtained their PhDs before changes in the science system had taken effect (PhD pre-1988), while the younger generation had obtained their PhDs when changes in the science system had already taken effect (PhD post-1987). Thus, group leaders from cohort old are educated as researchers in a different period than are group leaders from cohort young.

Three life-cycle phases can be distinguished. The starting phase includes academic leaders who have 5 or fewer years experience as leaders of their current groups. The departing phase includes academic leaders who are 57 years or older,[4] and are anticipating retirement. Experienced group leaders who have been leading their group for a long period (more than 5 years) and are not anticipating their retirement yet (younger than 57 years) are considered to be in the intermediary phase. Table 6.1 shows the distribution of the respondents among generations and life life-cycle phases.[5]

To test for *generation effects* on leadership, we compared academic leaders from cohort old (PhD pre-1988) with academic leaders from cohort young (PhD post-1987). Given the skewed distribution of our data (indicated by a Shapiro-Wilk test), we conducted Mann-Whitney U and chi-square tests to analyze differences in academic leadership. There are three possible *life-cycle effects*. We identified a 'starting effect' when new group leaders differ from the experienced and the departing peer group leaders in their academic leadership. A 'maturation effect' is detected when experienced group leaders differ from starting and departing group leaders in their academic

Table 6.1 Distribution of Respondent Among Generations and Life Cycle Phases

Generation	Mean	Starting group leaders	Experienced group leaders	Departing group leaders	Total
	N	*10*	*51*	*44*	*105*
Cohort old	Age	52.2	52.7	60.2	55.8
PhD pre-1988	Experience	2.9	12.8	17.4	13.8
	N	*12*	*18*		*30*
Cohort young	Age	44.3	45.1		44.8
PhD post-1987	Experience	4.0	8.7		6.8
	N	*22*	*69*	*44*	*135*
Total	**Age**	**48.0**	**50.7**	**60.2**	**53.4**
	Experience	**3.5**	**11.8**	**17.4**	**12.3**

leadership. Finally, a 'retirement effect' can be identified when departing group leaders differ from starting and from experienced research leaders. To analyze the differences in academic leadership for the three life-cycle phases, we also conducted Mann-Whitney *U* and chi-square tests. As some subsamples are very small (i.e., 10 starting group leaders in cohort old), we firstly tested generation and life-cycle effects on the total sample independently. However, we control generation effects for life-cycle effects and vice versa.

Dependent Variables

Our survey addressed items on the various components of academic leadership, namely, resource strategy, leadership, group management, and network management. The survey contained 49 questions in total (Van der Weijden, 2007). In Table 6.2, we describe the items that were used for this study and the descriptive statistics for each item.

RESULTS

In the following sections we report our findings. Results of the test for generation effects are summarized in Table 6.3, and those for the life cycle effects are summarized in Tables 6.4 through Table 6.6. In Table 6.7, we show how the results change when taking both independent variables into account simultaneously.

Table 6.2 Description and Descriptive Statistics of Dependent Variables

Variables	Description	Descriptive statistics				
		Mean	Median	Std dev	Minimum	Maximum
Resource strategy						
Funding sources	Percentage of total funding from:					
	–Institutional funding	41.7	40.0	24.5	0	100
	–External funding	58.3	60.0	24.5	0	100
	–Competitive funding	17.0	15.0	15.9	0	70
	–Contract funding	11.2	0	16.9	0	90
	–Charity funding	21.0	15.0	22.2	0	80
	–European funding	3.4	0	7.4	0	40
	–Other funding sources	5.8	0	11.4	0	66.7
Number of funding sources	Measure the total number of funding sources	3.5	3.0	1.1	1	6
Submitted research proposals	Dummy variable measuring whether or not group leaders had submitted research proposals in the period 1999–2001				Yes: 42.3%	No: 57.7%
Group size	Number of scientific and support staff (FTE)	14.8	14.0	7.0	3.0	38.0
Staff constellation	Percentage of total scientific staff:					
	–Professors	11.2	9.5	7.7	0	33.0
	–Senior staff	35.0	33.3	17.3	0	80.0
	–PhDs	47.2	50.0	17.4	0	91.0
	–Other scientific staff	6.7	0	11.3	0	67.0
	–Support staff (analysts and technicians)	24.0	24.1	15.4	0	57.0
	–Permanent staff	54.9	50.0	29.4	0	100

(Continued)

Table 6.2 (Continued)

Variables	Description	Descriptive statistics				
		Mean	Median	Std dev	Minimum	Maximum
Leadership						
Time allocation	Percentages of time spent on:					
	–research	45.6	43.4	16.1	11.0	83.5
	–group management	17.3	16.7	8.4	4.6	52.0
	–education	13.2	12.1	9.7	0	46.7
	–patient care	14.7	10.5	16.6	0	78.0
	–percentage of groups				Yes: 57.4%	No: 42.6%
	–external management tasks (network management)	9.2	6.6	8.2	0	52.0
Group management						
Intangible rewards	Dummy variable measuring whether or not leaders provide intangible rewards (i.e., special honours)				Yes: 53.3%	No: 46.7%
Quality control of research proposals	Measure importance of quality control for research proposals where 1 is totally disagree and 5 is totally agree: "Researchers are free to decide whether or not to implement comments of internal pre-evaluations in research proposals"	2.5	2.0	1.0	1	4
Research agenda	Dummy variables measuring whether or not group leaders value considerations to set the research agenda as important:					
	1) Research themes of international interest				Yes: 51.1%	No: 48.9%
	Theoretically challenged and innovative research				Yes: 68.9%	No: 31.1%
	themes				Yes: 30.8%	No: 69.2%

	Possibility for new research lines					
	4) PhDs interests and possibilities				Yes: 34.6%	No: 65.4%
	5) Opinion of international colleagues				Yes: 32.3%	No: 67.7%
	6) Interest of international research programmes				Yes: 25.0%	No: 75.0%
	7) Application possibilities in health care				Yes: 46.3%	No: 53.7%
	8) Possibility to obtain visibility in top journals				Yes: 40.6%	No: 59.4%
	Percentage of considerations that are valued as important to set the research agenda: 100 means that a group leader values all 8 considerations as important	46.0	42.9	15.2	7.1	78.6
Job assessments	Dummy variable measuring whether or not leaders evaluate their group members by means of job assessments				Yes: 94.9%	No: 5.1%
Network management						
Network activities	Intensity of network activities in days per year where network activities are the sum of days spent on:	42.7	40.0	21.6	10	130
	–lectures					
	–attending conferences					
	–participation in editorial boards					
	–participation in assessment committees					
Characteristics of the leader						
Age	Age of leader in 2002	53.5	55.0	6.5	37	68
Experience	Years of experience as group leader in current research group	12.2	11.0	6.8	1.0	37

Generation Effects on Academic Leadership

Based on major changes in research funding and research evaluation that had taken place in the Netherlands, we defined two cohorts of academic leaders. A generation effect occurs when the two cohorts of academic leaders differ in leadership, group, and network management, or in resource strategy. For example, the younger generation of academic leaders has significantly fewer permanent staff compared to the older generation (Figure 6.2).

Considering Hypothesis 1, we found evidence that the younger generation of academic leaders places a higher emphasis on research than does the older generation (Figure 6.3). The younger generation spends approximately half as much time on education than does the older generation. In contrast, the younger generation spends more time on research and group management activities (Table 6.3).

Given the high importance placed on obtaining external funding, we expected a higher time investment in network activities. Unexpectedly, we found that the younger generation spent less time participating in networking activities such as by holding memberships in committees and editorial boards (Table 6.3). However, one is invited for these types of activities based on reputation and network position, which the younger generation of academic leaders is still building. They do have less experience (Table 6.1), and therefore, we may measure here a starting effect and not a generation effect. We will come back to this later on in this chapter.

With respect to Hypothesis 2, our results show that the resource strategy of the younger generation of academic leaders has a considerable higher

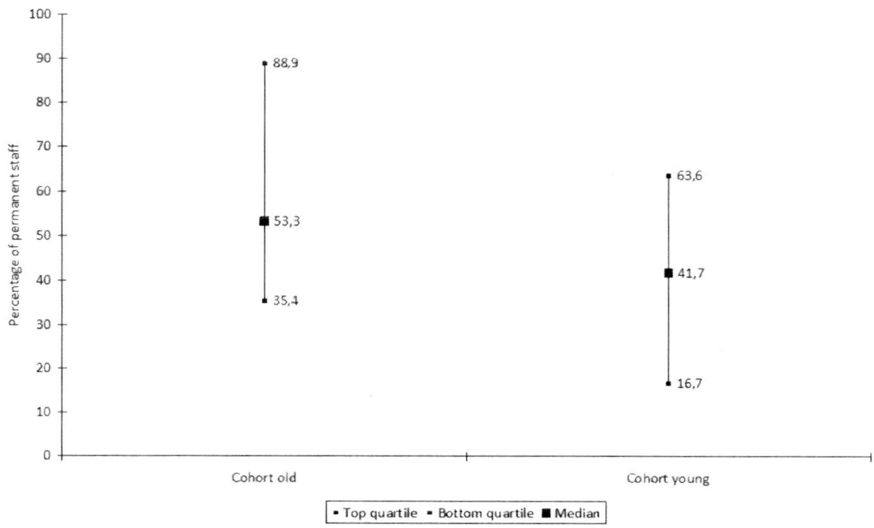

Figure 6.2 Generation effect of proportion of permanent staff

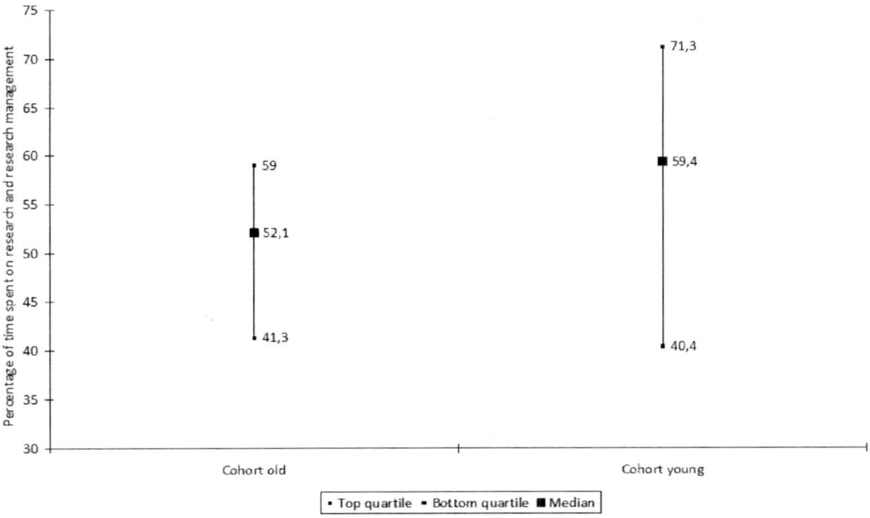

Figure 6.3 Generation effect of proportion of time spent on research and research management of the group

Table 6.3 Generation Effects on Academic Leadership

Academic leadership variables	Cohort old Median *(n)*	Cohort young Median *(n)*	Old versus young
Teaching	12.8 (104)	6.2 (30)	$r = -0.23$***
Research & research management	52.1 (104)	59.4 (30)	$r = -0.15$*
Networking	40.0 (104)	35.0 (30)	$r = -0.20$**
External funding	50.0 (99)	77.5 (28)	$r = -0.26$***
Number of funding sources	3.0 (99)	4.0 (28)	$r = -0.15$*
Permanent staff (%)	53.3 (99)	41.7 (27)	$r = -0.24$***
Professors (%)	10.0 (99)	7.7 (27)	$r = -0.18$**
PhDs interests and possibilities (% yes)	38.6 (101)	20.0 (30)	Odds ratio = 2.52*

Note: Cohort old = PhD before 1988; Cohort young = PhD after 1897.*$p < 0.10$.
**$p < 0.05$. ***$p < 0.01$.

[a] Reference category is cohort old.

focus on external funding (Table 6.3). The younger generation has a higher proportion of external funding (Figure 6.4), and they obtain funding from more sources. Furthermore, their group composition is different from the older generation. More specifically, the younger generation has substantially

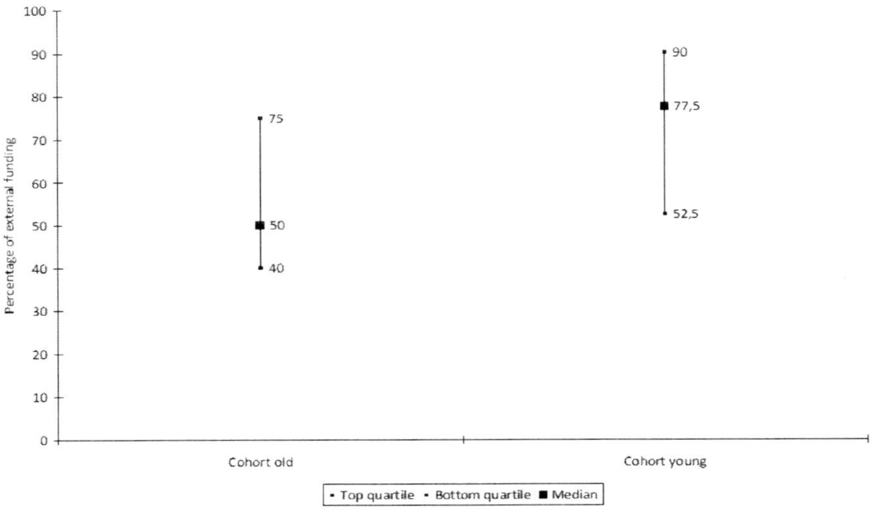

Figure 6.4 Generation effect of proportion of external funding

fewer permanent staff (Table 6.3). This is most probably an effect of the changing working conditions, where group leaders are increasingly expected to fund their group through external funding, resulting in mainly temporary positions. In short, when there is no guarantee of stable institutional funding, you cannot guarantee your staff permanent positions. The younger generation also has a lower proportion of professors (Table 6.3). But in next sections we will see that it concerns only the starting group leaders who have a lower percentage of professors in the group. This, therefore, seems to be a starting rather than a generation effect.

In addition, we found that the younger generation less often feels that PhD students should set their own research agenda (Table 6.3). This may be due to the fact that PhD students are generally paid through external funding, and therefore the research project is generally predefined by the group leaders' proposal.

In conclusion, Hypotheses 1 and 2 are supported by our test results: The younger generation of academic leaders focuses more on research and obtains their resources more from external sources.

Life-Cycle Effects

In the life cycle of an academic group leader, we distinguish between three phases. In the starting phase, an academic leader has to gain reputation for the work of his or her research group. In the maturity phase, an academic leader has a stable group and is not yet anticipating departing, and so focuses on strengthening the group vitality. In the departing stage, an academic leader is in the last phase of his or her career and anticipates retirement. We hypoth-

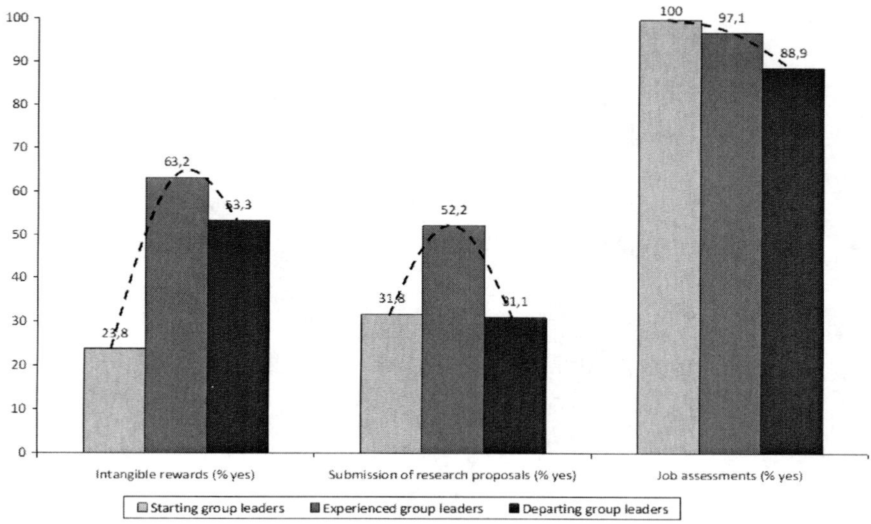

Figure 6.5 Starting effect of intangible rewards, maturity effect of submission of research proposal, and departing effect of job assessments

esized that academic leadership differs between these phases. We now report the starting, maturity, and departing effects on academic leadership.

A starting effect occurs when starting group leaders behave differently from experienced and departing group leaders. For example, starting group leaders less often provide intangible rewards such as special honors or nonfinancial prizes; a steep, increasing trend from starting group leaders to experienced group leaders is seen in Figure 6.5. A maturity effect occurs when experienced group leaders behave differently from starting and departing group leaders. For example, experienced group leaders often submit more research proposals to the medical research council; an inverse-U curve is seen in Figure 6.5 that represents less activity from starting and departing group leaders. A departing effect occurs when departing group leaders behave differently from starting and experienced group leaders. For example, departing group leaders often have fewer job assessments when evaluating their researchers; a declining trend is seen in Figure 6.5 from starting to departing group leaders.

Starting Effects

The starting effects are summarized in Table 6.4. As already noted in this chapter, starting group leaders spend significantly fewer hours on networking activities, such as participating in editorial boards, and participating in assessments committees. This may be due to the fact that starting group leaders are still building up their networks, and many of these activities are based on a positive reputation and the positions they hold within relevant networks.

Table 6.4 Starting Effects on Academic Leadership

Academic leadership Variables	Starting leaders	Experienced leaders	Departing leaders	*Starting vs. Experienced*		*Experienced vs. Departing*		*Starting vs. Departing*	
	Median (n)	Median (n)	Median (n)	Odds ratio	r	Odds ratio	R	Odds ratio	r
Networking	30.0 (22)	40.0 (69)	40.0 (43)		−0.30***		ns		−0.23*
Theoretically challenging	90.5 (21)	61.2 (67)	69.0 (42)	6.02**[a]		ns		4.26*[a]	
Intangible rewards (% yes)	23.8 (21)	63.2 (68)	52.3 (44)	5 50***[b]		ns		3.50**[b]	
Professors (%)	7.4 (18)	10.0 (67)	11.1 (41)		ns		ns		−0.32**

Note: Reference categories: a = starting group leaders and b = experienced group leaders. ns = not significant.

*$p < 0.10$. **$p < 0.05$. ***$p < 0.01$.

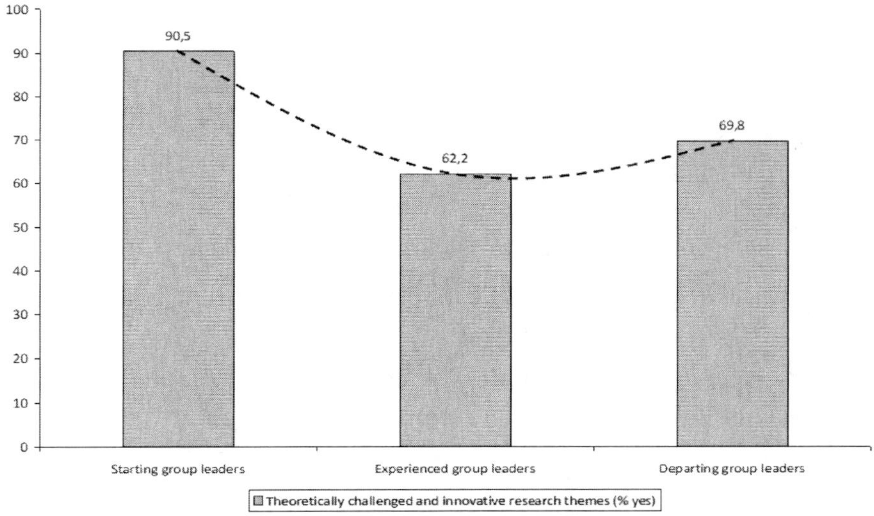

Figure 6.6 Starting effect of creativity

Another interesting starting effect is motivational. Starting research leaders emphasize their own research agenda; almost all starting academic leaders in our sample attach high value to challenging, theoretical, and innovative research themes. Apparently, starting group leaders consider creativity as highly important and necessary for establishing their reputation (Figure 6.6).

Concerning group management, starting leaders provide far fewer intangible rewards (Figure 6.5) than do experienced and departing group leaders. Perhaps less experienced group leaders place less emphasis on motivating their group members, because they need to promote their research line externally to build a reputation. Finally, starting group leaders have a smaller proportion of professors in their group, probably because some of these groups are not directed by full professors but by associate professors.

We found evidence for Hypothesis 3. Leadership practices of starting academic leaders focus more on developing a research agenda and building a reputation by finding an individual creative research niche, and they place less emphasis on internal management.

Maturity Effect

The maturity effects are summarized in Table 6.5. When an academic leader acquires reputation for his or her own new research line, the next phase is to guarantee survival of the group. Following reputation, the next obvious main goal is to obtain enough funding. Indeed, the results show that experienced group leaders acquire a significantly higher proportion of external funding compared to starting and departing group leaders (Figure 6.7).

Table 6.5 Maturity Effects on Academic Leadership

Academic leadership	Starting leaders	Experienced leaders	Departing leaders	*Starting vs. Experienced*		*Experienced vs. departing*		*Starting vs. departing*	
	Median (n)	Median (n)	Median (n)	Odds ratio	R	Odds ratio	R	Odds ratio	r
External funding	45.0 (22)	70.0 (64)	50.0 (42)		−0.22**		−0.20**		ns
Submitted research proposals (% yes)	31.8 (22)	52.2 (69)	31.1 (45)	2.34*[a]		2 34**[a]		ns	
Quality control of research proposals	3.0 (19)	2.0 (65)	3.0 (35)		−0.23**		−0.23**		ns

Note: ns = not significant.

*$p < 0.10$. **$p < 0.05$.

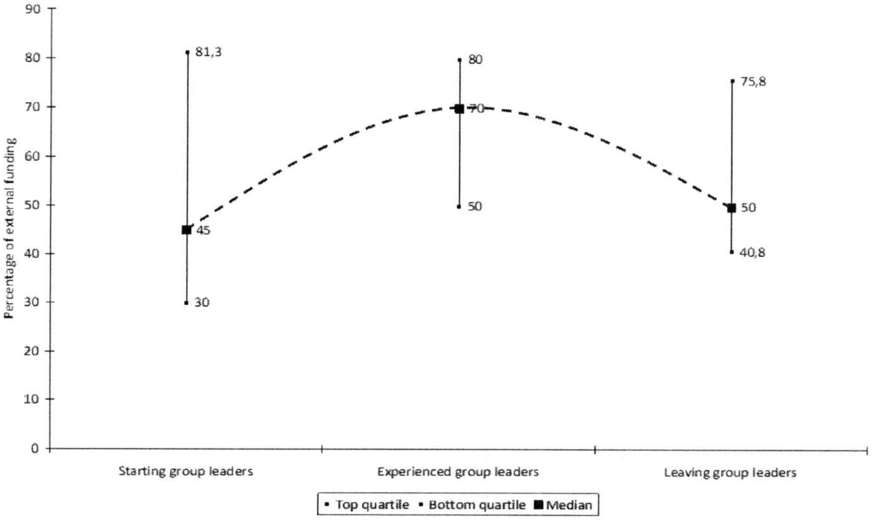

Figure 6.7 Maturity effect of the proportion of external funding

Also, the odds ratios show that experienced and departing group leaders are almost 2.5 times more likely to submit research proposals to the medical research council than starting group leaders (Table 6.5). Furthermore, they put considerably more emphasis on the quality control of research proposals. To conclude, Hypothesis 4 is confirmed, as leadership practice of academic leaders in their maturity phase aim to strengthen group vitality by obtaining new research funds.

Departing Effects

Several indicators demonstrate the departing effect (Table 6.6), referring to the activities of departing group leaders that are different from starting and experienced group leaders. They spent more time on teaching—that is, they spent twice as much time on teaching activities compared to the starting group leaders (Figure 6.8), while less time was spent on research activities and group management. It is plausible that departing group leaders develop a role that is more focused on education and less on research. That does not mean that departing group leaders are not devoted to research anymore; they still spend a considerable amount of their total work time on research and research management of the group (47.1 %, see Table 6.6).

This shift from research to education has an impact on the motivation to set the research agenda. Departing group leaders often find it more important that PhD students follow their own interests. At the end of their career, group leaders may put more emphasis on supervision, and—not to be hampered by group interests—they give young researchers more freedom to explore their individual interests. Producing high impact research is of less

Table 6.6 Departing Effects on Academic Leadership

Academic leadership	Starting leaders	Experienced leaders	Departing leaders	*Starting vs. Experienced*		*Experienced vs. departing*		*Starting vs. departing*	
	Median (*n*)	Median (*n*)	Median (*n*)	Odds ratio	*R*	Odds ratio	*R*	Odds ratio	*r*
External funding	45.0 (22)	70.0 (64)	50.0 (42)		–0.22**		–0.20**		ns
Submitted research proposals (% yes)	31.8 (22)	52.2 (69)	31.1 (45)	2.34*a)		2 34**		ns	
Quality control of research proposals	3.0 (19)	2.0 (65)	3.0 (35)		–0.23**		–0.23**		ns

a) Reference category is experienced group leaders. * p <0.10; ** p < 0.05.

Note: ns = not significant.

*$p < 0.10$. **$p < 0.05$.

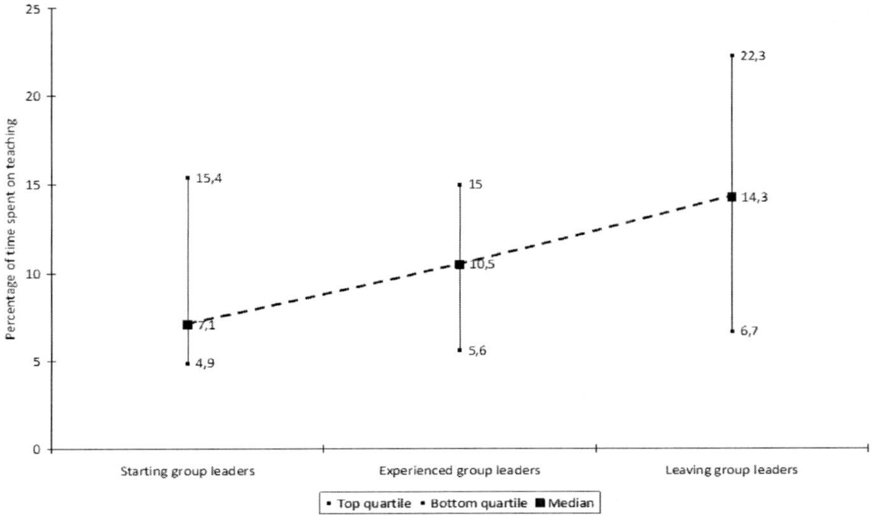

Figure 6.8 Departing effect of proportion of time spend on teaching

importance, and only a quarter of the departing group leaders attach value to obtaining visibility in top journals. In general, departing group leaders have a smaller focus on research. Finally, departing group leaders organize fewer job assessments; this task is possibly taken over by their successors (Figure 6.5).

Hypothesis 5 is also confirmed; the results show that academic leaders who are in their departing phase are relinquishing their roles as group leaders by shifting their focus from research to education and supervision.

Combined Effects

In the two previous sections, we investigated the effects of generation and life cycle stage on the total sample separately. However, some of the differences found in academic leadership practices may be the (additive) result of both generation and life-cycle processes. But other differences in academic leadership could be the result of either generation or life cycle processes. To test this, we control generation effects for life cycle and vice versa in this section.

In the results sections we already showed that the two generations differ in several variables of academic leadership. Some of the academic leadership variables were also influenced by life cycle processes. Table 6.7 shows the generation effects controlled for life cycle stage.

The *number of funding sources* and the *share of permanent staff* do not significantly differ between the three life cycle phases and are indeed solely generation effects. With respect to the *amount of external funding,* we found a generation effect as well as a life cycle effect. The amount of external

Table 6.7 Generation Effects Controlling for Life-Cycle Effects

Academic leadership variables	Generation	Starting leaders	*Starting vs. Experienced*	Experienced leaders	*Experienced vs. Departing*	Departing Leaders (old generation)	*Starting vs. Departing*
		Median (*n*)		Median (*n*)		Median (*n*)	
Number of funding sources	Cohort old	3.0 (10)		3.0 (48)		3.0 (41)	
	Cohort young	4.0 (12)		4.0 (16)			
	Total	3.5 (22)	ns	3.0 (64)	ns	3.0 (41)	ns
Permanent staff	Cohort old	42.7 (8)		53.1 (50)		58.3 (41)	
	Cohort young	47.5 (10)		35.7 (17)			
	Total	45.2 (18)	ns	50.0 (67)	ns	58.3 (41)	ns
External funding	Cohort old	30.5 (10)		60.0 (48)		50.0 (41)	
	Cohort young	55.0 (12)		80.0 (16)			
	Total	45.0 (22)	–0 .22**	70.0 (64)	–0 .20**	50.0 (41)	ns
Teaching	Cohort old	9.1 (10)		12.1 (50)		14.3 (44)	
	Cohort young	6.1 (12)		6.7 (18)			
	Total	7.1 (22)	ns	10.5 (68)	–0 .19**	14.3 (44)	–0 .23*
Research & research management	Cohort old	55.5 (10)		52.7 (50)		47.3 (44)	
	Cohort young	61.3 (12)		58.6 (18)			
	Total	58.7 (22)	ns	52.9 (68)	–0 .21**	47.3 (44)	–0 .21*
PhDs interests and possibilities (% yes)	Cohort old	20.0 (10)		34.7 (49)		47.6 (42)	
	Cohort young	25.0 (12)		16.7 (18)			
	Total	22.7 (22)	ns	29.9 (67)	2.13*	47.6 (42)	2.09*
Networking	Cohort old	32.5 (10)		50.0 (51)		40.0 (43)	
	Cohort young	30.0 (12)		35.0 (18)			
	Total	30.0 (22)	–0 .30***	40.0 (69)	ns	40.0 (43)	–0 .23*
Professors	Cohort old	9.4 (8)		9.1 (50)		11.1 (41)	
	Cohort young	6.7 (10)		11.1 (17)			
	Total	7.4 (18)	ns	10.0 (67)	ns	11.1 (41)	–0 .32**

Note: ns = not significant; # = odds ratio with reference category departing group leaders.

$^{*}p < 0.10$. $^{**}p < 0.05$.

funding that is acquired clearly represents a combined effect of generation and life cycle. Figure 6.9 illustrates that the proportion of external funding is the highest for experienced group leaders, who have as a main goal to secure vitality of the group. Figure 6.9 also shows that the newer generation of group leaders, in both the starting and the maturity life cycle phases, obtain much more external funding compared to the older generation. This generation comparison cannot be done for the departing leaders, as they all belong to the older generation.

Another variable that differs between generations (Table 6.7) is the shift from research to education. As Table 6.7 shows, here the departing age group differs from the two other groups. Given that the departing group is a subset of the older generation, the generation effect may be overestimated. As in both generations, the starting group leaders and the experienced group leaders differ from each other in the same direction; research and teaching are additively influenced by generation and life cycle. Both the older generation and the soon retiring group leaders shift their time allocation from research towards education. Also, departing leaders and leaders from the older generation seem to value more the own interests of PhD students.

In the sections on generation and life cycle effects, we already mentioned that the time spent on networking activities and the percentage of professors in the research group are both starting effects rather than a generation effect. This is seen in Table 6.7; the starting group leaders spend fewer days per year on networking activities and have a lower share of professors in their group. Table 6.7 shows no support for generation effects.

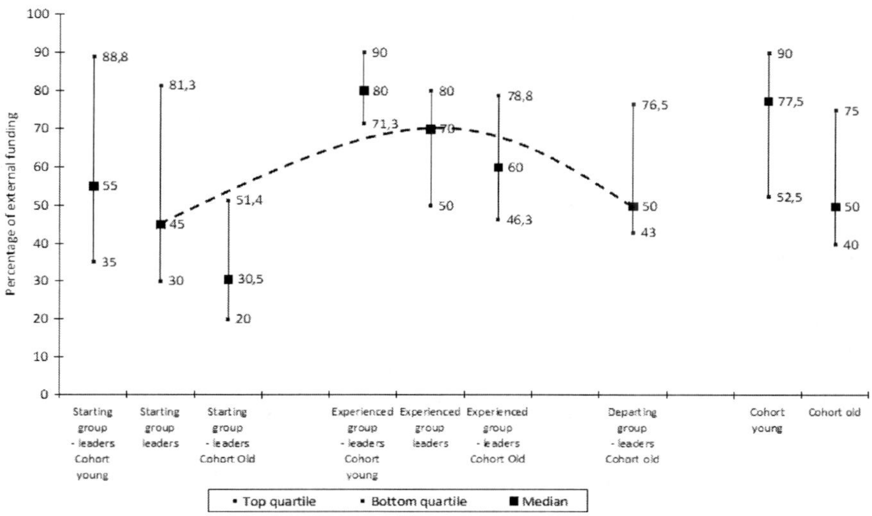

Figure 6.9 Maturity and generation effect of external funding

Table 6.8 Life-Cycle Effects Controlling for Generation Effects

Academic leadership Variables	Cohort Old	Cohort Young	*Cohort Old vs Cohort Young*	
Starting effects	Median (*n*)	Median (*n*)	Odds ratio	*r*
Theoretically challenging (% yes)	69.0 (100)	66.7 (30)	ns	
Intangible rewards (% yes)	52.9 (104)	55.2 (29)	ns	
Maturity effects				
Submitted research proposals (% yes)	39.0 (105)	53.3 (30)	ns	
Quality control of research proposals	2.0 (89)	2.0 (29)		ns
Departing effects				
Research agenda	37.5 (97)	50.0 (29)		ns
Job assessments (% yes)	94.3 (105)	96.7 (30)	ns	

Note: ns = not significant.

Finally, we checked whether the (previously not yet discussed) variables that showed differences between the life cycle groups differ between the generations. This is not the case (Table 6.8).

CONCLUSION AND DISCUSSION

The Study

In order to study the changes in academic leadership behavior, we distinguished between changes related to life cycle and changes related to generation differences. In the mid-1980s (until 1988), science policy implemented a national research evaluation procedure that assessed all medical and health science faculties and departments of non–university research institutes in the Netherlands. In the same period, a new funding system was introduced for universities. Finally, in this period, the share of competitive project funding grew fast. Based on these changes in research evaluation and funding, we defined two cohorts of academic leaders in our survey sample of biomedical and health group leaders. There is a new generation of group leaders who were socialized into the academic world after the implementation of these changes took effect, and there is an old generation of group leaders who were socialized before these changes had taken effect. As a result, we expected the younger generation to be more adapted to the new environment, which is reflected in their larger orientation on research and on external funding.

Considering life cycle effects on academic leadership, we distinguished three phases based on experience as a group leader and on age. The starting phase includes academic leaders with 5 or fewer years of experience. The maturity phase includes experienced academic leaders who are not anticipating retirement yet (younger than 57). The departing phase includes the oldest academic leaders (57 years or older) who are near retirement. We compared activities and attitudes between these three life-cycle phases and expected differences in goals, interests, and responsibilities.

Before summarizing our conclusions and presenting implications, we mention some limitations of our study. First, our sample is limited to the biomedical and health-care sciences, while cohort as well as life-cycle effects can vary between disciplines (Gonzalez-Brambila & Veloso, 2007; Simonton, 1988). However, the biomedical and healthcare sciences cover about 40% of all public research in the Netherlands. In addition, the biomedical and healthcare sciences consist of rather heterogeneous types of research, including fundamental preclinical research, application-oriented clinical research, and health-care research with a social science–oriented nature. Nevertheless, we do not disregard that the situation might be different in, for instance, technical sciences or humanities. The same holds for scientific disciplines where research is traditionally less organized in groups, and so academic leadership probably represents a slightly minor role. As a second limitation, the study is conducted in the Netherlands, which has some consequences for international generalization. Academic leadership practices may of course differ between countries, so comparative studies are certainly recommended. However, the Netherlands is not a small country in terms of research output and the country's academic culture is comparable with that of other Western countries. Finally, the sample was not large enough to conduct a more advanced statistical multivariate analysis.

Conclusion

In this study, we show differences in academic leadership practices influenced by generation and life-cycle phase. The institutionalization of research evaluations and the implementation of project funding in the Dutch science system in the middle of the 1980s have had consequences for the behavior of academic leaders. The young generation of academic leaders—socialized in a period wherein these changes had already taken effect—is more adapted to this new, research oriented, competitive environment later in their careers as group leaders. In comparison with the older generation of academic leaders, the younger generation spends more time on research activities and group management and they spend less time on education. Also, they acquire a higher proportion of external funding from more various sources. This higher dependence on external funds may have led to a lower proportion of permanent staff and it may determine the research topics of PhDs, giving young researchers less freedom to explore their own individual research interests.

As explained in literature section on life cycle effects, we expected that academic leaders who are in different stages of their careers have different goals, interests, and responsibilities. As a consequence they will show different academic leadership practices. Academic leaders who just started to lead their own research groups need to build up a reputation. This is confirmed by the finding that starting leaders emphasize a need for new and creative research themes. Their lower network activity mirrors their lower reputation; starting group leaders need more time to gain visibility in the scientific community before they get invited for, for example, assessment committees. The need to promote their research externally might result in less emphasis on internal group management, such as motivating their researchers with intangible rewards.

When reputation is obtained, the next step is obviously to retain it. Then, the main focus of experienced academic leaders is to preserve and extend group vitality. Our findings support this, as acquiring new research funds is the highest priority, and academic leaders in their maturity phase obtain more external funding than academic leaders in other life cycles. Furthermore, on average, experienced academic leaders are more active in submitting research proposals than starting and departing group leaders. Finally, they are more active in checking the quality of research proposals from their group members.

Academic leaders who are near retirement age were also expected to behave differently. In preparation of their leave, they are likely to reduce research activities and place more emphasis on education. More specifically, although departing academic leaders in our sample spend about half of their time on research and research management of their group, they spend a lower proportion of time on research activities and group management than starting and experienced academic leaders. Also, they spend a higher proportion of their time on education. On average, departing group leaders consider visibility in top journals as less important than do starting and experienced academic leaders. Finally, departing leaders (more than starting and experienced academic leaders) attach value to the autonomy of PhD students, which indicates a shift from a group leader's role to a supervisory role.

Implications

This study shows that if science policy changes incentives in the science system, researchers adapt their behavior. The period in which researchers become socialized determines their attitudes and activities in their further career. What does this imply for any mentioned current developments, such as stress for academic excellence, an increasing emphasis on the societal benefits of research, and the growing competition between researchers and research organizations for funding? With respect to the 1980s changes, academic leaders were forced into a more entrepreneurial attitude, indicated by a greater dependence on external funding and a stronger focus on research. More recent developments reinforce this tendency, and this entrepreneurial attitude is becoming important even earlier in research careers. For instance, the implementation of new grant schemes for young talented researchers[6] gives researchers in an early phase (immediately after defending their dissertation) the autonomy to elaborate their own research agenda. These grants provide great opportunities for young talented researchers to explore their own creative research niches. However, it also intensifies competition between researchers at an early stage in their career. At the same moment, it may create an 'obligatory point of passage' for entering an academic career. This is a potential risk, since the proportion of applicants that get funded is relatively low, and a lot of rejected applicants are good performers as well (Bornmann, Leydesdorff, & Van den Besselaar, 2010; Van den Besselaar & Leydesdorff, 2009). To summarize, the new developments, which force researchers to independence, will lead to a radicalization of competition. How this will affect scientific careers and research leadership in the future remains an open question.

The differences in goals, responsibilities, and interests in the different life cycle phases lead to differences in behavior. What can we learn from the observed role change over the life cycles of academic leaders? First, it should be noted that the roles of academic leaders who approach retirement age should not be underestimated. In fact, their roles move more from knowledge production to knowledge transfer (namely, educational and supervisory tasks). However, they remain spending a considerable amount of time on research, and they are still productive and thus do not lose touch with research practices. That is, they fulfill an important role in the education of the new generation of researchers. In particular, our results show that they are more sensitive for the autonomy, possibilities, and interests of PhD students and may therefore stimulate more strongly the creativity of young researchers. Given that researchers still fulfill a meaningful task in knowledge transfer at the end of their career, one could think about extending the scientific careers of researchers beyond the current retirement age of 65 in the Netherlands.[7]

Finally, a somewhat more speculative implication involves the size and composition of research groups. Looking at the different roles of starting, experienced, and departing academic leaders, one may think of the division of labor between leaders in various life-cycle stages. Then, a research group would consists of a less experienced group leader who develops his or her own creative research niche with support of a more experienced group leader that secures funding for the research group, and an older (highly experienced) leader who takes care of the supervision of young researchers. This would of course lead to larger teams, which may increase coordination costs. On the other hand, a bigger group size enables more variety of specialties and skills, and that is essential for addressing complex research problems (Borner et al., 2010; Stokols, Hall, et al., 2008; Stokols, Misra, et al., 2008). The development of larger research groups may also affect the vertical relations between the large research team and the research organization. Probably, larger research teams are more powerful players in the research organization, resulting in an expanded autonomy from the institute's management. This is also an interesting topic for future research.

ACKNOWLEDGMENTS

The collection of the data used in this study was supported by The Netherlands Organization for Scientific Research (NWO), grant no. 900–68–123. The authors would like to thank Dick de Gilder and Peter Groenewegen for their contributions in designing the questionnaire. Furthermore, we are grateful to Rosalie Belder and Pleun van Arensbergen for their contributions to the literature search, and to Eduard Klasen, Peter Groenewegen, Barend van der Meulen and two anonymous reviewers for comments on earlier versions of this chapter. Finally, this research was not possible without the support of the Rathenau Institute.

NOTES

1. Intermediary organization represents a dual alliance with both the government and the scientific community.
2. In Dutch Disciplineoverleg Medische Wetenschappen, the consultation body of the medical schools in the Netherlands. It has transformed into the NFU since 2004, the Dutch Association University Hospitals.
3. See VSNU, KNAW, and NWO (2009), the protocol for research assessment of public funded research in the Netherlands.
4. We choose 57 years as breakpoint because group leaders indicated in interviews that 62 is the age at which retirement can start and often starts. Legal retirement age is at 65. Similar to a maximum of 5 years starting phase, we also took a maximum of 5 years for phasing out.
5. Due to missing data about year of PhD obtainment, age, or years of experience as group leader, not all 137 respondents are included in the analyzes.
6. The Netherlands Organization for Scientific Research (NWO—Dutch research council) implemented the Innovational Research Incentives Scheme in 2000. These grants are meant for excellent researchers in different phases of their career; Veni (the youngest researchers who just obtained their PhD), Vidi (researchers who want to set up their own research line and appoint one or more researchers), Vici (senior researchers who want to start their own research group). Similar grants for young researchers are implemented, for example, in Germany, the Emmy Noether Programme, which aims to support young researchers in achieving independence at an early stage of their career, and in Sweden, the Project Grant Junior Researchers. Also, the European Research Council has funds that offer opportunities for young

investigators to develop independent careers through the ERC Starting Independent Research Grant.

7. Retirement age varies between countries due to their specific tax laws and/or pension rules.

REFERENCES

Aggarwal, R., Schirm, D., & Zhao, X. (2007). Role models in finance: Lessons from life cycle productivity of prolific scholars. *Review of Quantitative Finance and Accounting*, *28*, 79–100.

Aizenman, J., & Kletzer, K. (2008). *The life cycle of scholars and papers in economics—the "citation death tax"* (NBER Working paper No. 13891). Cambridge, MA: National Bureau of Economic Research.

Austin, A. E. (2002). Preparing the next generation of faculty—graduate school as socialization to the academic career. *Journal of Higher Education*, *73*(1), 94–122.

Babu, A. R., & Sing, Y. P. (1998). Determinants of research productivity. *Scientometrics*, *43*(3), 309–329.

Baldwin, R. G., Lunceford, C. J., & Vanderlinden, K. E. (2005). Faculty in the middle years: Illuminating an overlooked phase of academic life. *Review of Higher Education*, *29*(1), 97–118.

Bayer, A. E., & Dutton, J. E. (1977). Career age and research professional activities of academic scientists—tests of alternative nonlinear models and some implications for higher-education faculty policies. *Journal of Higher Education*, *48*(3), 259–282.

Blackburn, R. T., Behymer, C. E., & Hall, D. E. (1978). Research note—correlates of faculty publications. *Sociology of Education*, *51*(2), 132–141.

Bland, C. J., & Ruffin, M. T. (1992). Characteristics of a productive research environment: literature review. *Academic Medicine*, *67*(6), 385–397.

Bland, C. J., & Schmitz, C. C. (1986). Characteristics of the successful researcher and implications for faculty-development. *Journal of Medical Education*, *61*(1), 22–31.

Börner, K., Contractor, N., Falk-Krzesinski, H. J., Fiore, S. M., Hall, K. L., Keyton, J., . . . Uzzi, B. (2010). A multi-level systems perspective for the science of team science. *Science Translational Medicine*, *2*(49), 1–14.

Bornmann, L., Leydesdorff, L., & Van den Besselaar, P. (2010). A meta-evaluation of scientific research proposals: Different ways of comparing rejected to awarded applications. *Journal of Informetrics*, *4*(3), 211–220.

Brancati, F. L., Mead, L. A., Graves, P. L., Levine, D. M., & Klag, M. J. (1992). Personal predictors of career achievement among academic physicians. *Clinical Research*, *40*(2), A553-A553.

Cameron, S. W., & Blackburn, R. T. (1981). Sponsorship and academic career success. *Journal of Higher Education*, *52*(4), 369–377.

Carayol, N., & Matt, M. (2004). Does research organization influence academic production? Laboratory level evidence from a large European university. *Research Policy*, *33*(8), 1081–1102.

Cole, S., & Cole, J. R. (1967). Scientific output and recognition—study in operation of reward system in science. *American Sociological Review*, *32*(3), 377–390.

Costas, R., Van Leeuwen, T. N., & Bordons, M. (2010). A bibliometric classificatory approach for the study and assessment of research performance at the individual level: The effects of age on productivity and impact. *Journal of the American Society for Information Science and Technology*, *61*(8), 1564–1581.

Cronan-Hillix, T., Gensheimer, L. K., Cronan-Hillix, W. A., & Davidson, W. S. (1986). Students' views on mentors in psychology graduate training. *Teaching of Psychology*, *13*(3), 123–127.

D'Amato, A., & Herzfeldt, R. (2008). Learning orientation, organizational commitment and talent retention across generations: A study of European managers. *Journal of Managerial Psychology*, *23*(8), 929–953.

De Jong, S. P. L., Van Arensbergen, P., Daemen, F., Van der Meulen, B., & Van den Besselaar, P. (2011). Evaluation of research in context: an approach and two cases. *Research Evaluation*, *20*(1), 61–72.

Dennis, W. (1956). Age and productivity among scientists. *Science*, *123*(3200), 724–725.

Dietz, J. S., & Bozeman, B. (2005). Academic careers, patents, and productivity: industry experience as scientific and technical human capital. *Research Policy*, *34*(3), 349–367.

Dill, D. D. (1982). The management of academic culture—notes on the management of meaning and social integration. *Higher Education*, *11*(3), 303–320.

Dillman, D. (2000). *Mail and Internet surveys: The tailored design method*. New York, NY: John Wiley & Sons.

Dundar, H., & Lewis, D. R. (1998). Determinants of research productivity in higher education. *Research in Higher Education*, *39*(6), 607–631.

Etzkowitz, H., & Leydesdorff, L. (1995). The triple helix of university-industry-government relations: a laboratory for knowledge based economic developments. *EASST Review*, *14*(1), 11–19.

Gibbons, M., Limoge, S. C., Notwotney, H., Schwartzman, S., Scott, P., & Trow, M. (1994). *The new production of knowledge: the dynamics of science and research in contemporary societies*. London, England: Sage.

Gingras, Y., Lariviere, V., Macaluso, B., & Robitaille, J. P. (2008). The effects of aging on researchers' publication and citation patterns. *PLOS One*, *3*(12):e4048.

Göktepe-Hultén, D. (2008). Academic inventors and research groups: Entrepreneurial cultures at universities. *Science and Public Policy*, *35*(9), 657–667.

Gonzalez-Brambila, C., & Veloso, F. M. (2007). The determinants of research output and impact: A study of Mexican researchers. *Research Policy*, *36*, 1035–1051.

Goodwin, T. H., & Sauer, R. D. (1995). Life-cycle productivity in academic research—evidence from cumulative publications histories of academic economists. *Southern Economic Journal*, *61*(3), 728–743.

Göransson, B., Maharajh, R., & Schmock, U. (2009). New activities of universities in transfer and extension: Multiple requirements and manifold solutions. *Science and Public Policy*, *36*(2), 151–164.

Green, S. G. (1991). Professional entry and the adviser relationship—socialization, commitment, and productivity. *Group & Organization Studies*, *16*(4), 387–407.

Gursoy, D., Maier, T. A., & Chi, C. G. (2008). Generational differences: An examination of work values and generational gaps in the hospitality workforce. *International Journal of Hospitality Management*, *27*(3), 448–458.

Hansson, F., & Monsted, M. (2008). Research leadership as entrepreneurial organizing for research. *Higher Education*, *55*, 651–670.

Harvey, J., Pettigrew, A., & Ferlie, E. (2002). The determinants of research group performance: towards mode 2? *Journal of Management Studies*, *39*(6), 746–774.

Hornbostel, S., Böhmer, S., Klingsporn, B., Neufeld, J., & Von Ins, M. (2009). Funding of young scientist and scientific excellence. *Scientometrics*, *79*(1), 171–190.

Horta, H. (2009). Holding a post-doctoral position before becoming a faculty member: Does it bring benefits for the scholarly enterprise? *Higher Education*, *58*, 689–721.

Knorr, K. D., & Mittermeir, R. (1980). Publication productivity and professional position—cross-national evidence on the role of organizations. *Scientometrics*, *2*(2), 95–120.

Knorr, K. D., Mittermeir, R., Aichholzer, G., & Waller, G. (1979a). Individual publication productivity as a social position effect in academic and industrial research units. In F. M. Andrews (Ed.), *Scientific productivity. The effectiveness of research groups in six countries* (pp. 55–94). Cambridge, England: Cambridge University Press.

Knorr, K. D., Mittermeir, R., Aichholzer, G., & Waller, G. (1979b). Leadership and group performance: a positive relationship in academic research units. In F. M. Andrews (Ed.), *Scientific productivity. The effectiveness of research groups in six countries* (pp. 95–120). Cambridge, England: Cambridge University Press.

Koninklijke Nederlandse Akademie van Wetenschappen. (1988). *Disciplineplan Geneeskunde: samenhang, samenwerking en profilering* [*Foresight medical sciences: coherence, collaboration and focus*]. Amsterdam, the Netherlands: Koninklijke Nederlandse Akademie van Wetenschappen, Commissie Geneeskunde, Subcommissie integratie disciplineplan geneeskunde.

Kyvik, S. (1990). Age and scientific productivity—differences between fields of learning. *Higher Education*, *19*(1), 37–55.

Kyvik, S., & Olsen, T. B. (2008). Does the aging of tenured academic staff affect the research performance of universities? *Scientometrics*, *76*(3), 439–455.

Lepori, B., Van den Besselaar, P., Dinges, M., Póti, B., Reale, E., Slipersaeter, S., . . . van der Meulen, B. (2007). Comparing the evolution of national research policies: What patterns of change? *Science and Public Policy*, *34*(6), 372–388.

Levin, S. G., & Stephan, P. E. (1991). Research productivity over the life-cycle—evidence for academic scientists. *American Economic Review*, *81*(1), 114–132.

Long, J. S., Allison, P. D., & McGinnis, R. (1993). Rank advancement in academic careers: Sex differences and the effects of productivity. *American Sociological Review*, *58*(5), 703–722.

Long, J. S., & McGinnis, R. (1985). The effects of the mentor on the academic career. *Scientometrics*, *7*(3–6), 255–280.

Lord, R. G., & Hall, R. J. (2005). Identity, deep structure and the development of leadership skill. *The Leadership Quarterly*, *16*, 591–615.

Morrison, E., Rudd, E., Picciano, J., & Nerad, M. (2011). Are you satisfied? PhD education and faculty taste for prestige: Limits of the prestige value system. *Research in Higher Education*, *52*(1), 24–46.

Mumford, M. D., Scott, G. M., Gaddis, B., & Strange, J. M. (2002). Leading creative people: Orchestrating expertise and relationships. *Leadership Quarterly*, *13*(6), 705–750.

Mumford, M. D., Zaccaro, S. J., Harding, F. D., Jacobs, T. O., & Fleishman, E. A. (2000). Leadership skills for a changing world: Solving complex social problems. *Leadership Quarterly*, *11*(1), 11–35.

Paglis, L. L., Green, S. G., & Bauer, T. N. (2006). Does adviser mentoring add value? A longitudinal study of mentoring and doctoral student outcomes. *Research in Higher Education*, *47*(4), 451–476.

Pezzoni, M., Sterzi, V., & Lissoni, F. (2009). *Career progress in centralized academic systems: An analysis of French and Italian physicists* (Working paper n. 26/2009). Milano, Italy: Knowledge, Internationalization and Technology Studies.

Prpic, K. (1996). Characteristics and determinants of eminent scientists' productivity. *Scientometrics*, *36*(2), 185–206.

Rauber, M., & Ursprung, H. W. (2008). Life cycle and cohort productivity in economic research: The case of Germany. *German Economic Review*, *9*(4), 431–456.

Reskin, B. F. (1979). Academic sponsorship and scientists careers. *Sociology of Education*, *52*(3), 129–146.

Rip, A., & Van der Meulen, B. (1995). The patchwork of the Dutch evaluation system. *Research Evaluation*, *5*(1), 45–53.

Sessa, V. I., Kabacoff, R. I., Deal, J., & Brown, H. (2007). Generational differences in leader values and leadership behaviors. *The Psychologist-Manager Journal*, *10*(1), 47–74.

Simonton, D. K. (1988). Age and outstanding achievement: What do we know after a century of research? *Psychological Bulletin*, *104*(2), 251–267.

Simonton, D. K. (2004). *Creativity in science. Chance, logic, genius and zeitgeist.* Cambridge, England: Cambridge University Press.

Smith, C. G. (1971). Scientific performance and composition of research teams. *Administrative Science Quarterly*, *16*(4), 486–496.

Spaapen, J. (1995). *Research for society: Towards a method for the assessment of research in a policy context.* Amsterdam, the Netherlands: University of Amsterdam.

Spangenberg, J. F. A., Starmans, R., Bally, Y. W., Breemhaar, B., Nijhuis, F. J. N., & Vandorp, C. A. F. (1990). Prediction of scientific performance in clinical medicine. *Research Policy*, *19*(3), 239–255.

Stankiewicz, R. (1976). Research groups and the academic research organization. *Sociologisk Forskning*, *13*(2), 20–32.

Stephan, P. E., & Levin, S. G. (1993). Age and the Nobel-Prize revisited. *Scientometrics*, *28*(3), 387–399.

Stokols, D., Hall, K. L., Taylor, B. K., & Moser, R. P. (2008). The science of team science—overview of the field and introduction to the supplement. *American Journal of Preventive Medicine*, *35*(2), S77-S89.

Stokols, D., Misra, S., Moser, R. P., Hall, K. L., & Taylor, B. K. (2008). The ecology of team science—understanding contextual influences on transdisciplinary collaboration. *American Journal of Preventive Medicine*, *35*(2), S96-S115.

Sturman, M. C. (2003). Searching for the inverted U-shaped relationship between time and performance: Meta-analyses of the experience/performance, tenure/performance, and age/performance relationships. *Journal of Management*, *29*(5), 609–640.

Turner, L., & Mairesse, J. (2002). *Explaining individual productivity differences in public research: how important are non-individual determinants? An economic study of French physicists' publications (1986–1997).* Retrieved from http://www.opengrey.eu/item/display/10068/6105

Van Arensbergen, P. (2010). The Dutch system: Combining the best of both worlds?Internal report, Rathenau Instituut, Den Haag..

Van Balen, B. (2010). *Op het juiste moment op de juiste plaats. Waarom wetenschappelijk talent een wetenschappelijke carrière volgt* [At the right moment at the right spot: Why scholarly talent decides for an academic career]. The Hague, the Netherlands: Rathenau Instituut.

Van den Besselaar, P., & Leydesdorff, L. (2009). Past performance, peer review, and project selection: A case study in the social and behavioral sciences. *Research Evaluation*, *18*(4), 273–288.

Van der Meulen, B. (2009). Science policy in the Netherlands. In S. Hornbostel, A. Knie, & D. Simon (Eds.), *Handbuch Wissenschaftspolitik* (pp. 214–228).Wiesbaden, Germany: VS Verlag fur Sozialwissenschaften.

Van der Weijden, I. (2007). *In search of performance. Research management within the Dutch public medical and health sector.* Amsterdam, the Netherlands: Vrije Universiteit.

Van der Weijden, I., De Gilder, D., Groenewegen, P., & Klasen, E. (2008). Implications of managerial control on performance of Dutch academic (bio)medical and health research groups. *Research Policy*, *37*, 1616–1629.

Van der Weijden, I., Verbree, M., Braam, R., & Van den Besselaar, P. (2009). *Management en prestaties van onderzoeksgroepen* [*Management and performance of research groups*] (SciSA Report 0913). The Hague, the Netherlands: Rathenau Instituut.

Van Leeuwen, T. N., Visser, M. S., Moed, H. F., Nederhof, T. J., & Van Raan, A. F. J. (2003). The holy grail of science policy: Exploring and combining bibliometric tools in search of scientific excellence. *Scientometrics*, *57*(2), 257–280.

Van Rijnsoever, F. J., Hessels, L. K., & Vandeberg, R. L. J. (2008). A resource-based view on the interactions of university researchers. *Research Policy, 37*, 1255–1266.

Van Steen, J. (2011). *Feiten en cijfers: Overzicht Totale Onderzoek Financiering (TOF) 2009–2015 [Facts and figures: Public research funding 2009–2015]*. The Hague: Rathenau Instituut.

VSNU, NWO, & KNAW. (2009). *Standard evaluation protocol 2009–2015: Protocol for research assessment in the Netherlands.* Retrieved from http:// www.knaw.nl/sep

Versleijen, A., Van der Muelen, B., Van Steen, J., Boneschansker-Kloprogge, P., Braam, R., Mamphuys, R., & Van den Besselaar, P. (2007). *Dertig jaar publieke onderzoeksfinanciering in Nederland 1975–2005 [Thirty years of public research funding in the Netherlands 1975–2005]* (SciSA report 0703). The Hague: Rathenau Instituut.

Westerheijden, D. F. (1997). A solid basis for decisions. *Higher Education, 33*, 397–413.

7 Time to Create

Pathways to Earlier and Later Creative Discoveries in Nobel Prize Winners

Dawn L. Eubanks, Michael E. Palanski, Juani Swart, Michelle M. Hammond, and Joy Oguntebi

Creative discoveries are increasing in importance across domains such as medicine, pharmacology, and technology. With millions of dollars being spent on high-risk creative endeavors, it is important to understand the processes involved with highly successful innovative initiatives. While recent noteworthy strides have been made in this area (e.g., Hunter, Bedell, & Mumford, 2007; Mumford, Hunter, Eubanks, Bedell, & Murphy, 2007), there is still much room for advancement. Gone are the days when companies achieved profits solely by streamlining a process or cleverly advertising a product. Clearly, these elements are important; however, true competitive advantage often comes through creative discovery (Amabile, 1988, 1996), and it does not come cheap. Because implementable creative ideas are rare, are embedded with high levels of risk, and thus often fail, the process is incredibly expensive (Mumford, Scott, Gaddis, & Strange, 2002). Therefore, an increased understanding of what leads to a creative discovery is extremely valuable.

The review article by Hennessey and Amabile (2010) distinguishes between everyday and eminent creativity, defining everyday creativity as "daily problem solving and the ability to adapt to change" (p. 572) and eminent creativity as "relatively rare displays of creativity that have a major impact on others" (p. 572). Based on these definitions, the current study is focused on eminent creativity because we are interested in ideas that have a major impact. For example, Marie Curie had a large scientific impact with her discovery of polonium and radium. Specifically, we focus on the more pragmatic kind of creativity—that which involves engaging with complex, ill-defined problems where the solutions are unclear (Ford, 2000; Mumford et al., 2002; Reiter-Palmon & Illies, 2004).

Leadership plays a critical role in the success of a creative discovery (Mumford, Peterson, & Robledo, this book; Puccio, Mance, & Zacko-Smith, this book). The role of leadership can take two forms. First, in the case of early-career researchers the role of the supervisor or mentor as a leader is important. It is the supervisor that shapes work habits, teaches norms in the field, provides ethical guidelines, and so on. This is an important relationship because in some sense, it teaches and shapes future work habits. In a study by Mumford, Murphy, et al. (2007), the authors identified the following variables as having a noteworthy influence over decision making in one's early career: professional leadership, poor coping, lack of rewards, limited competitive pressure, and poor career direction. Many of these variables can be influenced by the mentor or supervisor. Because of this, the role of the mentor as a leader plays a significant role early in one's career. Second, in the case of later career researchers, they must spend more time managing demands (time, allocating resources, managing relationships, etc.) that come their way and acting in a mentoring or supervisory role to less experienced researchers. In both career stages, leadership plays a key role with a different focus either as being effectively led or as being an effective leader (Mumford, Peterson, & Robledo, this book).

Because of the aforementioned high costs and large risks in the creative endeavor, organizational decision makers are interested in knowing how creativity occurs. The romantic notion of the artist at his easel creating a masterpiece or the scientist in her lab making a discovery is belied by a growing body of evidence that demonstrates that many factors beyond individual creative genius influence the creative process, factors such as job characteristics, organizational resources, support, and education (Shalley & Gilson, 2004). Given the multitude of factors that affect creativity, an important question begins to emerge: Which conditions might support a creative breakthrough early in one's career versus later? One way to approach this question is by examining the age or career stage at which a creative person (in the current study, a scientist) makes an important creative discovery.

Research from multiple fields has supported the notion that creativity tends to decline with age and that groundbreaking discoveries tend to occur earlier in life (e.g., McCrae, Arenberg, & Costa, 1987; Mumford & Gustafson, 1988). Indeed, such findings seemed to have crystallized into something of an axiom that creativity happens early in life. Such thinking is supported by examining the lives of the most famous scientists and artists; for example, Einstein was only 26 when he developed the theory of relativity, and Bill Gates was only 20 when he left Harvard and founded Microsoft. Yet, despite the research and popular romantic ideas that support the 'creativity-occurs-at-a-young-age' notion, other research suggests that creativity need not cease as one approaches mid-life and beyond. For example, Lindauer (1993) showed that some artists produced masterpieces much later in life. Other research has shown that creativity can be enhanced by having a deep and robust knowledge of a particular field and its accompanying issues (Mumford & Gustafson, 1988; Shalley & Gilson, 2004), knowledge which is built over the course of a career.

Because some individuals make a creative discovery early and others make creative discoveries later in their careers, an important question to ask is whether there are different processes in place for early or later career success, and what types of factors may lead to success earlier or later, respectively, in one's career? In other words, are there different paths for creativity, one leading to relatively early career success and the other leading to later career success? For example, some individuals may have a great idea and then for some reason fail to generate more creative discoveries. Others may endure and flourish as expertise becomes refined.

One way to address these questions is by examining the multilevel factors that could play a role in creativity, particularly scientific creativity. Explicit consideration of levels of analysis is important for two primary reasons (Yammarino, Dionne, Chun, & Dansereau, 2005). First, failure to consider levels of analysis leaves theory building and theory testing incomplete and may lead to incorrect conclusions. For example, Silvia (2007) has suggested that much of the early creativity research may be flawed because it typically did not examine the possible effects of having a nested data structure (e.g., individual level phenomenon may be embedded in group- or organizational-level variables, etc.) Second, consideration of different levels of analysis may provide important insights. For example, Yammarino et al. (2005) pointed out that a revolution occurred in physics when physicists proposed, and subsequently demonstrated, that quantum mechanics operate at a level *lower* than the atomic level. Within the context of innovation, preliminarily support exists to suggest that individual level factors may be more important for idea generation, whereas collective-level factors may be more important for the implementation of such ideas (Axtell et al., 2000; Hammond, Neff, Farr, Schwall, & Zhao, 2011). For the current study, we examine the influences at multiple levels of analysis in the lives of 59 Nobel Prize winners in physics, chemistry, medicine, and economics to attempt to investigate whether there may be different paths leading to early versus later career success.

LITERATURE REVIEW AND CONCEPTUAL DEVELOPMENT

Background in Creativity Research

Much research has taken place on the topics of creativity and innovation. There is noteworthy research that investigates the process of creativity, including studying the role of logic and computer simulations (Langley, Simon, Bradshaw, & Zythow, 1987; Shrager & Langley, 1990) or individual characteristics (Cattell, 1963; Eysenck, 1995). Although studying personality gives us useful insights into what allows a person to be creative, it is necessary to study creativity at multiple levels because of the strong influence of team and organizational level variables (Hennessey & Amabile, 2010). Research and development initiatives in organizations do not take place with individuals working in isolation; rather, they generally occur in teams with a variety of expertise (Crane, 1972). In our attempt to understand what allows an individual to achieve noteworthy creative discoveries early or later in his or her career, we relied on the wealth of existing literature on these topics (e.g., Mumford et al., 2002; Shalley & Gilson, 2004;

Shalley, Zhou, & Oldham, 2004). Although there are certain dispositional characteristics appearing consistently among creative scientists that support the investigation of alternatives under conditions of uncertainty (Mumford & Gustafson, 1988), there are other variables that become more or less important throughout the life span of the creative scientist.

It is no surprise that context plays an important role in creativity. For example, Denti and Hemlin (this book) discuss the importance of designing a creative knowledge environment. In the meta-analysis by Hunter, Bedell, and Mumford (2007), the authors analyzed the work from 42 studies investigating situational variables and various measures of creativity. They found that situational variables are particularly important to consider in high-pressure, turbulent environments. This study examined variables from a multilevel perspective, finding that there are many contradictions between what is called for based on the level examined. The current study will add to this effort in looking at multilevel variables among a highly creative and successful group of people. In particular we are interested in how the emphasized importance of these variables changes between early and later career.

We focused upon Nobel Prize winners in this study because winning a Nobel Prize demonstrates a particularly eminent creative discovery. In studying the careers of Nobel Prize winners, we can see that there are differences in the ways that these individuals went about their work. Similarly, while creativity can take the form of either incremental improvements or major breakthroughs (Mumford & Gustafson, 1988), the focus of this study is on individuals who achieved a major breakthrough for which they were then awarded the Nobel Prize.

Earning a Nobel Prize is an indication that the individual has successfully engaged in structuring a problem and generating a solution. In order to determine the importance of different variables based on career stage, we recorded the age at which the individual published his or her most cited first- or sole-authored piece of work. Although Nobel Prize winners generally work in academic institutions, several have also an applied component to their work (in pharmaceutical companies, government agencies, etc.; see www.nobelprize.org). In fact, many indicated in their autobiographies or interviews that they received funding from industry sources.

We have considered the career stage at which laureates produced their Nobel Prize–winning work as a variable of interest in terms of understanding the different ways that a career can be structured. The work by Simonton (1991) explored the idea of creative output and age. In this, Simonton discussed differences in productivity that may arise due to the fact that scientific fields have different average ages for beginning a career. However, Simonton's study did not look at contextual influences that could lead to these differences. As such, we selected winners from different fields in order to control for this possible effect. Therefore, the purpose of this study is to identify the multilevel contextual factors that can influence whether a scientist achieves eminence in early or later career. Work by Agars, Kaufman, and Locke (2008) and Shalley et al. (2004) have outlined some of the key contextual characteristics that influence creative output.

Early Creative Success

For the purposes of this study, we conceptualized 'early career' as the time when the winners were in graduate training through to their first full-time position. This can be viewed as an apprenticeship period. 'Later career' included the time from their second job through to the end of their career. Although these timeframes are not equal in terms of number of years, they illustrate the career stages that most people in the study experienced. The early career stage is characterized by a learning relationship with the mentor and the graduate school cohort and includes the first major transition experienced by the scientist as he or she assumes a level of autonomy and moves to establish a research path. The later career timeframe is characterized by bringing in research grants and generally building a research group at the home institution. This later career stage also includes research leadership.

Diverse Perspectives

Divergent thinking, or the ability to look at problems from different perspectives and to consider a wide range of possible solutions, is critically important for creativity (Shalley & Gilson, 2004). As Amabile (1988) points out, individuals who categorize knowledge in 'wide' as opposed to 'narrow' categories are better equipped to see possible relationships between seemingly disparate pieces of information. Research also indicates that having both breadth and depth of knowledge is linked to creativity (Shalley & Gilson, 2004). Evidence of this can be observed by the prevalence of major granting agencies that are providing funds for interdisciplinary efforts (Reiter-Palmon, de Vreede, & de Vreede, this book). Other research indicates that highly creative scientists engage in frequent and widespread communications with peers and read more vigorously and broadly than other scientists (Pruthi & Nagpaul, 1978). Thus, we would expect that such diverse thinking may enable researchers to 'see' possible solutions soon after entering a field because they have 'fresh eyes' and are not yet prone to acceptance of the dominant paradigm in a field. However, as one's career advances, one generally needs to become specialized and more narrowly focused in order to maintain this level of expertise in a particular topic area.

Certainly conducting a careful examination of each winner's background may provide evidence for a particular winner having an unusually broad perspective or the ability to categorize information in divergent areas. However, a more straightforward and objective indicator is a winner having obtained a university degree in a field that is different from the one in which that person won his or her award, thus indicating multiple areas of expertise. Such a criterion provides evidence that the winner has been exposed to at least a modest level of breadth of knowledge outside of his or her field of expertise. Therefore, we predict the following:

> *Hypothesis 1:* Laureates who earn degrees in multiple fields will tend to have success earlier in their careers.

Trying New Things

While individual-level factors are clearly important to earlier creative success, the expanding body of research about contextual variables at higher levels of analysis (i.e., organizational level) indicates that other factors may be important. Although we surmise that most of these contextual factors will encourage success later in one's life, there is one notable exception: the opportunity and the ability to try new things.

The action of trying new things (e.g., trying new ideas, trying new processes, and seeking different goals) is, along with taking risks, the sine qua non of creativity (Shalley & Gilson, 2004) because, by definition, creativity involves generating novel solutions. In their interviews and autobiographies, many winners mentioned the importance of being in an organization where members were free to try new ideas, and these new ideas often came about as a result of discussing and refining ideas in a collaborative manner. Given the importance of freedom and working with diverse colleagues (Mumford et al., 2002), it is not surprising that the work team itself has an important impact on creativity as identified by Madjar (2005). Although having such freedom to try new things is likely to be important at all stages of one's creative life, Mumford and Gustafson (1988) described how creative scientists had early career experiences in a less structured environment than did noncreative scientists. As such, we suggest the following:

> *Hypothesis 2:* Laureates who were members of organizations in their early careers where there was freedom to try new things will tend to have success earlier in their careers.

Later Creative Success

Leadership

Later creative success is typically marked by a deep knowledge of the field and the issues, opportunities, and methods of that field (Mumford & Gustafson, 1988; Shalley & Gilson, 2004). Moreover, research has indicated that the technical skills held by a leader are a robust predictor of creativity and innovation (Andrews & Farris, 1972; Barnowe, 1975), and this sort of learned expertise is typically developed over the course of a career. In the later stages of one's career, scientists often take on more of a leadership role after achieving a level of expertise in order to guide those around them (Mumford et al., 2002). It is not only individual- and team-based interpersonal factors that need to be considered in creative environments, but the structure-based tactics as well (Mumford et al., 2002). Oftentimes as individuals become known in a particular field, they move into leadership positions where they need to manage resources including equipment, time, and people. In order to do this effectively, individuals need to develop strategies to minimize distractions, strategies such as building team cohesion, finding ways of reducing outside demands, and ensuring that they and their teams have enough time and resources; in essence, they need to practice leadership (Mumford et al., this book). To be effective, they need to have a strong understanding of contextual factors (Mumford et al., 2002) because later career success is facilitated by controlling resources and the environment (Cole, 1979). In particular, previous research (Mumford et al., 2002) suggests that providing optimal (i.e., neither too much nor too little) time, resources, and access to expertise through leadership actions is critical to success. As such, we suggest the following:

> *Hypothesis 3:* Laureates who create, through leadership actions, optimal time, resources, and access to expertise will tend to have success later in their careers.

Sufficient Cognitive Capacity and Time

Research indicates that minimizing distractions and allowing plenty of time to explore ideas are particularly important later in one's creative career, for it is during this time that individuals spend so much time developing their expertise that they often leave little time for other aspects of life (Feldman, 1999; Weisburg, 1999). Similarly, Amabile (1983) indicates that the ability to concentrate for long periods is particularly important in the creative process. Thus, one of the most important things that a scientist can do to facilitate creativity is to create a team environment that is relatively free from distractions, allowing an optimal amount of cognitive capacity to be used for creative endeavors.

> *Hypothesis 4:* Laureates who have cognitive capacity free from extraneous concerns will tend to have success later in their careers.

METHOD

In order to study the complex subject of creative discovery, multilevel influences in early and later career, we decided to use a historiometric approach. This analysis of historical documents allows for evaluation and analysis of the multilevel phenomenon at play among highly creative individuals, in this case, Nobel Prize winners. Because of the high-level nature of this sample of scientists, online interviews and autobiographies were judged to be the best source materials. This approach has been used in research studying various behaviors or cognitions of high-level leaders (e.g., Bedell-Avers, Hunter,

Angie, Eubanks, & Mumford, 2009; Eubanks et al., 2010). There are many complexities involved in studying the careers of highly creative individuals that make laboratory studies difficult, if not impossible. For example, the longitudinal nature of the data would pose many challenges as subject pools disperse and research priorities progress. Using a sample of Nobel Prize winners provides an interesting sample for the study of creativity, particularly because we were able to study these creative individuals across the domains of economics, medicine, chemistry, and physics. We were able to gain access to this otherwise elusive population through historical documents in the form of interviews and autobiographical articles. Furthermore, interviews and autobiographical information were particularly useful in this study because we were interested in reflections on early career experiences and affects from teams and organizations.

This historiometric approach is a hybrid qualitative and quantitative design involving the content analysis of historical records, as described by Simonton (1991, 2003), allowing for examination of these complex interactions and relationships in their historical context. For example, a person's career may have been influenced by a major event such as a banking crisis or war that would provide an important context for that person's experience. Historiometric research within the domain of leadership has been used successfully for some time now (Ballard, 1983; Hermann, 1980; House, Spangler, & Woycke, 1991; Mumford, 2006; O'Connor, Mumford, Clifton, Gessner, & Connelly, 1995; Winter, 1993; Zullow & Seligman, 1990), allowing a wide range of research questions to be addressed. In order to design a rigorous historical study, the sample and data source must be carefully considered. Thus, a carefully designed selection plan for the sample and data source must be developed and implemented.

Sample and Data Source

Sample

In this study, we considered relevant criteria related to creativity and Nobel Prize winners and engaged in the following process. Initially, we identified 100 individuals from economics, medicine, physics, and chemistry who won Nobel Prizes between the years of 1939 and 2007. We decided on these fields because they were all scientific in nature, yet still diverse. Other areas such as Peace or Literature were not included because we wanted to only include the creative processes that might be applied to a research and development setting. This list was then reduced based on the length of the interview (some were overly long or short) and the availability of the markers of interest (i.e., in some interviews the winner only discussed technical topics specific to his or her scientific field). Finally, a list of 59 Nobel Prize winners was established (see Table 7.1 for a complete list) based on the following criteria: (a) he or she won a Nobel Prize in economics, medicine, physics, or chemistry; (b) there was an autobiography in an

Table 7.1 Nobel Prize Winner Information

Name	Field	Year of Nobel Prize	Country of Birth	Reference
Steven Chu	Physics	1997	USA	Smith, A. (Interviewer) & Chu, S. (Interviewee). (2008). Retrieved from Nobelprize.org Web site: http://nobelprize.org/mediaplayer/index.php?id=1079
Roy J. Glauber	Physics	2005	USA	Smith, A. (Interviewer) & Glauber, R. J. (Interviewee). (2008). Retrieved from Nobelprize.org Web site: http://nobelprize.org/mediaplayer/index.php?id=1103
Theodor W. Hnsch	Physics	2005	Germany	Smith, A. (Interviewer) & Hnsch, T. W. (Interviewee). (2008). Retrieved from Nobelprize.org Web site: http://nobelprize.org/mediaplayer/index.php?id=1076
Heinrich Rohrer	Physics	1986	Switzerland	Smith, A. (Interviewer) & Rohrer, H. (Interviewee). (2008). Retrieved from Nobelprize.org Web site: http://nobelprize.org/mediaplayer/index.php?id=979
Gerardus 't Hooft	Physics	1999	Netherlands	Carlson, P. (Interviewer) & Hooft, G. (Interviewee). (1999). Retrieved from Nobelprize.org Web site: http://nobelprize.org/mediaplayer/index.php?id=455
Donald A. Glaser	Physics	1960	USA	Brny, A. (Interviewer) & Glaser, D. A. (Interviewee). (2000). Retrieved from Nobelprize.org Web site: http://nobelprize.org/mediaplayer/index.php?id=710
Martinus J.G. Veltman	Physics	1999	Netherlands	Carlson, P. (Interviewer) & Veltman, M. (Interviewee). (1999). Retrieved from Nobelprize.org Web site: http://nobelprize.org/mediaplayer/index.php?id=456
Robert C. Richardson	Physics	1996	USA	Griehsel, M. (Interviewer) & Richardson, R. C. (Interviewee). (2004). Retrieved from Nobelprize.org Web site: http://nobelprize.org/mediaplayer/index.php?id=715
Nicolaas Bloembergen	Physics	1981	Netherlands	Griehsel, M. (Interviewer) & Bloembergen, N. (Interviewee). (2004). Retrieved from Nobelprize.org Web site: http://nobelprize.org/mediaplayer/index.php?id=886
Douglas D. Osheroff	Physics	1996	USA	Griehsel, M. (Interviewer) & Osheroff, D. D. (Interviewee). (2004). Retrieved from Nobelprize.org Web site: http://nobelprize.org/mediaplayer/index.php?id=714

Ivar Giaever	Physics	1973	Norway	Griehsel, M. (Interviewer) & Giaever, I. (Interviewee). (2004). Retrieved from Nobelprize.org Web site: http://nobelprize.org/mediaplayer/index.php?id=713
Carl E. Wieman	Physics	2001	USA	Smith, A. (Interviewer) & Wieman, C. E. (Interviewee). (2007). Retrieved from Nobelprize.org Web site: http://nobelprize.org/mediaplayer/index.php?id=985
Antony Hewish	Physics	1974	United Kingdom	Smith, A. (Interviewer) & Hewish, A. (Interviewee). (2009). Retrieved from Nobelprize.org Web site: http://nobelprize.org/mediaplayer/index.php?id=1144
George F. Smoot	Physics	2006	USA	Smith, A. (Interviewer) & Smoot, G. F. (Interviewee). (2008). Retrieved from Nobelprize.org Web site: http://nobelprize.org/mediaplayer/index.php?id=1083
Brian D. Josephson	Physics	1973	United Kingdom	Griehsel, M. (Interviewer) & Josephson, B. D. (Interviewee). (2004). Retrieved from Nobelprize.org Web site: http://nobelprize.org/mediaplayer/index.php?id=977
George J. Stigler	Economics	1982	USA	Stigler, G. J. (1995). George J. Stigler. In W. Breit & R. Spencer (Eds.), Lives of laureates-thirteen Nobel economists (pp. 95–112), Cambridge, MA: The MIT Press.
Sir Arthur Lewis	Economics	1979	United Kingdom	Lewis, A. (1995). W. Arthur Lewis. In W. Breit & R. Spencer (Eds.), Lives of laureates -thirteen Nobel economists (pp. 1–19), Cambridge, MA: The MIT Press.
Lawrence R. Klein	Economics	1980	USA	Klein, L. R. (1995). Lawrence R. Klein. In W. Breit& R. Spencer (Eds.), Lives of laureates-thirteen Nobel economists (pp. 21-40), Cambridge, MA: The MIT Press.
Milton Friedman	Economics	1976	USA	Friedman, M. (1995). Milton Friedman. In W. Breit& R. Spencer (Eds.), Lives of laureates-thirteen Nobel economists(pp. 79–93), Cambridge, MA: The MIT Press.

(*Continued*)

Table 7.1 (Continued)

Name	Field	Year of Nobel Prize	Country of Birth	Reference
Paul A. Samuelson	Economics	1970	USA	Samuelson, P. A. (1995). Paul A. Samuelson. In W. Breit & R. Spencer (Eds.), Lives of laureates- thirteen Nobel economists (pp. 59–76), Cambridge, MA: The MIT Press.
Douglass C. North	Economics	1993	USA	North, D. C. (1995). Douglass C. North. In W. Breit& R. Spencer (Eds.), Lives of laureate -thirteen Nobel economists(pp. 251–267), Cambridge, MA: The MIT Press.
James M. Buchanan	Economics	1986	USA	Buchanan, J. H. (1995). James M. Buchanan. In W. Breit& R. Spencer (Eds.), Lives of laureates- thirteen Nobel economists (pp. 165–181), Cambridge, MA: The MIT Press.
Myron S. Scholes	Economics	1997	Canada	Smith, A. (Interviewer) & Scholes, M. S. (Interviewee). (2008). Retrieved from Nobelprize.org Web site: http://nobelprize.org/mediaplayer/index.php?id=1134
Daniel L. McFadden	Economics	2000	USA	Smith, A. (Interviewer) & McFadden, D. L. (Interviewee). (2008). Retrieved from Nobelprize.org Web site: http://nobelprize.org/mediaplayer/index.php?id=1132
Edmund S. Phelps	Economics	2006	USA	Bergstrm, R. (Interviewer) & Phelps, E. S. (Interviewee). (2006). Retrieved from Nobelprize.org Web site: http://nobelprize.org/mediaplayer/index.php?id=85
Robert C. Merton	Economics	1997	USA	Griehsel, M. (Interviewer) & Merton, R. C. (Interviewee). (2004). Retrieved from Nobelprize.org Web site: http://nobelprize.org/mediaplayer/index.php?id=443
William F. Sharpe	Economics	1990	USA	Sharpe, W. F. (1995). William F. Sharpe. In W. Breit & R. Spencer (Eds.), Lives of laureates-thirteen Nobel economists (pp. 205–225), Cambridge, MA: The MIT Press.
John F. Nash Jr.	Economics	1994	USA	Griehsel, M. (Interviewer) & Nash Jr., J. F. (Interviewee). (2004). Retrieved from Nobelprize.org Web site: http://nobelprize.org/mediaplayer/index.php?id=429

James A. Mirrlees	Economics	1996	United Kingdom	Griehsel, M. (Interviewer) & Mirrlees, J. A. (Interviewee). (2004). Retrieved from Nobelprize.org Web site: http://nobelprize.org/mediaplayer/index.php?id=440
Kenneth J. Arrow	Economics	1972	USA	Arrow, K. J. (1995). Kenneth J. Arrow. In W. Breit & R. Spencer (Eds.), Lives of laureates-thirteen Nobel economists(pp. 47–57), Cambridge, MA: The MIT Press.
Robert W. Fogel	Economics	1993	USA	Griehsel, M . (Interviewer) & Fogel, R. W. (Interviewee). (2004). Retrieved from Nobelprize.org Web site: http://nobelprize.org/mediaplayer/index.php?id=689
Fritz Lipmann	Medicine	1953	Russia	Lipmann, F. (1984). A long life in times of great upheaval. Annual Review of Biochemistry, 53, 1–31.
Gertrude B. Elion	Medicine	1988	USA	Elion, G. B. (1993). The quest for a cure.Annual Review of Pharmacology and Toxicology, 33, 1–22.
Arthur Kornberg	Medicine	1959	USA	Smith, A. (Interviewer) & Kornberg, A. (Interviewee). (2006). Retrieved from Nobelprize.org Web site: http://nobelprize.org/mediaplayer/index.php?id=84 Kornberg, A. (1989). Never a dull enzyme. Annual Review of Biochemistry, 58, 1–29.
Robert F. Furchgott	Medicine	1998	USA	Furchgott, R. F. (1995). A research trail over half a century. Annual Review of Pharmacology and Toxicology, 35, 1–27.
Julius Axelrod	Medicine	1970	USA	Axelrod, J. (1988). An unexpected life in research. Annual Review of Pharmacology and Toxicology, 28, 1–23.
Konrad Bloch	Medicine	1964	Germany	Bloch, K. (1987). Summing up. Annual Review of Biochemistry, 56, 1–19.
Severo Ochoa	Medicine	1959	Spain	Ochoa, S. (1980). The pursuit of a hobby. Annual Review of Biochemistry, 49, 1-30.
Alan L. Hodgkin	Medicine	1963	United Kingdom	Hodgkin, A. L. (1983). Beginning: Some reminiscences of my early life. Annual Review of Physiology, 45, 1–15.

(Continued)

Table 7.1 (Continued)

Name	Field	Year of Nobel Prize	Country of Birth	Reference
Rolf M. Zinkernagel	Medicine	1996	Switzerland	Smith, A. (Interviewer) & Zinkernagel, R. M. (Interviewee). (2007). Retrieved from Nobelprize.org Web site: http://nobelprize.org/mediaplayer/index.php?id=982
Christian de Duve	Medicine	1974	Belgium	Orrenius, S. (Interviewer) & Duve, C. (Interviewee). (2000). Retrieved from Nobelprize.org Web site: http://nobelprize.org/mediaplayer/index.php?id=726
Rita Levi-Montalcini	Medicine	1986	Italy	Smith, A. (Interviewer) & Levi-Montalcini, R. (Interviewee). (2008). Retrieved from Nobelprize.org Web site: http://nobelprize.org/mediaplayer/index.php?id=1101
Eric R. Kandel	Medicine	2000	Austria	Smith, A. (Interviewer) & Kandel, E. R. (Interviewee). (2008). Retrieved from Nobelprize.org Web site: http://nobelprize.org/mediaplayer/index.php?id=984
Alfred G. Gilman	Medicine	1994	USA	Smith, A. (Interviewer) & Gilman, A. G. (Interviewee). (2007). Retrieved from Nobelprize.org Web site: http://nobelprize.org/mediaplayer/index.php?id=895
Arvid Carlsson	Medicine	2000	Sweden	Smith, A. (Interviewer) & Carlsson, A. (Interviewee). (2008). Retrieved from Nobelprize.org Web site: http://nobelprize.org/mediaplayer/index.php?id=902
David Baltimore	Medicine	1975	USA	Pettersson, R. (Interviewer) & Baltimore, D. (Interviewee). (2001). Retrieved from Nobelprize.org Web site: http://nobelprize.org/mediaplayer/index.php?id=172
Gerhard Herzberg	Chemistry	1971	Germany	Herzberg, G. (1985). Molecular spectroscopy: A personal history. Annual Review of Physical Chemistry, 36, 1–30.
Frederick Sanger	Chemistry	1958	United Kingdom	Rose, J. (Interviewer) & Sanger, F. (Interviewee). (2001). Retrieved from Nobelprize.org Web site: http://nobelprize.org/mediaplayer/index.php?id=360
Leopold Ruzicka	Chemistry	1939	Switzerland	Ruzicka, L. (1973). In the borderland between biorganic chemistry and biochemistry. Annual Review of Biochemistry, 42, 1–20.

Arne Tiselius	Chemistry	1948	Sweden	Tiselius, A. (1968). Reflection from both the sides of the counter. Annual Review of Biochemistry, 37, 1–22.
Luis Leloir	Chemistry	1970	Argentina	Leloir, L. (1983). Far away and long ago. Annual Review of Biochemistry, 52, 1–15.
Johann Deisenhofer	Chemistry	1988	Germany	Smith, A. (Interviewer) & Deisenhofer, J. (Interviewee). (2007). Retrieved from Nobelprize.org Web site: http://nobelprize.org/mediaplayer/index.php?id=893
Walter Kohn	Chemistry	1998	Austria	Griehsel, M. (Interviewer) & Kohn, W. (Interviewee). (2004). Retrieved from Nobelprize.org Web site: http://nobelprize.org/mediaplayer/index.php?id=843
Sidney Altman	Chemistry	1989	Canada	Forsn, S. (Interviewer) & Altman, S. (Interviewee). (2000). Retrieved from Nobelprize.org Web site: http://nobelprize.org/mediaplayer/index.php?id=731
John C. Polanyi	Chemistry	1986	Canada	Forsn, S. (Interviewer) & Polanyi, J. C. (Interviewee). (2000). Retrieved from Nobelprize.org Web site: http://nobelprize.org/mediaplayer/index.php?id=729
Manfred Eigen	Chemistry	1967	Germany	Jrnvall, H. (Interviewer) & Eigen, M. (Interviewee). (2000). Retrieved from Nobelprize.org Web site: http://nobelprize.org/mediaplayer/index.php?id=879
Kary B. Mullis	Chemistry	1993	USA	Griehsel, M. (Interviewer) & Mullis, K. B. (Interviewee). (2005). Retrieved from Nobelprize.org Web site: http://nobelprize.org/mediaplayer/index.php?id=428
Gerhard Ertl	Chemistry	2007	Germany	Smith, A. (Interviewer) & Ertl, G. H. (Interviewee). (2007). Retrieved from Nobelprize.org Web site: http://nobelprize.org/mediaplayer/index.php?id=767
Aaron Ciechanover	Chemistry	2004	Israel	Smith, A. (Interviewer) & Ciechanover, A. (Interviewee). (2007). Retrieved from Nobelprize.org Web site: http://nobelprize.org/mediaplayer/index.php?id=990
George A. Olah	Chemistry	1994	Hungary	Brny, A. (Interviewer) & Olah, G. A. (Interviewee). (2000). Retrieved from Nobelprize.org Web site: http://nobelprize.org/mediaplayer/index.php?id=733

Note: N = 59. Boldfaced values on the diagonal represent the square root of the average variance extracted (AVE). EC = early career; LC = later career.

annual review piece for an academic journal or there was a video interview conducted by nobelprize.org; and (c) the article or interview was an appropriate length (approximately 30 minutes or 20 pages, respectively). Roughly equal numbers of winners in economics, medicine, physics, and chemistry were identified to allow us to observe differences based on field of research. Winners were selected from several countries to account for cultural differences that may occur in the creative process. Similarly, winners from different periods were selected.

Data Source

Autobiographies were selected from annual review articles appearing in academic journals. These were found by conducting a search through an electronic library system. Because this approach was unable to yield a sufficient number of individuals winning economics prizes, we also used a book called *Lives of Laureates—Thirteen Nobel Economists* (Breit & Spencer, 1995). When an autobiography was not available for an individual, interviews were viewed on nobelprize.org to ensure an adequate sample size. The criteria used to select the articles or interviews for inclusion in the study were as follows: (a) the articles and interviews took place within the past 50 years (our article and interview sample spanned the years 1961 to 2009) and (b) the article or interview contained a discussion of the relevant markers. The entire article or interview was read or viewed to evaluate the relevant markers developed for this study.

Predictors and Criteria

After conducting a literature review on the relevant aspects of creativity and innovation and an initial reading of the source material, two academics in business schools created event markers to be evaluated on a 5-point Likert scale. The literature review coupled with appraisal of autobiographies and interviews allowed us to include markers capturing components both at early and later career as well as individual, group, and organizational levels. These markers were written to assess observable behaviors that consistently appeared in the autobiographies or interviews. Therefore, our markers included those that related to (a) individuals (e.g., "To what extent does the scientist display inquisitiveness related to their research area in childhood?"), (b) leadership—later career (e.g., "To what extent does the scientist through leadership actions, create optimal resources to complete tasks?"), (c) role of the mentor (e.g., "To what extent does the scientist have a high-quality exchange relationship with his or her mentor?"), (d) team or work group—early career (e.g., "To what extent does the scientist have task interdependence with group members?"), (e) team or work group—later career (e.g., "To what extent does the scientist have diversity in his or her network?"), (f) organizational—early career (e.g., "To what extent does the scientist have opportunities to try new things?"), (g) organizational—later career (e.g., "To what extent is the scientist free from extraneous concerns?"). A total of 75 predictor markers were developed. Approximately 15 items were written to assess each level and stage of career. The criterion measure was age at which the winner's most cited first- or solo-authored publication was published. This variable was used as a proxy to understand whether individuals were achieving the height of success early or later in his or her career. For the current study, we focused on only a subset of the available markers as warranted by extant research relating to the age at which creativity occurs. Other marker subsets will be used to address other research questions.

Controls

In addition to the predictors and criteria, controls were developed to account for variables that may otherwise influence the dependent variables of interest. The information to assess the controls came from nobelprize.org or the annual review articles. Several control variables were selected following the historiometric research methodology conducted by Mumford (2006) that were anticipated to account for variation among the variables of interest. For example, variables about the autobiography or interview such as the length of the article and nature of the publication media (i.e., interview or autobiographical article) were used. Control items such as length of interview or autobiography, country of origin, country where they are living, and domain were included. Research by Simonton (1991) considered the age that creators produce their first, last, and best contributions based on domain and the associated inherent cognitive requirements of the field. Simonton's findings indicated that mathematicians produced their best work at a young age, although physicists and chemists are even younger. Those in medicine achieve their career peak at a later age and those in earth sciences reach their peak even later. Therefore, we also controlled for the domain of each winner (physics, chemistry, medicine, and economics).

Rating Procedures

The ratings included five judges, all academics in business schools studying organizational behavior. They engaged in a 40-hour training program via the videoconferencing software Skype because of the geographic dispersion of the team. In this training, they were asked to assess predictor and criteria items based on the creativity literature on a 5-point Likert scale. Next, the markers for 14 winners were evaluated by all five raters. After being exposed to this training, the average interrater agreement coefficients for these items was adequate with interclass correlation coefficients reaching reasonable levels (ICC = .79) using the procedures suggested by Shrout and Fleiss (1979). Each of the remaining winners was evaluated by at least four raters. Because the interrater agreement coefficient was at an acceptable level, individual ratings were averaged to provide a single score for each of the 59 laureates for every marker. Another judge, a graduate student in business, assessed the control items as well as the criterion item (age when the most cited piece of work was published). This information was collected and reported by only one individual as these items were objective in nature (e.g., length of time between winning award and writing autobiography/giving interview, number of words in autobiography/interview, etc).

Analyses

We first conducted preliminary analyses (examining correlations and running regressions) for the various control variables. Results from both approaches revealed that none of the controls had a statistically significant effect on the outcome; therefore, we excluded them from the test of the theoretical model. For testing of both the measurement model and the theoretical model, we used partial least squares (PLS) analysis with SmartPLS (Ringle, Wende, & Will, 2005). PLS is widely used for exploratory data testing and has several advantages over other techniques (Chin & Newsted, 1999). PLS does not require multivariate normal distribution and is especially suitable for the analysis of small samples, as was the case in this study. Moreover, PLS can help to reduce measurement error by weighing the individual indicators of a multi-indicator variable (Sosik, Kahai, & Piovoso, 2009). Other forms of path modeling, such as covariance-based structural equation modeling, are generally used in confirmatory model testing and may be susceptible to error in situations in which there is a low construct-to-sample-size ratio, as was the case in this study. PLS also has the ability to test both the measurement model and theoretical model simultaneously. This ability makes PLS preferable to multiple regression analysis in which the measurement model and theoretical models must be tested independently.

The test of the measurement model includes three primary parts: (a) individual item reliability, (b) internal consistency, and (c) discriminant validity. Individual item reliability was assessed by examining the factor loadings of each measure on its corresponding construct. In this study, most of the variables were assessed with a single item, which is the average of the five individual ratings. For these items, factor loadings, internal consistency and average variance extracted (AVE) are all 1. The lone exception is leadership, which consists of three items (i.e., optimal time, resources, and people). Fornell and Larcker (1981) suggest accepting items that have more explanatory power than error variance. In practice, the generally accepted cutoff is .70 or greater, and all items met this standard.

Construct internal consistency may be assessed by composite internal scale reliability, which is similar to Cronbach's alpha. Fornell and Larcker (1981) suggest a cutoff of .70 for internal consistency. A second way to measure internal consistency is with AVE, which is a measure of variance accounted for by the underlying construct. Fornell and Larcker (1981) suggest a cutoff of .50 for AVE. Internal scale reliability (.86) and AVE (.68) were both acceptable for the leadership variable.

Table 7.2 Means, Standard Deviations, Reliabilities, and Correlations

	Variables	Mean	SD.	1	2	3	4	5
1	Age at most-cited work	49.11	15.12	**1.00**				
2	Degree in different field†	0.61	.69	–.28*	**1.00**			
3	Try new things–EC	3.63	.42	–.26*	–.06	**1.00**		
4	Leadership - LC	3.48	.30	.27*	–.23	.35**	**.86**	
5	Free cognitive capacity - LC	3.50	.34	.36**	–.08	.06	.48**	**1.00**

* $p < 0.05$. ** $p < 0.01$.

† 0 = no; 1 = yes.

Discriminant validity in PLS is assessed in two ways. First, each item should load higher on the construct that it is supposed to measure than on any other construct (Carmines & Zeller, 1979). All items in the study met this criterion. Second, each construct should share more variance with its items than with any other construct in the model (Barclay, Higgins, & Thompson, 1995). This criterion is usually assessed similarly to a multitrait/multimethod approach. Specifically, the square root of the AVE of a construct should be greater than the construct's correlation with any other construct in the model. An examination of Table 7.2 (in which the square root of the AVE is located on the diagonal) demonstrates that this criterion was also met.

Results

Results of the test of the theoretical model are shown in Figure 7.1. The standardized beta coefficient for each path in the model was obtained from the PLS algorithm in SmartPLS. The statistical significance of each path in the theoretical model was determined by the *t*-value for a given bivariate relationship based on a bootstrapping technique with 200 iterations.

The results showed that having a degree in a different field was inversely related to the age at which the Nobel Prize winner published his or her most cited work ($b = -.21$, $p < .05$); thus, Hypothesis 1 was supported. This then indicates that having a diversity of knowledge is important to early career success. This stands to reason as early in one's career idea exploration is important while later in one's career specialization becomes important. Similarly, the results showed that having freedom to try new things was inversely related to the age at which the Nobel Prize winner published his or her most cited work ($b =-.38$, $p < .01$); thus, Hypothesis 2 was supported. This is in line with the previous statement in that freedom of ideas, knowledge, and

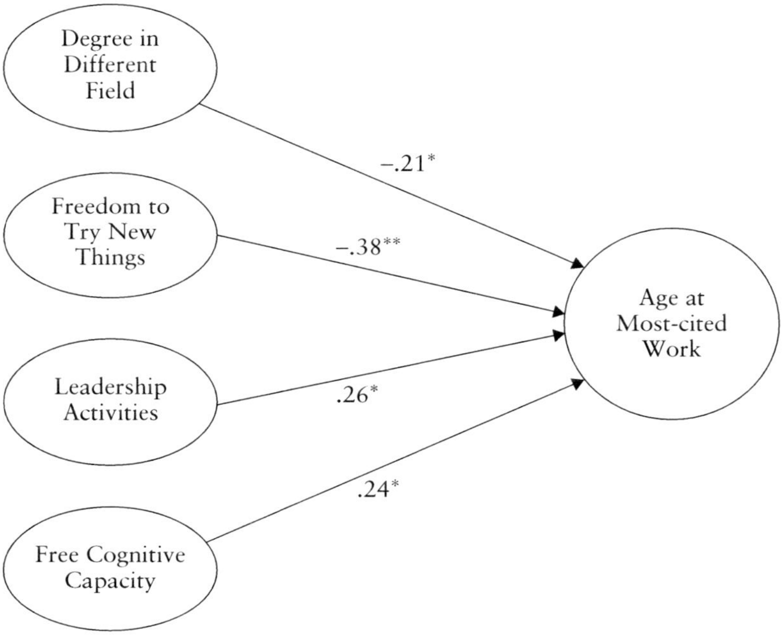

Figure 7.1 Results of partial least squares analysis. Results on path lines are standardized *b* weights. $^*p < .05$. $^{**}p < .01$.

work processes are particularly important early in one's career. In support of Hypothesis 3, there was a direct relationship between leadership activities and the age at which the Nobel Prize winner published his or her most cited work ($b = .26$, $p < .05$). Whereas Hypotheses 1 and 2were focused on characterization of success during early career, Hypothesis 3 is focused on the scientist as a leader. Leadership is particularly important in later career as one might expect, because the laureates need to determine how to ensure that they and their groups have optimal time, resources, and access to relevant expertise to achieve success. Finally, in support of Hypothesis 4, there was a direct relationship between having sufficient free cognitive capacity in the later career stage and the age at which the Nobel Prize winner published his or her most cited work ($b = .24$, $p < .05$). Again this hypothesis is relevant to the laureate as a leader. The support of this hypothesis indicates the importance for the laureate to allow sufficient space for thinking and working with team members in order to generate creative discoveries. Although related to Hypothesis 3, this is distinct because, although a leader may successfully carve out time, acquire resources, and develop relationships that can foster creative discovery, having sufficient cognitive capacity is another matter. Life is full of distractions and demands grow in scope and intensity as we progress in our careers. Being skilled at balancing these demands and focusing for long stretches of time appears to be particularly important for creative discovery later in one's career.

DISCUSSION

General Findings

In this study, we have outlined two paths to career success. This allows us to understand that not all scientists follow the same path to creative discovery. Although some achieve their height of success at an early career stage, others reach eminence much later. This appears to occur independently of field (economics, medicine, chemistry, or physics). Success at an early career stage was associated with holding a degree in a different field and having freedom to try new things, thus representing the importance of heterogeneous knowledge at an early career stage. Creative output at a later career stage tended to be characterized by exhibiting leadership and managing multiple demands. As one advances in his or her profession, a scientist is faced with increasing demands. It then becomes the challenge of the scientist to free up cognitive capacity to carve out time for research to continue. This can happen in several different ways by appropriately managing time, people, and resources.

Limitations

Before turning to the implications flowing from this study, we acknowledge some limitations. First, although using an exceptional sample such as Nobel Prize winners allows us to learn from the best, it also makes it difficult to generalize to a broader population of research scientists. Although this is an elite population, we believe that results from this study could be generalized to scientists working in research and development areas both in industry and in academia with industry affiliations. Because Nobel Prize winners and scientists working in research and development are both engaging in scientific discovery, there is certainly a similarity here. Second is the fact that the material used in this study was restrictive in the sense that the only information available was that provided in the autobiography or interview. Because of this, the sample was reduced to the autobiographies or interviews that contained a substantial discussion of information captured by our markers.

Next, autobiographical material or interviews often tend to contain an element of impression management meaning that the scientist may try to present him- or herself in a positive light. Although it is certainly the case that all information presented may not be factually accurate, we were interested in the individual's interpreted reality of events. Although there are certain events such as education and degrees earned that are unlikely to be invented because of the ability to verify such information, other events such as experience with mentors and advisors may be more subjective. However, an individual's perception of events may drive his or her current and future behavior. In essence, it may not matter what the objective truth is, but rather the interpretation of events that is important.

Finally, there is a potential for single source bias because there was overlap between the two individuals that designed the conceptual framework for the variables of interest in this study and conducted the ratings. Attempts were made to reduce this potential bias by having discussions about the variables of interest among all judges and providing several rater training sessions.

Theoretical Implications

The theoretical implications are twofold. First, we have managed to address the contradictory findings about having a breadth in knowledge and expertise in a particular domain. Breadth of knowledge appears to be particularly important for early career success. By identifying specific variables associated with each of these aspects, we have been able to further develop our understanding of how creativity and innovation intersect with success in early and late career. This in particular helps in understanding the important role of the leader in guiding creative discovery. This advances work that clearly identifies the important role of the leader, but does not address the issue of time, resource, and relationship management in such an explicit manner (Friedrich, Mumford, Vessey, Beeler, & Eubanks, 2007; Mumford, Eubanks, & Murphy, 2007; Mumford, Hunter, et al., 2007; Mumford, et al., this book). This study clearly demonstrates the important role of the leader both in terms of mentoring and setting research habits in early career and in managing time, resources, and relationships later in one's career.

Second, specific variables associated with success at an early or later career stage have been identified. Rather than having a general understanding of what allows a scientist to have a successful career, we now have a more specific view of what is important at an earlier and later stage in order to stimulate creativity. This study allowed us to identify the critical variables for success (diversity of thought and having the freedom to try new things) early in one's career versus an emphasis on leadership based around time, resource, and relationship management later on in one's career. In addition to this, the importance of having free cognitive capacity was identified as being important later in one's career. This finding is noteworthy because as researchers advance in their careers spare cognitive capacity becomes increasingly scarce due to growing external demands and expectations.

Practical Implications

As we have been able to identify specific variables that are key for earlier and later career stage groups, we may be able to inform scientists in these respective career stages of their relative importance. In addition, we may be able to structure environments to foster creativity more effectively. This research therefore extends some of the earlier work by Hemlin, Allwood, and Martin (2008), where the authors discuss their Creative Knowledge Environments framework. Understanding how career stage influences the way to encourage creative ideas can be considered when structuring training and development contexts (including doctoral programs). For instance, individuals early in their careers should be allowed to try new things and explore rather than imposing a strong vision on them. A structure in which this is allowed to take place should be established by the immediate supervisor. Next, when hiring early career researchers, it may be an advantage for them to have a degree in another field besides the one in which they are currently specializing. As demonstrated in this study, this will encourage diversity of thought and in turn enhance creative discoveries. Similarly, these findings may have implications for job design. When younger scientists join an organization, it is important to provide them with the proper flexibility and support to try new things. On the other hand, organizations need to make sure that these later career scientists can reach their potential by avoiding the temptation to place onerous management tasks on them. Instead, these individuals need to protect their cognitive capacity so that they have time, space, and necessary resources to solve complex problems. The problem then becomes 'Who will undertake these management tasks?'

Future Research Directions

Future research on individuals working in research and development will help to develop further our understanding of the various pathways to a successful scientific career. Although the current effort provides substantial information about early and later career success, conducting research on a broader population working in research and development would provide a more generalizable set of conclusions.

In sum, creative discoveries will continue to be an important part of human advancement. This study provides us with information about the characteristics that allow someone to engage in creative discovery early in his or her career as well as the contextual components that can lead to his or her success in later career.

REFERENCES

Agars, M. D., Kaufman, J. C., & Locke, T. R. (2008). Social influence and creativity in organizations: A multi-level lens for theory, research, and practice. In F. J. Yammarino, F. Dansereau (Series Eds.) & M. D. Mumford, S. T. Hunter, & K. E. Bedell-Avers (Vol. Eds.), *Research in Multi-Level Issues: Vol. 7. Multi-level issues in creativity and innovation* (pp. 3–61). Bingley, England: Emerald.

Amabile, T. M. (1983). The social psychology of creativity: A componential conceptualization. *Journal of Personality and Social Psychology*, *45*, 357–377.

Amabile, T. M. (1988). A model of creativity and innovation in organizations. *Research in Organizational Behavior*, *10*, 123–167.

Amabile, T. M. (1996). *Creativity in context*. Boulder, CO: Westview.

Andrews, F. M, & Farris, G. F. (1972). Time pressure and the performance of scientists and engineers: A five year panel study. *Organizational Behavior and Human Performance*, *8*, 185–200.

Axtell, C. M., Holman, D. J., Unsworth, K. L., Wall, T. D., Waterson, P. E., & Harrington, E. (2000). Shop floor innovation: Facilitating the suggestion and implementation of ideas. *Journal of Occupational and Organizational Psychology*, *73*, 265–285.

Ballard, E. J. (1983). Canadian prime ministers: Complexity in political crises. *Canadian Psychology*, *24*, 125–129.

Barclay, D., Higgins, C., & Thompson, R. (1995). The partial least squares (PLS) approach to causal modeling, personal computer adoption and use as an illustration. *Technology Studies*, *2*, 285–309.

Barnowe, J. T. (1975). Leadership and performance outcomes in research organizations. *Organizational Behavior and Human Performance*, *14*, 264–280.

Bedell-Avers, K. E., Hunter, S. T., Angie, A.D., Eubanks, D. L., & Mumford, M. D. (2009). Charismatic, Ideological, and pragmatic leaders: An examination of leader-leader interactions. *The Leadership Quarterly*, *20*, 299–315.

Breit, W., & Spencer, R. W. (1995). *Lives of laureates: Thirteen Nobel economists* (3rd ed.). Boston, MA: Massachusetts Institute of Technology.

Carmines, E.G., & Zeller, R.A. (1979). *Reliability and validity assessment*. Beverly Hills, CA: Sage.

Cattell, R. B. (1963). The personality and motivation of the researcher from measurements of contemporaries and from biography. In C. W. Taylor & F. Barron (Eds.). *Scientific creativity: Its recognition and development* (pp. 119–31). New York, NY: Wiley.

Chin, W. W., & Newsted, P. R. (1999). Structural equation modeling analysis with small samples using partial least squares. In R. Hoyle (Ed.), *Statistical strategies for small sample research* (pp. 307–341). Thousand Oaks, CA: Sage.

Cole, S. (1979). Age and scientific performance. *American Journal of Sociology*, *84*, 958–977.

Crane, D. (1972). *Invisible colleges*. Chicago, IL: University of Chicago Press.

Eubanks, D. L., Antes, A L., Friedrich, T. L., Caughron, J. J., Blackwell, L. V., Bedell-Avers, K.E., & Mumford, M. D. (2010). Criticism and outstanding leadership: An evaluation of leader reactions and critical outcomes. *The Leadership Quarterly*, *21*(3), 365–388.

Eysenck, H. J. (1995). Creativity and personality: Suggestions for a theory. *Psychological Inquiry*, *4*, 147–178.

Feldman, D. H. (1999). The development of creativity. In R. J. Sternberg (Ed.), *Handbook of creativity* (pp. 169–188). Cambridge, England: Cambridge University Press.

Ford, C. M. (2000). Creative developments in creativity theory. *Academy of Management Journal*, *46*, 618–630.

Fornell, C., & Larcker, D.F. (1981). Evaluating structural equation models with unobserved variables and measurement error. *Journal of Marketing Research*, *18*, 39–50.

Friedrich, T. L., Mumford, M. D., Vessey, B., Beeler, C., & Eubanks, D. L. (2010). Leading for innovation type and complexity. *International Studies of Management and Organization: Special Issue on Creativity and Innovation in Organization*, *40*, 6–29.

Hammond, M. M., Neff, N. L., Farr, J. L., Schwall, A. R., & Zhao, X. (2011). Predictors of individual level innovation at work: A meta-analysis. *The Psychology of Aesthetics, Creativity, and the Arts*, *5*(1), 90–105.

Hemlin, S., Allwood, C.M., & Martin, B.R. (2008). Creative knowledge environments, *Creativity Research Journal*, *20*(2), 196–210.
Hennessey, B. A., & Amabile, T. M. (2010). Creativity. *Annual Review of Psychology*, *61*, 569–598.
Hermann, M. G. (1980). Assessing the personalities of Soviet politburo members. *Personality and Social Psychology Bulletin*, 6, 332–352.
House, R. J., Spangler, W. D., & Woycke, J. (1991). Personality and charisma in the U.S. presidency: A psychological theory of leader effectiveness. *Administrative Science Quarterly*, *36*, 364–396.
Hunter, S. T., Bedell, K. E., & Mumford, M. D. (2007). Climate for creativity: A quantitative review. *Creativity Research Journal*, *19*(1), 69–90.
Langley, P., Simon, H. A., Bradshaw, G. L., & Zythow, J. M. (1987). *Scientific discovery.* Cambridge, MA: MIT Press.
Lindauer, M. S. (1993). The span of creativity among long-lived historical artists. *Creativity Research Journal*, *6*, 221–239.
Madjar, N. (2005). The contributions of different groups of individuals to employees' creativity. *Advances in Developing Human Resources*, *7*, 182–207.
McCrae, R. R., Arenberg, D., & Costa, P. T. (1987). Declines in divergent thinking with age: Cross-sectional, longitudinal, and cross-sequential analyses. *Psychology and Aging*, 2, 130–137.
Mumford, M. D. (Ed.). (2006). *Pathways to outstanding leadership: A comparative analysis of charismatic, ideological, and pragmatic leadership*. Mahwah, NJ: Erlbaum.
Mumford, M. D., Eubanks, D. L., & Murphy, S. T. (2007). Creating the conditions for success: Best practices in leading for innovation. In J. A. Conger & R. E. Riggio (Eds.), *The practice of leadership: Developing the next generation of leaders* (pp. 129–149). San Francisco, CA: Jossey-Bass.
Mumford, M. D., & Gustafson, S. B. (1988). Creativity syndrome: Integration, application, and innovation. *Psychological Bulletin*, *103*, 27–43.
Mumford, M. D., Hunter, S. T., Eubanks, D. L., Bedell, K. E., & Murphy, S. T. (2007). Developing leaders for creative efforts: A domain-based approach to leadership development. *Human Resource Management Review*, *17*, 402–417.
Mumford, M. D., Murphy, S. T., Connelly S., Hill J. H., Antes, A. L., Brown, R.P., & Devenport, L. D. (2007). Environmental influences on ethical decision making: Climate and environmental predictors of research integrity. *Ethics & Behavior*, *17*(4), 337–366.
Mumford, M. D., Scott, G. M., Gaddis, B. H., & Strange, J. M. (2002). Leading creative people: Orchestrating expertise and relationships. *The Leadership Quarterly*, *13*, 705–750.
O'Connor, J. A., Mumford, M. D., Clifton, T. C., Gessner, T. E., & Connelly, M. S. (1995). Charismatic leaders and destructiveness: An historiometric study. *The Leadership Quarterly*, *6*, 529–555.
Pruthi, S., & Nagpaul, P. S. (1978). Communication patterns in small R&D projects. *R&D Management*, *8*, 53–58.
Reiter-Palmon, R., & Illies, J. J. (2004). Leadership and creativity: Understanding leadership from a creative problem solving perspective. *The Leadership Quarterly*, *15*, 55–77.
Ringle, C.M., Wende, S., & Will, A. (2005). Smart PLS 2.0 [Computer software]. Hamburg, Germany: University of Hamburg. Retrieved from http://www.smartpls.de
Shalley, C. E., & Gilson, L. L. (2004). What leaders need to know: A review of social and contextual factors that can foster or hinder creativity. *The Leadership Quarterly*, *15*, 33–53.
Shalley, C. E., Zhou, J., & Oldham, G. R. (2004). The effects of personal and contextual characteristics on creativity: Where should we go from here? *Journal of Management*, *30*, 933–958.

Shrager, J., & Langley, P. (1990). *Computational models of scientific discovery and theory formation*. San Mateo, CA: Kaufmann.

Shrout, P. E., & Fleiss, J. L. (1979). Intraclass correlations: Uses in assessing rater reliability. *Psychological Bulletin*, *2*, 420–428.

Silvia, P.J. (2007). Another look at creativity and intelligence: Exploring higher-order models and probable confounds. *Personality and Individual Differences*, *44*, 1012–1021.

Simonton, D. K. (1991). Career landmarks in science: Individual differences and interdisciplinary contrasts. *Developmental Psychology*, *27*, 119–130.

Simonton, D. K. (2003). Scientific creativity as a constrained stochastic behavior: The integration of produce, person, and process perspectives. *Psychological Bulletin*, *129*, 475–494.

Simonton, D. K. (2009). Varieties of (scientific) creativity: A hierarchical model of domain-specific disposition, development, and achievement. *Perspectives on Psychological Science*, *4*, 441–452.

Sosik, J., Kahai, S., & Piovoso, M. (2009). Silver bullet or voodoo statistics? A primer for using the partial least squares data analytic technique in group and organization research. *Group & Organization Management*, *34*, 5–36.

Weisburg, R. W. (1999). Creativity and knowledge: A challenge to theories. In R. J. Sternberg (Ed.), *Handbook of creativity* (pp. 226–259). Cambridge, England: Cambridge University Press.

Winter, D. G. (1993). Power, affiliation, and war: Three tests of a motivational model. *Journal of Personality and Social Psychology*, *65*, 532–545.

Yammarino, F. J., Dionne, S. D., Chun, J. U., & Dansereau, F. (2005). Leadership and levels of analysis: A state-of-the-science review. *The Leadership Quarterly*, *16*, 879–919.

Zullow, H. M., & Seligman, M. E. P. (1990). Pessimistic rumination predicts defeat of presidential candidates, 1900 to 1984. *Psychological Inquiry*, *1*, 52–61.

Part III

Implications Section

8 Succession Planning for Scientific Positions

Identifying, Developing, and Retaining Leaders for Innovation

Ginamarie S. Ligon, Kate T. Dembroski, Robyn C. Mapp, and Bianca M. Zongrone

Planning who will lead for scientific innovation poses challenges on multiple fronts. First, leadership in scientific organizations is difficult—requiring substantial domain, organizational, and field-level expertise and identifying individuals who have potential and interest in such leadership roles is a challenge at best. Much evidence suggests that these leaders come from the ranks of scientists—individuals characterized by high levels of field commitment (vs. organizational commitment), need for autonomy (vs. need for affiliation), and intense levels of intellectual curiosity for the work (vs. for managing the work) (Mumford & Gustafson, 1988). Although these exact characteristics might help one innovate as an individual contributor, they may militate against what is required to lead teams and organizations in scientific pursuits (Mumford, Scott, Gaddis, & Strange, 2002). Thus, organizations are charged with a complex job of identifying these rare individuals early and preparing them for roles in which they may not succeed without substantial planning and development.

Second, although more work has occurred in the past decade with respect to what leaders of innovation do and think about (Byrne, Mumford, Barrett, & Vessey, 2009), less attention has been paid to levels of leadership involved with scientific innovation. For example, it is less clear what senior leaders, sometimes in roles such as CEOs or senior vice presidents (SVPs), are responsible for executing when compared to entry-level leaders of innovation. To date, no clear taxonomy exists that differentiates what might be necessary to lead in senior versus junior roles in scientific organizations.

Finally, the impending exodus of Baby Boomers coupled with rapid growth in emerging industries and countries has made finding and developing these leaders for scientific innovation urgent and important. As these roles are presently filled by individuals between the ages of 45 and 65 (i.e., baby boomers—those born between 1946–1964), the need to develop their successors and transmit what they know to them is increasingly cited as a top concern for organizations, and much more so in those companies where knowledge is the important commodity. Although more organizations have realized the import of succession planning for these roles, it is critical that leaders, and particularly leaders in scientific domains, be involved sooner rather than later with any type of planning in this domain.

Thus, the present chapter has three goals. First, we differentiate what it is that senior versus junior leaders for innovation do in scientific organizations. Reviewing the literature surrounding innovation leadership as well as job analysis material of exemplar positions, the first goal of this chapter is to bifurcate the tasks and focus of influence as seen in senior versus junior leaders of innovation. It should be noted that midlevel positions in scientific organizations (e.g., area coordinators, product managers) are equally critical to examine. However, the present effort is constrained to examine the upper and lower echelons of scientific leadership as these roles likely have the most differences in terms of developmental needs. For example, making the jump from individual contributor to entry-level manager requires a shift in perspectives—from self- to group focused (Ligon, Wallace, & Osburn, 2011), whereas shifting from mid- to senior-level leadership positions requires much more awareness at the organizational and field levels than lower level leadership might entail. In addition, it is imperative to focus on what is needed for senior-level leaders because so many incumbents in these roles plan on retiring in the upcoming years. Thus, the present chapter is focused solely on bifurcating current models of scientific leadership into senior- versus entry-level leaders.

As a second goal of this chapter, given these differences in job requirements, we outline a succession planning model with implications and suggestions for grooming leaders for both types of roles. This model has recommendations for the identification, development, and tracking efforts required for planning for future leaders of scientific innovation. Third, we intend to outline the theoretical and practical implications associated with succession planning for these two levels of scientific leaders, with particular emphasis on what needs to occur given the dynamic nature of their organizations and fields. Before turning our attention to these goals, it is important to discuss what about our current operating environment drives this imperative need for succession planning.

BUSINESS CASE FOR SUCCESSION PLANNING

Given recent changes in the nature of doing business, impending trends over the next decade and the current composition of the global workforce, succession planning has become a vital business necessity. In a survey conducted by the American Society for Training and Development in 2009, the most popular driver of succession planning in organizations was listed as the identification and preparation of future organizational leaders. This reason was followed closely by the goal of assuring business continuity. These results indicate the belief that succession planning is intended to ensure the future health of an organization. What the survey also found, however, is that 55% of responding organizations did not have a formal succession plan in operation (Paradise, 2010). Organizations are aware of the need for succession planning, but may not always establish the programs to fulfill that need.

Businesses must incorporate succession planning into their long-term strategies and corporate cultures. Creating a board committee devoted to succession planning is imperative to the plan's success, as noted by a large percentage of CEOs in a recent report on succession planning best practices (McKeown, 2010). The need for a continuous and integral succession plan is fueled by the global growth, competition, and recent economic collapses that have led companies to restructure more frequently and more rapidly. This hasty organizational restructuring leaves in its wake a large number of leadership gaps at all levels, from line managers to the executive suite. In this type of setting, sustained organizational performance and successful organizational restructuring can be achieved through the expansion of traditional succession planning models. In addition to identifying potential leaders, succession plans should develop and transition them into their new roles (Kleinsorge, 2010).

Succession planning is not only imperative in the leadership ranks, however, as it also plays an important role in technical positions. 'Technical succession planning' is critical for those positions whose success is based on specialized knowledge. This differs from the traditional managerial succession planning in that it focuses on transfer of knowledge from those who possess specialized knowledge gained from experience (Rothwell, 2010). A major driver behind the need for technical succession planning is the aging of the baby boomer population. This demographic currently comprises a significant portion of the labor force and has already started reaching retirement age (Hunt, Revell, & Rotenberg, 2007; Myron, 2005). Organizations stand to experience job vacancies and the loss of critical knowledge if Baby Boomers, who consist of one quarter of the workforce, were to retire without the proper succession plans in place (Stark, 2009; Young, Piktialis, & Rappaport, 2007). In creative scientific organizations, knowledge transfer from this age segment of employees to emerging leaders is particularly critical for future success. A significant portion of these Baby Boomers are considered key employees, as they have the knowledge, skills and potential to continue contributing to the success of their organizations (Towers Perrin, 2005). While global business and economic conditions mandate the need for succession planning (European Commission: Employment, Social Affairs, and Inclusion, 2011), the challenging work required—and the years required to develop such technical expertise (Ericsson & Charness, 1994)—to lead for scientific innovation may make succession planning particularly critical in positions that require managing organizations and teams. In the following section, we provide an outline of the nature of this creative leadership work, contrasting those in senior positions to those in more junior roles.

THE NATURE OF CREATIVE WORK

To understand what it is that creative leaders do, it is important to understand the context in which they work. Creative work has often been associated with domains and fields that are ill defined, unstructured, and afford multiple pathways for the generation of novel, useful products or services (Mumford & Gustafson, 1988). Positions that require such innovation are comprised of two separate but related activities—generation and implementation. Generation of ideas involves processes such as problem definition, information gathering, idea formation, and conceptual combination (Mumford, Baughman, & Sager, 2003). Implementation of those ideas is composed of idea evaluation, revision, and outcome forecasting (Mumford, Lonergan, & Scott, 2002). In scientific organizational settings, both are required to produce products, services, and social structures that are valued as novel and useful to stakeholders.

This overview of the processes associated with scientific organizational innovation brings to light four important observations about the nature of creative work and what is needed to execute it. First, creative work requires the active gathering and manipulation of ideas in a particular domain, activities that require substantial amounts of domain-specific knowledge. For example, in order to gather information about a potential biotechnical development, researchers must have sufficient field, process, and organizational expertise to know how and where to find data used for decision making. These types of expertise may take years to develop (Ericsson & Charness, 1994; Weisburg, 1999), and there is little doubt that such focused, domain-specific knowledge is critical for multiple components of scientific innovation.

Second, because the nature of the problems associated in creative positions is described as ill defined and complex, they often require the imposition of multiple types of expertise and collaboration among the individuals who hold each (Reiter-Palmon, Herman, & Yammarino, 2008). This issue is an important one, as it highlights the point that engaging in creative problem solving in an organization requires collaboration between multiple individuals—individuals with high levels of expertise, professionalism, and need for autonomy (Feist & Gorman, 1998). Thus, it is imperative that a leader be one who can have credibility with each of these constituencies, understand the nature of their knowledge structures, *and* foster effective communication among them.

Third, scientific innovation is uncertain and resource intensive. The nature of creative work is inherently risky, in that ideas and techniques with limited to no prior history of success are required, and failure of some of those new ideas and techniques is often considered the cost of doing business when it comes to creative endeavors. Groups of people devote time, effort, equipment, and other resources during the development and implementation stages, and often those resources are met with unexpected time complications and un-forecasted setbacks (Cardinal & Hatfield, 2000). Thus, leaders of scientific innovation must be skilled at politically managing such uncertainty to stakeholders. One example activity where this occurs can be found in meetings; leaders of innovation can reduce anxiety in stakeholders by presenting strategy and progress updates with confidence, even when the leader herself may be less certain due to the ambiguity inherent in innovative work. It also implies that leaders must be skilled at estimating such costs and securing resources, with particular adeptness at forecasting timelines and budget expenditures. These components mandate the need for leaders to have an acute understanding of the creative work they lead and the broader organizational context in which they work.

Finally, scientific innovation requires creative people to execute it. There is a substantial body of research that profiles these individuals (Feist & Gorman, 1998; Mumford & Gustafson, 1988). As noted earlier, one characteristic of creative people is high amounts of domain expertise (Ericsson, 2008). To develop this level of knowledge, creative people often are highly involved with their jobs and intrinsically motivated by the field (Collins & Amabile, 1999). This intense identification with the field serves as a buffer against the labor associated with creative problem solving, but also may result in less attention being paid to other activities such as relationship maintenance both in and outside of work (Mumford, Scott, et al., 2002). There are two important implications of this aspect of creative people. First, one implication of managing such a unique workforce is that leaders of creative efforts should be skilled at fostering a collaborative climate between experts who likely would rather engage in task-related work than in organizational activities or group-related exchanges. Second, such individuals already have high levels of field commitment, viewing professional development as an important component of their identities. Thus, although these highly professional, creative individuals may not always see their developmental activities as tied to the betterment of the organization (instead viewing development as a natural responsibility of someone who committed to his *field's* advancement), their inclination and participation in such activities is precisely what is needed to execute an effective succession planning framework. In other words, although attending conferences, reading the latest periodicals in their field, and staying abreast of competitors' developments is naturally part of what mid- and senior-level scientific leaders do, transmitting such techniques and developmental opportunities earlier and more systematically to younger field members is imperative for succession planning frameworks to have credibility and impact. Thus, any human resource initiative geared at succession planning for scientific leaders should make use of the developmental expertise of mid- and senior-level leaders early and often.

JOB REQUIREMENTS FOR LEADERS OF SCIENTIFIC INNOVATION

Given that the nature of creative work is complex, costly, uncertain, and resource-intensive, leading the people who can actually produce scientific innovation in the context of a larger organization requires a certain skill set. In the last decade, attention has shifted onto what is involved in and required for leading innovative efforts (Byrne et al., 2009; Cooper & Kleinschmidt, 2000; Hemlin, 2009; Mumford, Scott, et al., 2002). In a recent review, Byrne and colleagues (2009) synthesized this literature and identified 11 core functions associated with leading for innovation: (a) environmental analysis, (b) strategy formation, (c) leadership team building, (d) mission definition, (e) team construction, (f) climate definition, (g) idea generation, (h) idea evaluation, (i) obtain support and resources, (j) planning, and (k) process management.

Their taxonomy demonstrates that the challenges associated with leading for creativity revolve around the need to understand (a) the work itself, (b) the broader organizational context, (c) management of diverse functional teams, and (d) politics associated with buffering creative workers from the larger, mechanistic organization.

Because this literature is relatively neophytic in its development, it makes sense that further specification of this taxonomy is required. First, although some of the reviews focus on the importance of team building in general, none specifically highlights the critical need for these leaders to actively engage in succession planning. As argued earlier, planning for future leaders of scientific innovation requires both management and technical succession planning, and the impending exodus of the Baby Boomer leaders has made succession planning a central function for leaders. Thus, the first addition to Byrne and colleagues' taxonomy (2009) for the purposes of the present chapter is the addition of a 12th function: succession planning. Sample tasks and requisite knowledge, skills, abilities/other (KSAOs) are discussed in the following section and are displayed in Table 8.1.

Second, what Byrne et al. (2009) and other similar theories do not address is what differences in these functions are required across levels of organizational leadership. Moreover, although most of the literature converges around the importance of these functions, it is unclear whether the tasks associated with each of them are robust across senior level leaders (e.g., senior vice presidents of research and development [R&D]) *and* junior-level leaders (e.g., lab managers of creative project teams). For example, in Hemlin's (2009) study of 84 group leaders in the biotech field, recommendations for group composition were established; it is encouraged that leaders are careful to structure creative teams so that they are from the similar domains, but may have varying levels of expertise within that domain. This clearly falls under the function of *team construction,* but it seems most relevant with those (i.e., junior-level leaders) managing individuals directly engaged in creative work, versus those (i.e., senior leaders) concerned with leading larger creative functions such as R&D departments. In fact, if examined closely, the literature on creative leadership dips across levels of analysis more often than not, which has implications for the practical and theoretical implications of leading for scientific innovation across an organization. Thus, the first goal of the present chapter is to better define core functions required for junior and senior leaders when managing for scientific innovation.

SENIOR VERSUS JUNIOR LEADERS OF INNOVATION

The broader leadership literature is only recently beginning to articulate differences between levels of leadership (Chan & Drasgow, 2001; Lord & Hall, 2005). In all types of organizations, there appear to be differences

Table 8.1 Position Functions, Primary Focus of Influence, and Sample Tasks for Senior and Junior Leaders of Innovation

	Junior Leaders for Innovation		Senior Leaders for Innovation	
Core Functions	Primary Focus of Influence	Sample Tasks	Primary Focus of Influence	Sample Tasks
1) Environmental Analysis	External	1) Interpret market research 2) Consult with clients on needs 3) Read professional journals	External	1) Review customer feedback 2) Monitor industry competitor portfolios 3) Obtain feedback from Board Members
2) Strategy Formation	Internal	1) Provide feedback about strategy with regard to lab capabilities 2) Determine technology needs given defined strategy by senior leadership 3) Provide relevant questions to help in evaluating strategy against market constraints	External and Internal	1) Assess market saturation 2) Identify organizational strengths/ differentiators 3) Reference innovation strategy against other organizational projects and expenses
3) Mission Definition	Internal	1) Partner with senior leaders to understand how lab capabilities fit with firm strategy 2) Explain parameters of strategy to individual innovators 3) Articulate how strategy can be defined in his lab	External and Internal	1) Develop portfolio of projects across labs 2) Make changes based on market needs 3) Structure idea generative processes by introducing themes related to strategy

(*Continued*)

Table 8.1 (Continued)

	Junior Leaders for Innovation		Senior Leaders for Innovation |	
Core Functions	Primary Focus of Influence	Sample Tasks	Primary Focus of Influence	Sample Tasks
4) Leadership Teambuilding	Internal	1) Participate in collaborative discussions with internal leaders of other labs 2) Identify gaps in expertise in lab and feed information upwards for planning purposes	External and Internal	1) Coordinate lab leaders of innovation 2) Structure cross-disciplinary information sharing 3) Build team based on environmental scanning, composed of diverse experts
5) Team Construction	Internal	1) Design teams to have varying levels of expertise 2) Break larger teams into optimal size (5-7) 3) Monitor teams for increasing levels of cohesion	Internal	1) Mentor junior leaders on best practices for team composition and size
6) Climate Definition	Internal	1) Reinforce cooperation and task-based conflict 2) Publicly engage in open exchange and candid debate 3) Frame errors as learning opportunities, not mistakes	Internal	1) Send consistent message that innovation is valued 2) Emphasize importance of trust, cooperation norms 3) Role model ethical, creative leadership

7) Idea Generation	Internal	1) Stimulate idea exploration through seed projects 2) Role model conceptual combination in meetings 3) Structure team meetings for optimal idea generation and open communication of ideas	External and Internal	1) Gathering relevant information 2) Generating viable mission/framework 3) Constructing problems in a way that provides structure but affords multiple pathways for success
8) Idea Evaluation	External and Internal	1) Evaluate how well proposed idea fits into broader lab and organizational goals 2) Assess using operating criteria (e.g., costs) and innovative criteria (e.g., degree of novelty) 3) Identify conceptual strengths and weaknesses	External and Internal	1) Gauge viability of proposals against operating constraints (e.g., budget) 2) Evaluate merits of proposal from standpoint of political relationships 3) Assess risk against other organizational projects
9) Support and Resources	External and Internal	1) Obtain technical and monetary resources for lab 2) Provide compensatory feedback about ideas 3) Demonstrate viability of projects using small seed projects in order to garner organizational support	External and Internal	1) Identify project champions 2) Manage uncertainty in meetings with upper level leaders and outside stake holders 3) Negotiate vendor, distributor, and federal and state agency contracts
10) Planning	External and Internal	1) Monitor fielding of new products as they are brought "online" 2) Evaluate market reactions to initial releases of product 3) Test requisite technology needed for implementation	External and Internal	1) Ask for seed projects 2) Direct and implement policies to maximize ROI and competitive advantage 3) Review technical reports of managers to assess progress and make plan revisions

(*Continued*)

Table 8.1 (Continued)

Core Functions	Junior Leaders for Innovation		Senior Leaders for Innovation l	
	Primary Focus of Influence	Sample Tasks	Primary Focus of Influence	Sample Tasks
11) Process Management	Internal	1) Prepare project specifications for clients 2) Prepare budgets, bids, and direct negotiations of research contracts 3) Perform administrative tasks (e.g., review/approve expenditures, enforce rules, make purchase decisions) Internal	1) Create pathways for communication buffered from organization (e.g., wikis) 2) Review and approve budgets 3) Determine potential areas of cost reduction, program improvement, or policy change	
12) Succession Planning	External and Internal	1) Establish relationships with potential outside hires 2) Oversee experiential development and mentoring programs with individual contributors 3) Track performance of individuals in high potential group/ succession plan pipeline	External and Internal	1) Plan for leader positions for new markets 2) Identify external talent when internal bench strength is lacking in domain 3) Meet with HR to develop metrics for succession planning progress

required in KSAOs when examining the job content of senior-level leaders and junior-/entry-level leaders (Chan & Drasgow, 2001). It follows, then, that leaders of scientific innovation may also exhibit some of these differences.

Senior-level leaders are those charged with formulating policies and providing overall direction of R&D for an organization, and most frequently within guidelines set by a board of directors or a similar governing body (Occupational Information Network [O*Net], 2011). They plan, direct, and coordinate research activities at the highest level of management with the assistance of subordinate executives and staff managers. In short, they are *leaders of the leaders* of innovation, and they can be found in roles such as chief research officers or executive vice president for R&D. The focus of their influence is pulled between activities such as shaping the nature of the creative performance of their various departments (internal focus) and what the market needs for the next scientific innovation (external focus).

Junior-level leaders plan, coordinate, and direct activities in scientific fields. They guide the research and development of their field via the direct supervision and development of staff scientists engaged in the work. They manage the labs, the project teams, and the design groups. For most, it is their first experience as a leader, and given the nature of creative efforts they often have rose from the ranks of the labs they now manage (Hunter, Thoroughgood, Myers, & Ligon, 2011). The focus of their leadership efforts is often directed internally, or downward to their reports and across to their peers. However, some of the functions identified in Byrne's (2009) model suggest they may also have some external focus (e.g., environmental scanning), albeit through tasks different from those used by their senior leader counterparts.

In the following section, we delineate how the 12 core functions (Byrne et al., 2009) associated with leading for innovation vary in tasks and focus of influence between senior- and junior-level positions. This discussion can be augmented by reviewing Table 8.1, which has a taxonomy of the sample tasks and primary foci of influence associated with each of the 12 core job functions of senior and junior leaders for scientific innovation.

TWELVE CORE FUNCTIONS OF LEADING FOR INNOVATION

Environmental Analysis

Environmental analysis is one of the first steps in planning for scientific innovation—in order to innovate, it is critical that organizations understand the market, customers, and operating environment that will have an impact on receptivity to new products and services. For both senior and junior leaders, the focus of this function is external to the organization, but the tasks required for both roles differ. For senior leaders, environmental analysis entails a macro-view of the portfolio mixes of the competition (Hunter et al., 2011), with particular emphasis on gathering information at the firm level of performance (Benner & Tushman, 2002). For junior leaders, environmental scanning is more focused, composed of staying abreast in changes in the field but often those specific or closely related to their product lines. Sample tasks then might involve reading current journal articles, monitoring similar labs, and interpreting market research.

Strategy Formation

Developing a strategy for scientific innovation is a significant responsibility for leaders, and it requires them to balance a mix of projects across multiple people, timelines, and resource demands (Mumford, Scott, et al., 2002). Basing a strategy on a portfolio serves at least two purposes in an innovative organization. First, it helps to buffer the uncertainty associated with creativity—having multiple *similar but distinct* lines of research ensures that when some fail due to unforeseen causes, others are still in the pipeline that can generate revenue or market share. Second, having multiple projects increases the likelihood of conceptual combination among labs within an organization (Scott, Lonergan, & Mumford, 2004), and capitalizes on efficiencies among sister labs and researchers. These benefits can only be realized with the oversight of a leader who understands the overarching strategy and where synergies might occur. Thus, for senior leaders, both an external and internal focus of influence is required. Senior leaders must understand how their strategy for scientific innovation places them in the market at large, but they also must recognize the internal capabilities and projects of various labs and teams. Junior leaders, more proximal to the work being done, have more internally focused tasks when it comes to strategy. Junior leaders are charged with providing feedback upward—often to the senior leaders of innovation—with regard to new products and capacity, and they also need to interpret their own lab's strategic advantages in the context of the field and competitors (Mumford et al., 2007).

Mission Definition

Defining the mission in a way that allows projects to still have intrinsic value for innovators but that aligns with organizational strategy is critical to maintaining the personal investment of innovative individuals (Hunter et al., 2011). Missions provide an organizational framework that guides work and helps to evaluate ideas and proposals. For senior leaders, defining this mission in the context of the broader organizational strategy is key, and the focus of attention and influence is primarily internal. For this function, senior leaders must be able to have a strategic understanding of the entire firm, affording them the capacity to drive scientific innovation streams and take advantage of contrasting capabilities and structures (Benner &

Tushman, 2002). Effective senior leaders are those who can readily identify synergies across the firm and integrate teams to realize those strategic partnerships. For junior leaders, the primary task associated with mission definition is understanding what objectives at the lab and individual levels best exemplify that mission (Strange & Mumford, 2002). Junior leaders must be well versed in providing a framework for doing the work in the context of the organizational strategy, but not to the degree where individuals working on projects lose ownership and autonomy in design (Hunter et al., 2011). Articulation of the mission internally is critical for junior leaders.

Leadership Team Building

As creativity is complex and requires multiple types of expertise (Andrews & Farris, 1967), it follows that the leadership team in charge of managing that process is diverse, particularly in terms of domain expertise (Hunter et al., 2011). Given the need for these multiple types of domain expertise, multiple players are needed in creative efforts. Identifying and selecting these people is primarily a charge for senior leadership. For example, a head of R&D for a design firm may want to utilize a new computer technology as part of a differentiation strategy. As a result, it is the R&D leader's job to identify who might possess this expertise—either internally or externally—and attract and recruit him or her for this organization. Thus, a senior leader's focus is both internal and external, and teambuilding tasks may range from headhunting to structuring communication between experts once a team is in place. A junior leader's role may not require as much external networking and focus, as these more proximal managers are unlikely to be part of the teambuilding process except for playing supportive roles (e.g., identifying key industry players, participating in meetings set by senior leadership).

Team Construction

Team construction is related to teambuilding and centers on designing teams for optimal size and cohesion. This function may be more important for junior leaders (vs. senior leaders, who have more mentoring tasks associated with this function) as they have autonomy in structuring project teams. Junior leaders, upon defining the mission and understanding the nature of their workers, are charged with creating the project teams who can implement that mission. Thus, these individuals should engage in these tasks with an awareness of optimal team size (Byrne et al., 2009), cohesion (Abra, 1994), and communication best practices (Mumford, Scott, et al., 2002). Orchestrating expertise (Mumford, Scott, et al., 2002) is central to this effort, and junior leaders must actively monitor and make adjustments to team composition in light of gathered feedback (Keller, 2003).

Climate Definition

Individual perceptions of the work environment, or climate, have been reviewed extensively with regard to innovation (Ellison, James, & Carron, 1970; Mumford & Gustafson, 1988; Nystrom, 1990). What is important to bear in mind is that both senior and junior leaders are critical lynchpins in shaping and transmitting climate to followers (Schneider & Reichers, 1983). In a meta-analysis of climate variables and innovative performance, Hunter, Bedell, and Mumford (2007) identified 14 dimensions of optimal climate for creativity. It is reasonable to assume that senior and junior leaders play important roles in transmitting this climate, albeit in markedly different ways. Senior leaders have a larger sounding board as they have opportunity to emphasize the importance of scientific innovation, focus on development, and encouragement of risk taking at a firm level (Hunter, Bedell, & Mumford, 2006). This has implications for others, even those in noninnovative, supportive roles, to encourage and accept innovation. Junior leaders, although their focus is more directed at their teams, are responsible for encouraging climate of open communication, information sharing, learning from errors (rather than avoiding them), and having fun at work (Hemlin, 2009). In essence, senior leaders can have an impact on the organizational climate and can foster lab climate for scientific innovation, and junior leaders may have more direct influence on lab climate for scientific innovation.

Idea Generation

There is substantial evidence that all leaders play an important role in idea generation (Andrews & Farris, 1967; Barnowe, 1975). For senior leaders, this likely occurs through information gathering tasks, as they often structure missions and define problems based on external and internal scanning (Mumford et al., 2007). In addition, they engage in conceptual combination activities in their capacity to see connections between seemingly unrelated strains of work in their organizations (Scott et al., 2004). Junior leaders influence idea generation through two mechanisms—structuring interactions to encourage it and generating ideas themselves. Leaders of the individuals charged with doing the innovative work shape idea generation by gathering and sharing information to be used in problem construction (Reiter-Palmon et al., 2008), as well as by role modeling generative processes through directly working with creative teams and individuals (; Mumford et al., 2007; Scott et al., 2004; Shalley & Gilson, 2004).

Idea Evaluation

Leaders for scientific innovation engage in idea evaluation by imposing criteria against which ideas are reviewed, as well as determining how an idea will fit within the context of the broader organizational and field context

(Mumford et al., 2007). Both senior and junior leaders exercise internal and external focus for this function, though admittedly in different ways. Senior leaders, with their understanding of the operating constraints of the entire organization, must evaluate proposals against other organizational plans. For example, if new infrastructure is being purchased to increase efficiency goals, senior leaders may need to table expensive R&D projects in a given quarter. In addition, senior leaders must be skilled at evaluating how ideas will be received by others—both internal and external to the organization. Given that scientific innovation requires an extant amount of risk, senior leaders must also be able to evaluate what components will be most aversive to more cautious organizational stakeholders (e.g., board members) and identify ways to overcome those objections. In short, senior leaders need a fair amount of technical expertise when evaluating the merits of proposals as well as political/interpersonal expertise when assessing what concerns will be voiced by others. Junior leaders, however, are more focused on evaluating the technical merits of ideas, because they often are on the initial stages of idea development and plan revision (Lonergan, Scott, & Mumford, 2004). Thus, junior leaders engage in tasks such as assessing the costs, novelty, and viability of specific plans, and they often are charged with creating small seed demonstrations to show the benefits of their team's proposal to senior management.

Support and Resources

As mentioned earlier, scientific innovation is a resource-intensive endeavor for an organization to pursue (Hemlin, 2009; Hunter et al., 2011). Support for creativity can be conceptualized in three main ways: (a) support for ideas, (b) support for the work, and (c) social support for the employee (Byrne et al., 2009). Although senior and junior leaders are responsible for providing all three of these support and resource mechanisms, they go about obtaining it in markedly different manners. Senior leaders need a fair amount of organizational knowledge, an understanding of resources and their capabilities, and knowledge of the political landscape. This is key to leveraging relationships for the attainment of resources (Cooper & Kleinschmidt, 2000) and senior meetings within the organization is one venue where this can be seen. Although senior leaders lack the certainty that an idea or project will actually yield results, they must present information convincingly and in a way that allays the concerns of others (Boehlke, 2008). In a sense, senior leaders need to run meetings with confidence, even when all of the information is not available.

Junior leaders, although also responsible for vying for resources for their teams, may be heavily involved with providing the support for ideas and social support needed for scientific innovation. Much research suggests that this can be identified in the form of how leaders provide feedback to their team members (Ligon, Graham, Edwards, Osburn, & Hunter, 2012; Lonergan, Scott, & Mumford, 2002; Zhou & Oldham, 2001). Moreover, junior leaders must take care to provide compensatory feedback about the ideas of their team members, which requires a certain degree of expertise for both operational and creativity criteria used to assess the merits of ideas.

Planning

After ideas have been developed and evaluated, innovative leaders then move into an extensive planning and coordination stage (Mumford et al., 2007). In a review of the planning that is required of innovative organizations, Hunter, Bedell, and Mumford (2007) developed a planning model based on incrementalism and consisting of five related processes. The tasks associated with each of these involve both internal and external focus of influence and work harmoniously between senior and junior leaders. For example, senior leaders are often charged with gauging how product prototypes are received by various organizational stakeholders, and junior leaders may then propose ways to respond to such feedback. In short, as the planning phase is multifaceted and complex, it requires constant reciprocal information sharing among all levels of the innovation team.

Process Management

Although the discussion thus far has focused on the functions meant to encourage innovation, it is important to remember that innovation occurs in the context of organizations—organizations that are often large, bureaucratic, and highly mechanistic (Benner & Tushman, 2002). Thus, a central function of a leader for scientific innovation at any level is to serve as a buffer for the creative individuals working within a broader organizational context. Tasks associated with process management can be grouped into two domains: (a) ensuring that effective communication occurs in organizations that are large, hierarchical, and often slow and (b) shouldering much of the administrative burden so that their team members can focus on the generative and evaluative aspects of their jobs rather than budgets and reporting. Thus, for both senior and junior leaders, process management appears to be an internally focused function, and it may require a substantial amount of organizational and administrative expertise (Boehlke, 2008; Hunter et al., 2011).

Succession Planning

Although the preceding 11 functions have been historically supported for leaders of innovation, less attention has been devoted to their roles in succession planning, or the planned development of future leaders of innovation. However, given the complex, multifaceted, and intensive functions required in these positions, it follows that current incumbents be charged with the primary responsibility of filling their own roles in the future—as they tend to know what is required more than others understand. As succession planning is a lengthy process that entails a fair amount of expertise in processes such as human resource planning, selection, development, and performance appraisal, the next section of this chapter is devoted to specific best practices for succession planning for senior and junior leaders of scientific innovation.

IMPLICATIONS OF 12 FUNCTIONS, VARYING TASKS, AND FOCUS OF INFLUENCE

Four important conclusions can be drawn from Table 8.1. First, leaders of scientific innovation have it tough. They are challenged to manage highly complex work, difficult individuals, and buffer overly mechanistic and often bureaucratic organizations (Hunter et al., 2011). There are at least 12 core functions associated with these jobs, and likely many more tasks underlying each one. Leaders of scientific innovation must be competent at executing each one.

Second, although the 12 functions appear robust across senior and junior levels of leadership, the requisite tasks are likely to differ depending on how close a leader interacts with the individuals charged with doing the innovative work. Moreover, on reviewing the literature that has examined what it is that such leaders do, it became clear that some tasks are more suited to senior leaders (e.g., R&D SVPs) whereas others are more likely taken on by junior leaders (e.g., lab managers working with teams of 5–7 innovators). This finding has significant implications for both the importance and type of skills required to execute each position. To date, most literature on leadership development in general has focused on broad competencies of leadership (for exceptions, see Day, Harrison, & Halpin, 2009; Lord & Hall, 2005). The bifurcation of tasking identified in Table 8.1, however, indicated that we may be thinking of leaders of scientific innovation much too simplistically. Rather, we may need to focus on developing these two types of leaders differentially, and we may need to consider that development from one role to the next may not always be linear nor always certain (Mumford et al., 2007).

Third, the two types of leaders tend to focus on different constituents and types of information when exercising these two functions. Specifically, it appears that senior leaders are much more (but not solely) focused on factors external to the organization, whereas junior leaders tend to exercise their influence more often (but not always) internal to the organization. In addition, senior leader tasks appear more cross-department (i.e., macro-organizational) in scope, whereas junior leaders may be more team based (i.e., micro-organizational) in their influence focus. For developmental purposes, it is clear that attention be paid to how individuals develop these relationships and domain expertise. This also has implications for how these individuals in both types of leadership positions develop the many and varied kinds of expertise—technical, organizational, and domain-specific expertise—that can take years of deliberate practice and focused instruction, suggesting that mentoring play a particularly important role in the development of scientific leaders (Ericsson, 2008).

Finally, the developmental challenges of becoming a senior leader or a junior leader of scientific innovation vary (Mumford et al., 2007), and the motivation to lead may be markedly different (Chan & Drasgow, 2001). Junior leaders, often assuming their first leadership role, face the new challenge of seeing themselves as a leader versus solely an individual contributor. Because domain expertise is essential for leading an innovative team, they most likely were once team members who had developed high levels of such technical knowledge and creative problem solving skills. In addition, individuals in innovative positions tend to identify much more with the work and field than with the organization in which they execute (Mumford, Scott, et al., 2002). Thus, a central challenge that must be addressed in any developmental effort is helping these individuals to see the value of leadership, and to begin incorporating such roles into their personal identities.

Senior leaders, also requiring high levels of domain expertise, often rise through the ranks of mid-level laboratory leadership prior to obtaining pinnacle leadership positions. They face the challenge of having to understand the nature and people involved with the technical work, but also understand the broader organization and field-level stakeholder concerns so that they can successfully navigate them. The capacity to build organizational and stakeholder alliances seems paramount for those most senior positions, and unfortunately, is often not developed until they enter these roles. In addition, because senior leaders often set the overall innovation direction and mission of the organization, these leaders must be versant in emerging new technologies and products so that they can capitalize on trends and stay ahead in their own innovations. In order to do this, they often rely on a diverse set of contacts in a loosely related network, a network that can take years to develop and nurture (Perry-Smith, 2008). Thus, the need to develop expertise in such technologies and collaborative networks is also requisite for these senior leaders.

Given these implications, the urgency for development, and the lack of attention in the innovation literature, we detail how both levels of leader might benefit from succession planning to develop the capacities requisite to execute these 12 functions. In addition, drawing from best practices in succession planning, we outline a framework to stimulate ideas on how technical and managerial competencies might be developed in both senior and junior leaders of scientific innovation.

SUCCESSION PLANNING: SENIOR AND JUNIOR LEADERS OF SCIENTIFIC INNOVATION

As mentioned earlier, succession planning is a change effort intended to prepare individuals for promotions or new positions within an organization by emphasizing the identification, selection, and development of internal talent

(Rothwell, 2010). For leaders of scientific innovation, succession planning revolves around both the need to meet promotion goals (i.e., typical management succession planning) and the need to meet knowledge transfer goals (i.e., technical succession planning). In addition, given the differing roles as delineated in Table 8.1 between senior and junior leaders, a one-size-fits-all approach cannot be developed in haste for this important set of activities. For the purposes of the present chapter, we focus on the very different roles of senior and junior level leaders, but acknowledge that much of the development required to move between these positions will be found in mid-level positions. Although numerous skills overlap (e.g., creative problem solving skills; Mumford, Scott, et al., 2002), we argue that succession planning for junior and senior leaders differ in at least two important ways. First, it is critical that junior leaders make the cognitive and self-concept leap of seeing themselves as leaders and no longer solely as individual innovators (Lord & Hall, 2005). To this end, succession planning activities revolve around encouraging high-potential individuals to see the value in leading a team, even when they may prefer to do the work themselves. In addition, when it comes to identifying individual scientists who have the potential to develop into leaders, it is imperative that motivation to lead be assessed and nurtured prior to investment in broader succession planning activities (Chan & Drasgow, 2001).

Second, activities of senior leaders require the development of organizational and field level expertise, as they serve as both the buffers and lynchpins between these entities and the individuals engaged in the innovative work. Thus, in addition to creative problem-solving skills and planning capacity, the identification of individuals with interpersonal skills associated with managing political relationships and building alliances is key for senior leaders. In addition, providing access to such developmental activities may help strengthen these skills prior to them assuming roles where performance needs to occur in real time.

Succession planning for leaders of scientific innovation can be broken into five related but distinct processes (Haworth, 2005; Hollington, 2007), and each must be shared between current scientific innovation leaders and individuals working in human resource functions: (a) strategy formation, (b) identification and selection, (c) development, (d) tracking and validation, and (e) placement into leadership position. Figure 8.1 shows the how each of these steps relates to planning for senior and junior positions and offers sample activities for current leaders to undertake in these efforts.

Strategy Formation

As the innovation strategy of the firm is developed, the strategy for the requisite people to conduct and lead that work also needs attention. From a leadership perspective, issues such as what new product lines will be introduced, current skill/expertise gaps, and impending retirements should be considered as part of the workforce planning initiative (Murphy, 2006). Some

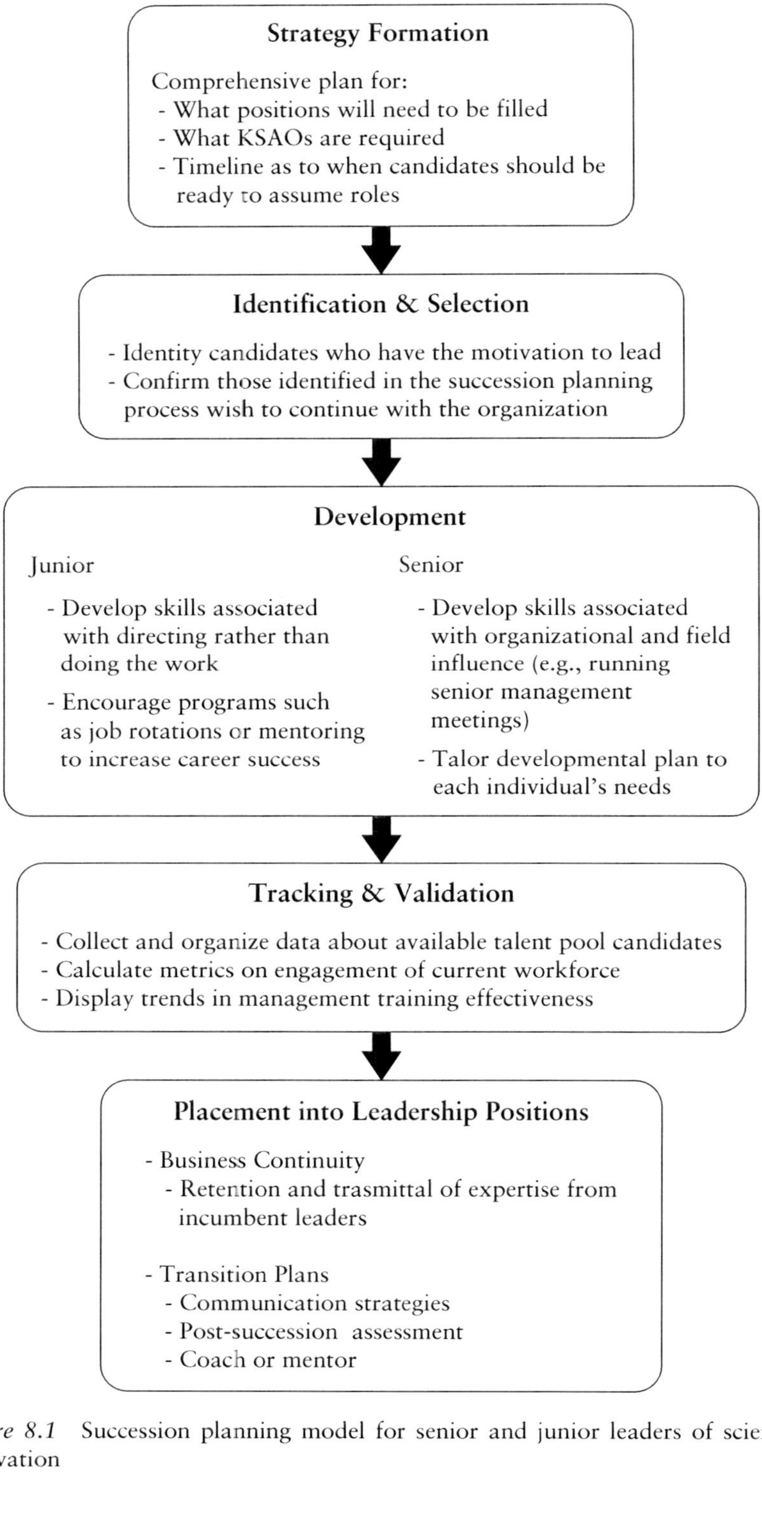

Figure 8.1 Succession planning model for senior and junior leaders of scientific innovation

scientific organizations are far more advanced at this. For example, each spring, managers at General Electric (GE) work with human resources to put together a rigorous succession plan for each position (Hoffman, 2005). They take time to define the critical success factors for each position, which results in an optimal KSAO profile needed for successful execution. The final product of strategy formation should be a comprehensive plan for what positions will need to be filled, what KSAOs are required to fill them, and a tentative timeline as to when candidates should be ready to assume such roles.

Identification and Selection

Once position needs are prioritized, leaders must work to identify and select individuals who might be groomed to develop into such roles. As much has been written about the competencies required for innovation (Feist & Gorman, 1998) and the leadership of innovation (Hunter, Bedell, & Mumford, 2007; Mumford, Baughman, & Sager, 2003), we focus on what might be required above and beyond these competencies to benefit from a formal succession planning initiative: (a) motivation to lead and (b) commitment to the organization. First, Chan and Drasgow (2001) have drawn a distinction about motivation to lead, indicating that because innovation leadership requires such different activities than individual innovation requires, it is difficult to identify those who might still be motivated by the work if further removed from it in a leadership role. They also draw distinction between junior- and senior-level leaders, indicating that the activities required from senior leaders are so markedly different from what is needed by junior leaders, identification of who can do what may require examining different two different individuals. In short, they argue that development might not be linear from junior to senior leadership positions. For leaders of scientific innovation, however, expertise in the domain at hand is so critical for performance that it is unclear how this finding may generalize to this particular group of leaders. In scientific organizations, we argue that senior leaders are indeed found in the ranks of junior leaders, but not all junior leaders will have the motivation or the skill sets to assume senior leadership. Thus, identifying who may have the motivation for leadership is key for succession planning.

Second, because succession planning is expensive, time-consuming, and focused, it follows that those identified be part of the process for it to actually continue with the organization. This may be particularly tough to identify in creative workers, as they typically have more field commitment than they have organizational commitment (Mumford et al., 2007). However, making certain that they are interested in being part of a succession planning initiative is important. At the Marriott organization, information and technology (IT) workers are provided with opportunities to take part in rotational assignments as part of a succession planning initiative. Entrance into this program is both by nomination and by self-selection, and both the leadership and incumbents need to agree on the succession plan (Hoffman, 2005).

Finally, in a review of best practices for succession planning, Garman and Glawe (2004) indicate that identifying a talent pool is optimal to identifying specific candidates for specific positions. This is a challenge for scientific leaders for at least two reasons. First, it seems that we are looking for a unique individual—one who can do the technical work yet is also willing to lead it in the current organization. Finding a 'pool' of many of these individuals may seem too lofty a goal. Second, because leading labs requires specific domain expertise within that field, organizations are limited by the number of individuals even available for identification. Thus, identification often may need to also search outside of the organization, particularly for the junior-level leaders early in their careers.

Development

Once positions and candidates are identified, developing them for future positions can occur. Work on how to develop leaders for creative efforts generally draws on a combination of activities—from formal training programs to experiential development programs (Ligon et al., 2012; Mumford et al., 2007). When bifurcating developmental activities for senior versus junior leaders, however, it is important to understand what different skills each will be required to develop. First, to revisit junior leader needs, these individuals are encouraged to not only begin to see themselves as leader (vs. individual contributors) but to also begin to develop the skills associated with directing the work rather than doing the work. In a review conducted on shaping individuals for entry-level innovation leadership roles, having a mentor was identified as an important developmental event in the trajectory associated with moving into leadership. Mentors might serve not only as transmitters of norms, skills, and best practices of leadership; they also may serve as role models for incorporating leadership into individual identities (Lord & Hall, 2005). One study examined career experiences that shaped high potential innovators, and it was also found that rotations in a high-performing lab and early mentoring were indicative of later career success (Mumford et al., 2007). As so many rote new skills are also required (e.g., evaluation of ideas, providing feedback), it makes sense that formal training programs have also been found effective for increasing innovation (Scott, Leritz, & Mumford, 2004).

For senior leaders, developmental interventions are also effective if outlined correctly. In a review by Groves (2007), 30 CEOs and human resource (HR) executives were asked to identify best practices for succession planning and development of senior leaders. One practice that garnered support for shaping senior leaders was establishing organization-wide forums and project-based learning activities so that would-be senior leaders become exposed to various parts of the organization. This kind of exposure and early networking activities may aid them in later alliance building initiatives needed to understand and obtain resources for scientific innovation.

Hicks (2000) also recommended the use of job rotations and other experiential development as part of a package to develop senior leadership skills, though Mumford and colleagues (2007) emphasize the importance of making these voluntary or optional rather than mandatory assignments because not all individuals will favor moving into a senior leadership role at a given time.

Overwhelmingly, for both junior and senior leaders, it is recommended that developmental plans be tailored to each individual's specific constellation of developmental needs given organizational strategy and anticipated growth (Byham, Smith, & Paese, 2002; Conger & Fulmer, 2003). Although there is a risk of overburdening the process if such plans become too administratively intensive (Conger & Fulmer, 2003; Garman & Glawe, 2004), individual developmental plans aid in communicating what is important and serve as anchors for gauging success.

Tracking and Validation

The validation literature on succession planning is sporadic at best (Garman & Glawe, 2004). This may be due to the complexity associated with the whole process itself, to the fact that the process is newer in most organizations, or to the point that it is difficult to tie development to firm-level outcomes. Regardless of the availability of peer-reviewed literature on how effective succession planning is, there are numerous vendors willing to assist organizations in tracking their succession planning programs (Worklogix, 2009). Two widely used providers for global organizations' succession planning data tracking are SAP and PeopleSoft (Oracle, 2011; SAP, 2004). Each of these for-profit organizations offers processes available to meet three stated goals: (a) collect and organize data about available talent pool candidates, (b) provide metrics on engagement of the current workforce, and (c) display trends in management training effectiveness. Software providers often promise to provide data on what has been termed 'gap analysis', but it is unclear how such metrics are derived other than inputting ratings on skills and established developmental objectives. Although little is known about which of the numerous vendors of succession planning tracking systems provide the best data, what is more clear is that they offer a way to track candidate progress through developmental initiatives and provide a framework for making succession decisions. Much more research is needed as to what organizational performance indicators will be affected by and should be measured in succession planning initiatives.

Placement Into Leadership Positions

Business continuity and transition plans are key for placing individuals in innovative leadership roles (Rothwell, 2010). Transition plans may involve communication strategies, both to internal and external stakeholders depending on the level of leader (Garman & Glawe, 2004). Transition plans should also include postsuccession assessments (e.g., 360-feedback surveys) in order to address problems early on (Tyler & Biggs, 2001). Finally, it is encouraged that new leaders for scientific innovation be provided with a coach or seasoned mentor to help navigate the process and learn from new experiences (Ligon & Hunter, 2010).

CONCLUSION

The present chapter was focused on developing leaders for scientific innovation using a succession planning framework. It offered a taxonomy of 12 core job functions, illustrating how they differ between senior and junior leaders of scientific innovation. We also reviewed five steps in succession planning and how they might be applied to plan for leaders of scientific innovation. Before turning to the theoretical and practical limitations of our model, it is important to consider the limitations to the present chapter.

Limitations

First, the findings and recommendations in this chapter are based on a review of the literature surrounding scientific leadership and succession planning, but no primary empirical data were collected for this chapter. However, in order to drive theory development of scientific innovation leadership, our extension and specification of Byrne's (2009) model of core functions is imperative to drive empirical assessment of these tasks and focus of influence.

Second, the literature on succession planning is scant at best. Most of the recommendations are based on case studies and other qualitative examinations of succession planning. In addition, none of these focused on succession planning for scientific leadership specifically. However, the lack of current empirical work in this area does not make it any less critical for those charged with leading for scientific innovation, and certainly not given the impending labor crisis in innovative fields. We view this piece as an important first step in this discussion.

Finally, our review focused on two types of leaders: entry level/junior and senior. It is quite possible that mid-level managers (e.g., area coordinators, division heads) may also have distinct tasks associated with their roles. Individuals charged with leading departments, product lines, and other mid-level positions between frontline leadership and the top of R&D may be especially taxed, particularly as organizations become flatter and those in mid-level positions are charged with more diverse labs and projects. We selected the two extremes of leadership for demonstrative purposes to illustrate that scientific leadership is not a unitary construct and should not be developed using a one-size-fits-all approach, but future research should also include important mid-level managers to complete our understanding.

Theoretical Implications

Despite these limitations, there are at least three important theoretical implications of the present chapter. First, it is clear that this model needs empirical testing: are there measurable differences between junior and senior leaders of scientific innovation? In addition, are there differences in how the 12 functions should be arrayed in a taxonomic form—are some functions more important to one level than to the other? Understanding the nature of the different tasks will ultimately shape our understanding of the varied KSAOs required for each role, and may also shed light on how to develop them. One way to obtain such data would be to conduct job analyses on a sampling of such roles across various innovative and noninnovative organizations.

Next, more work is needed on the exact nature of the developmental activities associated with preparing individuals to assume these roles. Specifically, while a variety of educational techniques is encouraged to develop these leaders, it is unclear which activities (e.g., formal training versus informal mentoring) are linked to the development of which skills (e.g., networking skills, process management skills). As a best practice in succession planning is to define individual development plans, it is critical that we have an understanding of which developmental interventions shape the numerous and varied knowledge and skills required for leading for scientific innovation. In addition, much of the development required for innovative leadership may be occurring informally already. Because a defining characteristic of innovative individuals is their high levels of professionalism and intellectual curiosity, they often view development in their field as a professional responsibility rather than one that is to be incentivized at the organizational level. The key here, then, is to provide organizational support for these developmental initiatives to occur, through resources (e.g., time for mentoring activities, conference travel) and through expressed acceptance (e.g., recognition as part of a formal job description to attend such developmental activities and spend time in formal succession planning). In addition, from a planning perspective, human resource professionals could do a better job at involving senior leaders of scientific domains early in succession planning activities, as they hold the most expertise in evaluating who might be ready for such development and how to best track their progress.

Finally, establishing a clearer link between performance indices and succession planning is needed to understand how this lengthy, expensive process works. Empirically, this is quite a challenge as firm-level performance metrics are (a) difficult to obtain, (b) often unrelated to interventions at the group level (e.g., junior leaders), and (c) fraught with measurement error and unreliability. Despite these challenges, efforts must be made to link creativity outcomes to the development of leaders for those efforts.

Practical Implications

In addition to these theoretical implications for social science investigators, there are at least two practical implications for application in an organizational setting. First, the proposed model adds the core function of succession planning to both junior and senior leader positions. In doing so, it adds a great deal of work to their roles and to an already convoluted process. In asking these leaders to perform more work processes, organizations must ensure that they are measured and compensated for such efforts. Succession planning cannot simply be viewed as an HR process alone. Rather, it must be incorporated into the stated goals and measured outcomes of a leader's job. In addition to expecting this as a core function of their roles, organizations must provide adequate training and development in the succession planning arena so that leaders can strategically and properly execute such efforts.

Next, a large amount of succession planning for these positions requires the transfer of knowledge—both organizational knowledge and of the scientific domain in which the innovation is taking place. Thus, organizations need to rely heavily on the baby boomers within those roles because they are the leaders with this acquired knowledge. Transition plans must be put in place to get this information to the new cadre of leaders upon the retirement of baby boomers. In addition, perhaps identifying phased retirement plans and other nontraditional work arrangements may be key in retaining these baby boomers who can then play vital roles in succession planning. As this process is a multifaceted one in response to an increasingly urgent issue, organizations reliant on both junior and senior leaders for innovation management are encouraged to look begin shoring up their current talent and making effort to develop expertise and skills for the future.

REFERENCES

Abra, J. (1994). Collaboration in creative work: An initiative for investigation. *Creativity Research Journal*, *8*, 205–218.

Andrews, F. M., & Farris, G. F. (1967). Supervisory practices and innovation in scientific teams. *Personnel Psychology*, *20*, 497–515.

Barnowe, J. T. (1975). Leadership and performance outcomes in research organizations. *Organizational Behavior and Human Performance*, *14*, 264–280.

Benner, M. J., & Tushman, J. L. (2002). Exploitation, exploration and process management: The productivity dilemma revisited. *Academy of Management Review*, *28*, 238–256.

Boehlke, S. (2008). The politics of creativity: Four domains for inquiry and action by leaders in R&D. *Creativity and Innovation Management*, *17*(1), 77–87.

Byham, W. C., Smith, A. B., & Paese, M. J. (2002). *Grow your own leaders*. Upper Saddle River, NJ: Prentice Hall.

Byrne, C. L., Mumford, M. D., Barrett, J. D., & Vessey, W. B. (2009). Examining the leaders of creative efforts: What do they do, and what do they think about? *Creativity and Innovation Management*, *18*(4), 256–268.

Cardinal, L. B. & Hatfield, D. E. (2000). Internal knowledge generation: The research laboratory and innovative productivity in the pharmaceutical industry. *Journal of Engineering & Technology Management*, *17*, 247–272.

Chan, K. Y., & Drasgow, F. (2001). Toward a theory of individual differences and leadership: Motivation to lead. *Journal of Applied Psychology*, *86*, 481–498.

Collins, M. A., & Amabile, T. M. (1999). Motivation and creativity. In R. J. Sternberg (Ed), *Handbook of creativity* (pp. 297–312). Cambridge, England: Cambridge University Press.

Conger, J., & Fulmer, R. (2003). Developing your leadership pipeline. *Harvard Business Review*, *81*(12), 76–90.

Cooper, R. G., & Kleinschmidt, E. J. (2000). New product performance: What distinguishes the star products. *Australian Journal of Management*, *25*, 17–46.

Day, D. V., Harrison, M. M., & Halpin, S. M. (2009). An integrative approach to leader development. *Connecting adult development, identity, and expertise*. New York: Routledge.

Ellison, R. L., James, L. R., & Carron, T. (1970). Prediction of R & D performance criteria with biographical information. *Journal of Industrial Psychology*, *5*, 37–57.

Ericsson, K. A., & Charness, W. (1994). Expert performance: Its structure and acquisition. *American Psychologist*, *49*, 725–747.

European Commission: Employment, Social Affairs, and Inclusion. (2011). *European Commission for active aging and solidarity between generations*. Retrieved from http://ec.europa.eu/social/keyDocuments.jsp?type=0&policyArea=970&subCategory= 0&country=0&year=0&advSearchKey=&mode=advancedSubmit&langId=en

Feist, G. J., & Gorman, M. E. (1998). The psychology of science: Review and integration of a nascent discipline. *Review of General Psychology*, *2*, 3–47.

Garman, A. N., & Glawe, J. (2004). Succession planning. *Consulting Psychology Journal: Practice and Research*, *56*(2), 119–128.

Groves, K. S. (2007). Integrating leadership development and succession planning best practices. *Journal of Management Development, 26*(3), 239–260.

Haworth, M. (2005). Tips for better succession planning. *The Journal for Quality and Participation*, *28*(3), 13–15.

Hemlin, S. (2009). Creative knowledge environments: An interview study with group members and group leaders of university and industry R&D groups in biotechnology. *Creativity and Innovation Management, 18*(4), 278–285.

Hicks, S. (2000). Succession planning. *Training and Development*, *54*(3), 79–90.

Hoffman, T. (2005). Grooming the next generation. *Management*, *39*(21), 29–41.

Hollington, S. (2007). Have you planned . . . for your replacement? *Engineering Management*, *17*(1), 17–18.

Hunt, D. A., Revell, J., & Rotenberg, J. (2007). What US workers don't know about retirement. (McKinsey Quarterly Web exclusive) http://www.mckinseyquarterly.com/What_US_workers_dont_know_about_retirement_1898.

Hunter, S. T., Bedell, K. E., & Mumford, M. D. (2006). Dimension of creative climate: A general taxonomy. *International Journal of Creativity and Problem Solving*, *15*, 97–116.

Hunter, S. T., Bedell, K. E., & Mumford, M. D. (2007). Climate for creativity: A quantitative review. *Creativity Research Journal*, *19*, 69–90.

Hunter, S. T., Thoroughgood, C., Myers, A., & Ligon, G. S. (2011). Managing the paradoxes of leading for innovation. *Psychology of Aesthetics, Creativity and the Arts – Special issue: Innovation in Organizations*, *5*(1), 54–66.

Keller, T. (2003). Parental images as a guide to leadership sensemaking: An attachment perspective on implicit leadership theories. *The Leadership Quarterly*, *14*(2), 141–160.

Kleinsorge, R. (2010). Expanding the role of succession planning. *Training and Development*, *64*(4), 66–69.

Ligon, G. S., Graham, K., Edwards, A., Osburn, H. K., & Hunter, S. T. (2012). Managing performance for innovation. In M.D. Mumford (Ed.), *The handbook of organizational creativity* (pp. 633–666). San Diego, CA: Elsevier.

Ligon, G. S., & Hunter, S. T. (2010). Putting the development into experiential development. *Journal of Industrial and Organizational Psychology: Perspectives on Science and Practice*, *3*, 28–32.

Ligon, G. S., Wallace, J. H., & Osburn, H. K. (2011). Experiential development for leaders of innovation. *Advances in Developing Human Resources*, *13*(3), 297–318.

Lonergan, D.C., Scott, G.M., & Mumford, M.D. (2004). Evaluating aspects of creative thought: Effects of idea evaluation and revision strategies. *Creativity Research Journal*, *16*, 231–246.

Lord, R. G., & Hall, R. J. (2005). Identity, deep structure and the development of leadership skill. *Leadership Quarterly*, *16*, 591–615.

McKeown, E. (2010). Turbulent times highlight the need for succession planning. *Training and Development*, *64*(1), 18–19.

Mumford, M. D., Baughman, W. A., & Sager, C. E. (2003). Picking the right material: Cognitive processing skills and their role in creative thought. In M. A. Runco (Ed.), *Critical creative processes* (pp. 19–68). Cresskill, NJ: Hampton.

Mumford, M. D., & Gustafson, S. B. (1988). Creativity syndrome: Integration, application, and innovation. *Psychological Bulletin*, *103*, 27–43.

Mumford, M. D., Hunter, S. T., Eubanks, D. L., Bedell, K. E., & Murphy, S. T. (2007). Developing leaders for creative efforts: A domain-based approach to leadership development. *Human Resource Management Review*, *17*, 402–417.

Mumford, M. D., Lonergan, D.C., & Scott, G. (2002). Evaluating creative ideas: Processes, standards, and context. *Inquiry: Critical Thinking Across the Disciplines*, *22*, 21–30.

Mumford, M. D., Mobley, M. I., Uhlman, C. E., Reiter-Palmon, R., & Doares, L. (1991). Process analytic models of creative capacities. *Creativity Research Journal*, *4*, 91–122.

Mumford, M. D., Scott, G. M., Gaddis, B., & Strange, J. M. (2002). Leading creative people: Orchestrating expertise and relationships. *Leadership Quarterly*, *13*, 705–750.

Murphy, S. (2006). Executive development and succession planning: Qualitative evidence. *International Journal of Police Science and Management*, *8*(4), 253–265.

Myron, D. (2005). Bye-bye boomers, welcome Yers. *Customer Relationship Management*, *9*(11), 12.

Nystrom, H. (1990). Organizational Innovation. In M. S. West & J. L. Farr (Eds.) *Innovation and creativity at work: Psychological and organizational strategies* (pp. 143–162). New York, NY: Wiley.

Occupational Information Network. (2011). *Sciences managers*. Retrieved from http://www.onetonline.org/link/summary/11–9121.00/

Oracle. (2010). PeopleSoft enterprise succession planning. Retrieved from http://www.oracle.com/us/products/applications/peoplesoft-enterprise/hcm/061844.html

Paradise, A. (2010). Many barriers inhibit success of succession planning. *Training and Development*, *64*(6), 60–61.

Perry-Smith, J. (2008). When being social facilitates creativity: Social networks and creativity within organizations. In J. Zhou & S. E. Shalley (Eds.), *Handbook of organizational creativity* (pp. 189–210). New York, NY: Erlbaum.

Reiter-Palmon, R., Herman, A., E., & Yammarino, F. J. (2008). Creativity and cognitive processes: Multilevel linkages between individual and team cognition. In M. D. Mumford, S. T. Hunter, & K. E. Bedell-Avers (Eds.), *Multilevel issues in creativity and innovation: Volume VII* (pp. 203–267). Oxford, England: Elsevier.

Rothwell, W. (2010). The future of succession planning. *Training and Development, 64*(9), 50–54.

SAP. (2004). My SAP ERP human capital for the public sector. Retrieved from http://www.sap.com/canada/industries/publicsector/pdf/BWP_ERP_HCM_Public_Sector.pdf

Schneider, B., & Reichers, A. E. (1983). On the etiology of climates. *Personnel Psychology, 36*, 19–39.

Scott, G., Leritz, L. E., & Mumford, M. D. (2004). The effectiveness of creativity training: A quantitative review. *Creativity Research Journal, 16*, 361–388.

Scott, G. M., Lonergan, D.C., & Mumford, M. D. (2005). Contractual combination: Alternative knowledge structures, alternative heuristics. *Creativity Research Journal, 17*, 21–36.

Shalley, C. E., & Gilson, L. L. (2004). What leaders need to know: A review of social and contextual factors that can foster or hinder creativity. *The Leadership Quarterly, 15*(1), 33–53.

Stark, E. (2009). Lost in a time warp: How age stereotypes impact older baby boomers who still want to work. *People & Strategy, 32*(4), 58–64.

Strange, J. M., & Mumford, M. D. (2002). The origins of vision: Charismatic versus ideological leadership. *The Leadership Quarterly, 13*(4), 343–377.

Towers Perrin (2005). *The business case for workers age 50+: Planning for tomorrow's talent needs in today's competitive environment.* Washington, DC: AARP.

Tyler, J. L., & Biggs, E. (2001). *Practical governance.* Chicago, IL: ACHE Management Series/Health Administration Press.

Worklogix, Inc. (2009). Succession management: Track and develop the right talent to execute your strategy. Retrieved from http://www.worklogix.com/productlist/succeed.htm

Young, M. B., Piktialis, D., & Rappaport, A. (2007). *Gray skies, silver linings: How companies are forecasting, recruiting, and managing a mature workforce.* New York, NY: The Conference Board.

Zhou, J., & Oldham, G. R. (2001). Enhancing creative performance: Effects of expected developmental strategies and creative personality. *Journal of Creative Behavior, 35*, 151–167.

9 Leading Interdisciplinary Creative Teams

Challenges and Solutions

Roni Reiter-Palmon, Triparna de Vreede, and Gert-Jan de Vreede

The use of teams in organizations in general and interdisciplinary teams in particular is on the rise (Galbraith, 2009). The term interdisciplinary teams is used to denote teams that include team members from multiple disciplines or departments in an organization working together. The term *interdisciplinary teams* is contrasted with multidisciplinary teams. Much of the literature indicates that multidisciplinary teams include members of multiple disciplines but focuses on complementary contributions, whereas interdisciplinary teams require going beyond complementary contributions and to integrative work (Fiore, Hoffman, & Salas, 2008; Korner, 2010).

The last few decades have seen the rise of matrix structures, cross-functional teams, and interdisciplinary teams to address the complex issues that organizations face. The rapid change and adaptation required to ensure organizational success and survival have led organizations to move away from a hierarchical and static structure to a more dynamic, team-oriented, and interdisciplinary structure (Rosen et al., 2011). In addition, the problems that organizations face today are generally too complex for one individual or a team from one discipline to solve (Kozlowski & Bell, 2008).

This movement toward the use of interdisciplinary teams can be seen in a wide range of organizations with various types of teams. Research and development (R&D) teams have typically included team members from several disciplines. However, it is not uncommon for organizations now to also include individuals who will be responsible for implementing the innovation from production to marketing (Galbraith, 2009). Granting agencies such as the National Science Foundation (NSF) and the National Institutes for Health (NIH) recognize the importance of interdisciplinary teams and include references to this in their guidelines to that effect. In health care, interdisciplinary teams that include doctors from different disciplines, as well as nurses and other health-care professionals, have been introduced to improve care and to reduce medical errors and cost (A. Rosen & Callaly, 2005). Finally, crisis-management and disaster-relief efforts such as those seen in response to weather (e.g., Hurricane Katrina, one of the five deadliest hurricanes in the history of United States that killed more than 1,800 people) or man-made crises (e.g., Deep Horizon, the oil spill in the Gulf of Mexico that could not be contained for 3 months, making it the largest marine oil spill in the history of petroleum industry) have focused on the interdisciplinary nature of the management teams charged with mitigating the effects (Crichton, Lauche, & Flin, 2005).

Although many organizations utilize interdisciplinary teams, teams that focus on creativity and innovation such as R&D and scientific teams have been at the forefront of this trend (Blansett, 2003). The use of interdisciplinary teams has increased as organizations have realized the advantages of using teams for creative endeavors; however, the use of such teams is not without problems (Gooch, 2005). The purpose of this chapter is to provide an overview of the benefits and challenges of using interdisciplinary creative teams and to discuss some possible solutions to the difficulties and challenges identified. We first discuss briefly the benefits and challenges of teamwork for creativity and interdisciplinary creative teams, more specifically. The role of the leader in addressing these challenges and the facilitation of effective collaboration and teamwork in interdisciplinary creative teams are the focus on this chapter.

TEAMS AND CREATIVITY: BENEFITS AND CHALLENGES

In recent years, interest in team creativity has increased (Henessey & Amabile, 2010; Reiter-Palmon, Wigert, & de Vreede, 2011). As creativity and innovation are viewed as necessary for organizational survival, organizations are looking for ways to increase creativity and innovation (Shalley, Zhou, & Oldham, 2004; West, Hirst, Richter, & Shipton, 2004).

Early research on team creativity and innovation has focused on the comparison between teams and individuals, or teams and nominal groups, in which the results of individuals working alone are pooled together. These early investigations concluded that nominal groups typically outperform teams (Mullen, Johnson, & Salas, 1991). This somewhat surprising finding, dubbed productivity loss or process loss, led to a stream of research focusing on the factors that affect process loss and on ways to overcome them (Diehl & Stroebe, 1987). Both social and cognitive factors have been suggested to influence process loss. Individuals in teams may not speak up and offer ideas because of evaluation apprehension. Process loss can also be inherent in the process of communication between team members. When working face-to-face, team members cannot all speak at the same time. As a result, process loss may occur as team members wait to take turns. This wait can result in less time to present ideas, as well as in idea loss. The need to wait, and the resulting process loss, increases as team-size increases (de Vreede, Briggs, & Reiter-Palmer, 2010). Other factors investigated include social loafing and social pressure (Nijstad, Diehl, & Stroebe, 2003). However, these studies have focused on one type of task, that of idea generation or of brainstorming in a laboratory setting (Brophy, 2006). In addition, most

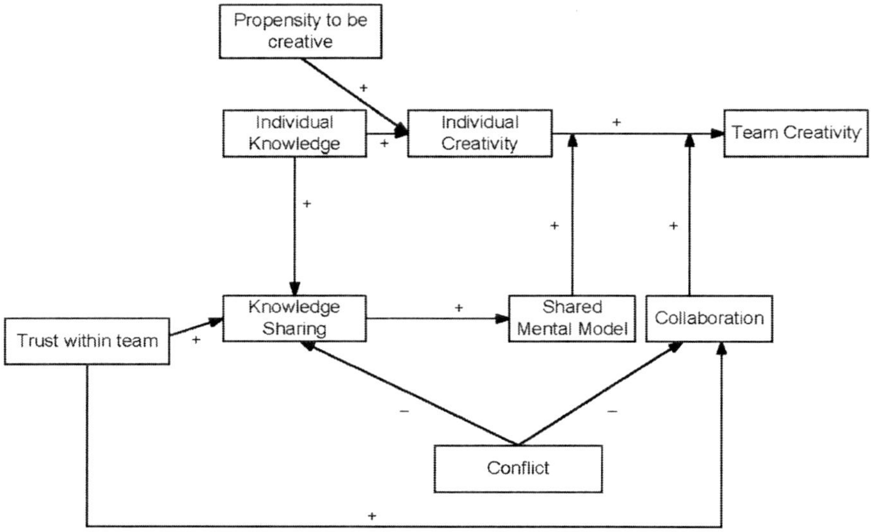

Figure 9.1 Model for creative interdisciplinary teams

of these studies have focused on one main outcome, fluency, or the number of ideas generated.

Research that has focused on teams in the field or multifaceted creative problem-solving activities paints a more complex picture regarding team creativity and innovation (Brophy, 2006; Reiter-Palmon et al., 2011). Team composition or team diversity, social processes, and team cognition have all been identified as important categories of antecedents of team creativity and innovation (Reiter-Palmon et al., 2011). Although multiple constructs within each category have been studied in relation to team creativity and innovation, in this chapter we focus on those constructs that are particularly important for interdisciplinary creative teams. We first review the research available regarding these constructs and the effect on creative teams and interdisciplinary creative teams. We then present a model (see Figure 9.1) linking the different constructs with team creative performance in interdisciplinary teams, and specifying the hypothesized relationships between these constructs. Finally, in the last section, we review the literature on leading creative teams as it relates to this model, and provide suggestions on what team leaders can do to facilitate creativity in interdisciplinary teams.

TEAM KNOWLEDGE DIVERSITY

Team composition, or team diversity, has been studied as an antecedent of team creativity and innovation (Reiter-Palmon et al., 2011). Early work that focused on demographic diversity suggests that demographic diversity does not influence team creativity and innovation (Hulsheger, Anderson, & Salgado, 2009). Another aspect of diversity that has been found to have a pronounced effect is functional diversity (Milliken, Bartel, & Kutzberg, 2003). Functional diversity refers to differences based on attributes that are relevant to job performance, such as diversity in education, function in the organization, and job-relevant knowledge, skills, and abilities (Hulsheger et al., 2009). Most research evaluating functional diversity found positive effects for functional diversity, suggesting that teams comprised of members from diverse functional backgrounds outperform homogeneous teams in terms of creativity and innovation (Fay, Borrill, Amir, Haward, & West, 2006; Keller, 2001; Shin & Zhou, 2007). Functional diversity is of particular importance to interdisciplinary teams. One reason for the creation of such teams is to capitalize on that functional diversity.

One important effect that group functional diversity has is that team members will have different knowledge and expertise. It has been suggested that these differences are what allows an interdisciplinary team to be more creative. As team members have access to different knowledge and information, and possess a variety of skills, abilities, and expertise, problems may be evaluated from multiple perspectives, and a more complete and creative solution can be developed.

Education and past experiences have a significant effect on how individuals understand problems and the approaches they take to try to solve them (Mumford, Reiter-Palmon, & Redmond, 1994). Problem construction and identification is a cognitive process in which individuals identify that there is a problem that needs to be solved, and the important parameters (such as goals, constraints, key information, and procedures) that would guide the problem-solving effort (Mumford et al., 1994). Problem construction has been found to be an important antecedent to creativity, in which individuals who are able to construct problems more creatively are also able to solve problems in a more creative fashion (Reiter-Palmon & Robinson, 2009). Problem construction and identification has been found to be influenced by past experiences (Mumford et al., 1994; Reiter-Palmon, Mumford, & Threlfall, 1998). It is therefore likely that members of functionally diverse teams will have different understanding and conceptualizations of the problem, and will adopt different approaches to try to solve the problem (Cronin & Weingart, 2007; Reiter-Palmon, Herman, & Yammarino, 2008). These various conceptualizations can be resolved in different ways, some of which may be more likely to lead to a creative problem construction and therefore a creative solution. Reiter-Palmon et al. (2008) speculated that if diverse problem constructions are integrated in such a way that the new problem construction capitalizes on the diversity of individual constructions, it is likely that the resulting solution will also be more creative. If, however, the team fails to capitalize on the diversity of problem constructions, then the team will not be able to also capitalize on the diversity of the knowledge of its members when developing a solution, thereby leading to a less creative solution.

This view was supported by Weingart, Todorova, and Cronin (2010), who studied interdisciplinary teams responsible for product development. They found that heterogeneity in constructions, that is, when team members differed in how they understood the problem, was positively related to the quality of the new product developed, but only in teams that exhibited task conflict and managed the conflict well.

The preceding discussion suggests that the benefits afforded by team functional diversity and team member creative potential are not straightforward. It is not enough to assemble teams that include creative individuals and that are diverse or interdisciplinary. Teams must be able to capitalize on these advantages. De Drue, Nijstad, Bechtoldt, and Baas (2011) have suggested that both cognitive and social processes play a critical role in team creativity and innovation. Specifically, based on the Motivated Information Processing model, they have suggested that although cognitive aspects associated with diversity may lead to increased creativity, social processes must be in place to ensure that teams can take advantage of the cognitive diversity and cognitive processes of the team.

COLLABORATION

It is not surprising that team social processes and interactions have been emphasized in all models that evaluate team effectiveness and team performance (Mathieu, Maynard, Rapp, & Gilson, 2008). One important factor for effective team performance is collaboration. Team collaboration refers to the joint effort of team members in achieving the team goal. Team collaboration has been found to be especially important for team functioning when tasks are complex, ambiguous, and dynamic, all of these being markers of creative tasks (Burke, Stagl, Salas, Pierce, & Kendall, 2006). Several studies have evaluated the relationship among team collaboration, team creativity, and innovation and have found that, when team goals emphasized collaboration, teams produced more new ideas and team innovation improved (Drach-Zehavy & Somech, 2001; Mitchell, Boyle, & Nicholas, 2009; Pearce & Ensley, 2004). Team collaboration was found to be particularly important in the early stages of the creative process (Hoegl, Weinkauf, & Gemueden, 2004).

However, collaboration can be time-consuming and requires significant effort. As a result, team members typically do not engage in this process unless they see the need, are trained to do so, or are forced by circumstances to do so (Marta, Leritz, & Mumford, 2005; Salas, Nichols, & Driskell, 2007). In many cases, when collaboration activities are performed in less than optimal ways, they increase unnecessary communication or add to the task load when the team is already overloaded (Salas et al., 2007). Research on team communication provides additional support for this notion, because it has been found that the amount of communication is negatively related to team creativity and innovation (Kratzer, Leenders, & van Engelen, 2004). Salas et al. found that team training in effective collaboration strategies was the best training strategy in terms of its effect on team performance.

Another reason why collaboration is difficult to carry out in the context of creative teams is rooted in the creative individual. Creative individuals have been identified as independent, introverted, individualistic, and achievement oriented (Barron & Harrington, 1981; Feist, 1998). In addition, creative individuals tend to be domineering, arrogant, and competitive (Mumford & Hunter, 2005; Silvia, Kaufman, Reiter-Palmon, & Wigert, 2011). These traits suggest that creative individuals may tend to work better independently, with limited supervision (George & Zhou, 2001; Oldham & Cummings, 1996). As a result, working in teams, in general, and on tasks requiring coordination and collaboration may be more difficult for creative individuals (Mumford & Hunter, 2005).

TRUST AND PSYCHOLOGICAL SAFETY

One important factor that has been found to influence collaboration in diverse teams is trust (Barczak, Lassk, & Mulki, 2010). Creative teams are characterized by high trust among team members and a sense of psychological safety (Burningham & West, 1995; Edmondson, 2004; West & Anderson, 1996). Psychological safety and trust have both been linked to team member willingness to discuss information openly and freely (Salas, Sims, & Burke, 2005). In addition, psychological safety has been found to encourage employees to take initiative, to make suggestions, and to facilitate the implementation of innovation (Burke et al., 2006; Edmondson, 2004). These findings suggest that team trust and psychological safety influence the motivation of team members to engage in behaviors that lead to creativity and innovation (Carmeli, Reiter-Palmon, & Ziv, 2010). Similarly, the research on climate for creativity has found that a climate of trust and psychological safety facilitates the creative performance of teams and individuals (Amabile & Gryskiewicz, 1989; Ekvall, 1996; Hunter, Bedell, & Mumford, 2007).

Team trust and psychological safety also influence communication patterns and the behavior of team members. Research indicates that when feeling psychologically unsafe, individuals are more likely to develop a defensive orientation and are less likely to display creativity and innovative behavior at work (Nicholson & West, 1988; West & Richter, 2008). In addition, when trust is low, disagreements and misunderstandings over ambiguous information are more likely to be interpreted in a negative way, resulting in negative responses from team members and often leading to conflict (Curseu & Schruijer, 2010; Salas et al., 2005). This may be particularly important for interdisciplinary creative teams, as diversity may lead to increased incidents of misunderstandings, disagreement, and difficulty in developing trust. Research suggests that trust is more easily developed in teams that are homogenous (Curseu & Schruijer, 2010). Furthermore, in a longitudinal study, Curseu and Schruijer have found that trust develops early in the team life, and lack of trust early in team life was predictive of team conflict later.

CONFLICT

Team conflict can be defined as either task-oriented (disagreements about what needs to be done or how to do it) or relational (with a focus on the person, personal characteristics, and personal dislike; Jehn, 1997). Early work on conflict and creativity did not distinguish between these two factors, and found that conflict is linked to lower creativity, in part as a result of reduced trust and a lower level of psychological safety (Carnevale & Probst, 1998). Others have suggested that task-related conflict can be important for enhancing creativity and innovation. Disagreement among team members on the task or on how to perform it will result in different views being discussed, multiple points of view that can be incorporated, and the resolution of differences in approaches (Kurtzberg & Amabile, 2001; Mannix & Neale, 2005). Jehn (1995) found that the effect of task conflict on team performance was dependent on the type of task. When tasks were routine, task conflict was detrimental to group functioning. In contrast, when teams performed nonroutine tasks, task conflict either had no effect or was beneficial for group functioning. Interviews conducted as part of the study revealed that when task conflict in groups performing nonroutine tasks was discussed openly, resulting in a reevaluation of the tasks, alternative ways were found to achieve team goals, and more creative ideas resulted.

However, other work has suggested that task conflict may not facilitate team creativity and innovation. Mortensen and Hinds (2001), in a study of product development teams, found that both relationship and task conflict were negatively related to creative performance. Furthermore, a study by De Dreu (2006) examining the impact of task conflict on creativity indicated that there was a curvilinear inverted-U relationship between task conflict and creativity in postal service teams. A moderate level of task conflict facilitated creative performance, but too much or too little task conflict stifled creativity. These results may be explained by the fact that task and relationship conflicts are not always easy to distinguish. Criticism about a suggested approach may be viewed as a personal criticism by the person who made the suggestion. Indeed, Jehn and Mannix (2001) found that task conflict often breeds interpersonal conflict.

Other work on conflict suggests that not only does the type of conflict matter, but also when the conflict occurs in the life of the project. In a study based on R&D teams, Kratzer, Leenders, and van Engelen (2006) found that team conflict had an overall negative relationship with team creativity. Further investigation found that the relationship between team conflict and team creativity differed depending on the phase of the project. During project conceptualization, an inverted-U shaped relationship was found, suggesting that moderate levels of conflict were beneficial for creativity. However, during later phases of the team creative process, conflict had a negative effect on creativity. The results of the studies presented earlier paint a complex picture of the relationship between conflict and team creativity and innovation, especially regarding task conflict. Although task conflict may be beneficial under some circumstances, it is not always helpful, and may even hinder team creativity and innovation.

In creative teams, especially interdisciplinary teams, there are many opportunities for conflict and disagreements to develop. Creative individuals, as indicated, tend to be independent and highly intelligent and to have strong personalities and opinions (Barron & Harrington, 1981; Silvia et al., 2011). Therefore, these individuals will be more likely to voice their opinion and be willing to disagree with members of their team. When teams include members from different disciplines, the differences due to the educational background, experiences, and knowledge can result in additional conflict. Not only would team members develop ideas and solutions that are based on their own educational background and knowledge; communicating these ideas and solutions can also be difficult as the vocabulary of the different disciplines may differ (Cronin & Weingart, 2007).

Trust, psychological safety, and conflict have also been shown to influence knowledge sharing in a team environment (Edmondson, 1999; Rank, Pace, & Frese, 2004). The concept of knowledge sharing is of particular importance for interdisciplinary creative teams. Interdisciplinary teams are typically assembled when organizations believe that no one individual possess the knowledge required to solve a problem, and therefore a team of experts from different disciplines is required (Bantel & Jackson, 1989; Cummings, 2004; West et al., 2004). In addition, expertise and domain knowledge have been found to be important contributors to individual creativity (Ericcson, 1999; Vincent, Decker, & Mumford, 2002). As a result, it is not surprising that knowledge sharing has emerged as an important variable for team creativity.

KNOWLEDGE SHARING

Knowledge sharing has been positively related to team creativity and innovation (Damanpour, 1991; Howell & Shae, 2006). Knowledge sharing provides a team with information that was not otherwise available, increasing the likelihood that the ideas generated are more creative. Knowledge or information sharing can take place both with members outside the group (external knowledge sharing) and within the group (internal knowledge sharing). External knowledge sharing focuses on the role of team members in obtaining knowledge that is not available to the team internally. It has been suggested that interdisciplinary teams will have access to more diverse knowledge resources and, as a result, access to more information (Ancona & Cladwell, 1992). Internal knowledge sharing focuses on sharing of knowledge within the team, such that team members inform each other.

Studies that have evaluated the role of external knowledge sharing directly, have found that it influences team creativity (Ancona & Cladwell, 1992; Howell & Shae, 2006). Further, the importance of knowledge sharing with external sources has been investigated in studies of weak ties. Studies by Perry-Smith and Shalley (2003) as well as by Baer (2010) found that network ties with members outside the group are related to increased creativity and innovation among the team members. It is expected that interdisciplinary teams will have a wider and more diverse external network, resulting in increased access to knowledge.

Focusing on the role of knowledge-sharing within the team, meta-analyses by Damanpour (1991) and Hulsheger et al. (2009) found that internal communication was positively related to creativity and innovation. De Dreu and West (2001) suggested that, when diverse groups (such as interdisciplinary teams) share information and task-relevant knowledge, they have a higher opportunity to integrate their possibly conflicting viewpoints and to generate more creative ideas. Similarly, Drach-Zehavy and Somech (2001) found that sharing information was positively related to team innovation.

Research on information sharing, which is closely related to knowledge sharing, typically uses the hidden profile paradigm, which identifies the conditions under which team members will be more likely to share information. These studies suggest that shared information, in other words, information that is available to all team members prior to team discussion, is more likely to be discussed, whereas information that is not available to all or most team members is less likely to be discussed (Mesmer-Magnus & DeChurch, 2009). Furthermore, a study on the effect of functional diversity on team performance in business unit management teams (Bunderson & Sutcliffe, 2002) found that having a degree of overlap in experiences and functions between team members facilitated information sharing and team performance. Bunderson and Sutcliffe argued that the overlap in information provides a common ground that facilitates information sharing by allowing for easier communication.

Teams that are specifically asked to discuss information prior to reaching a decision are more likely to discuss previously unshared information (Stewart & Stasser, 1995). In addition, social processes that promote open discussion and cooperation allow team members to overcome the tendency to share only information that is already known to all team members (Mesmer-Magnus & DeChurch, 2009). Finally, group members that were designated as experts tended to discuss more information, including previously unshared information, especially when the group was aware of their expert status (Franz & Larson, 2002; Stasser & Stewart, 1992). It has been suggested that one reason for this finding is that experts may feel more confident in speaking up in the group setting and when group members know that someone is an expert they are more likely to seek his or her input. It is possible that one reason members of interdisciplinary teams are able to communicate diverse information is because all members are believed to have their own area of expertise and, therefore, are more likely to communicate unshared information.

However, it must be noted that the nature of the knowledge sharing, whether collaborative or contentious, is critical (Lovelace, Shapiro, & Weingart, 2001). The results of Lovelace et al.'s study indicated that collaborative communication—that is, communication that is positive and intended to find mutually beneficial solutions—was related to higher creativity and innovation. On the other hand, contentious communication—that is communication that is less positive and focuses on win-lose situations—was related to lower creativity and innovation. Further, the effect of contentious communication was found to be particularly detrimental when frequency of information exchange is high. Supporting these results, Kratzer, Leenders, and van Engelen (2004) found that frequent information exchange resulted in reduced creativity and innovation as it led to cognitive overload and production blocking. However, other studies have shown that these drawbacks can be overcome by the use of electronic meeting technologies (e.g., Santanen, Briggs, & de Vreede, 2004; Valacich, Dennis, & Connolly, 1994). These results regarding internal knowledge sharing suggest that although overall knowledge sharing is important for the creative performance of teams, the conditions in which such knowledge exchange takes place cannot be ignored.

SHARED MENTAL MODELS

In addition to the contribution of knowledge sharing to team creativity directly, knowledge sharing may also be important as it contributes to the development of shared mental models (SMMs). SMMs are representations of knowledge structures that are shared among team members (Cannon-Bowers, Salas, & Converse, 1993; Klimoski & Mohammed, 1994). These shared structures enable the team members to find a common ground to describe, explain, and predict the events that occur in their environment (Mathieu, Goodwin, Heffner, Salas, & Cannon-Bowers, 2000). SMMs can be divided primarily into four areas: (a) knowledge about equipment and tools; (b) knowledge about the team task, goal, and performance requirements; (c) knowledge about other team members' abilities, knowledge, and skills; and (d) knowledge about appropriate team interactions (Cannon-Bowers et al., 1993).

The majority of research on SMMs has focused on one of the preceding dimensions or has consolidated them into two dimensions, namely, task and team shared mental models. Research shows a strong relationship between team dynamics and shared mental models. For example, many studies show a positive relationship between team effectiveness and shared mental models (Marks, Sabella, Burke, & Zacarro, 2002; Mathieu et al., 2000; Salas, Cooke, & Rosen, 2008). Other research shows that shared mental models play an important role in team adaptability (Burke et al., 2006) and team performance (Cannon-Bowers et al., 1993). The presence of shared mental models allow team members to understand each other's needs faster, identify performance gaps, and engage in effective forms of communication. In addition, SMMs have a positive influence on social processes like coordination, communication, and backup support, allowing teams to be adaptable and flexible to change.

Even though the research on SMMs and team creativity and innovation is limited, existing empirical studies suggest that SMMs have a positive relationship with creativity both at the individual and team level. For example, Mumford, Feldman, Hein, and Nagao (2001) found that when shared mental models are developed through sharing the same training program, they facilitate higher levels of creative idea generation within the team. Similarly, Pearce and Ensley (2004) found shared vision to be positively related to product and process innovation ideas in teams. In a related study, Gilson and Shalley (2004) found that increasing the number of shared goals in a team not only resulted in an increase in the creativity but also increased the engagement of the team members in the creative process.

However, it has also been noted that the presence of SMMs may not always result in increased creativity in teams. In fact, some researchers have argued that SMMs may lead to too much similarity and conformity, which would in turn, may result in less creativity among team members (Cannon-Bowers et al., 1993). This is illustrated in a study by Kellermanns, Floyd, Pearson, and Spencer (2008). Their results indicated an interaction between constructive confrontation norms and shared mental models in decision-making teams. Specifically, they found that reducing the similarity of SMMs resulted in higher quality decisions in teams that had norms supporting discussion of disagreement and constructive confrontation. In addition, it was found that a high degree of SMM contributed to decision quality only when the teams were not able to share information openly. This led Kellermanns et al. to hypothesize that teams that are able to effectively communicate and exchange information due to the presence of appropriate norms can take advantage of diversity (where the sharedness of the mental models are low). Thus, it seems likely that shared mental models influence team creativity positively under some circumstances but negatively in others.

While overall it appears that SMMs, especially those that include models of team functioning that facilitate discussion and constructive confrontation, are helpful for creativity, the development of SMMs is less well documented. McComb (2007) suggested that SMM development takes time and involves a continuous process occurring throughout the life the team.

Mental models are continuously revised based on new information resulting from team interactions and feedback. The development of SMMs may be facilitated when teams have common information and a common language to describe events, tasks, and behavior (McComb). Those are more likely to occur in teams that are homogeneous. As a result, diverse teams, such as interdisciplinary creative teams, may have difficulty especially in the early stages of SMM development and convergence. It is also possible that because of limited shared knowledge, difficulty in communication across disciplines, or different perceptions of goals or tasks, interdisciplinary teams will not be able to achieve a degree of sharedness that is necessary for effective functioning.

SUMMARY

In this section, we have described various factors that benefit team creativity and the unique challenges facing interdisciplinary creative teams. The model presented in Figure 9.1 depicts the constructs that are important for interdisciplinary creative teams, and provides suggestions for how these constructs are interrelated and lead to creative performance.

Teams offer significant benefits for problem solving, in general, and for creative problem solving, specifically. Teams offer the benefit of multiple perspectives, access to additional and varied knowledge, and the ability to share the work. Interdisciplinary teams offer the added benefits of more variety in perspectives and knowledge that can bring about a truly creative idea. However, creative teams also suffer from distinctive drawbacks. Working in a team requires trust so that team members can share knowledge. The type of knowledge that is most likely to be important for interdisciplinary creative teams is the knowledge that is unique to each team member based on his or her area of expertise. The benefit of an interdisciplinary creative team is realized only when that knowledge is shared. However, research indicates that the process of sharing knowledge that is not common is difficult. In addition to trust, conflict may influence knowledge sharing and make sharing more difficult. Creative teams are more likely to suffer from conflict because creative individuals are more likely to be independent, arrogant, competitive, and domineering (Feist, 1998; Mumford & Hunter, 2005; Silvia et al., 2011). Further, interdisciplinary creative teams are more likely to include experts in different areas, and therefore these individuals will be less likely to be able to understand the point of view of others in the team (Jassawalla & Sahittal, 2006). However, research suggests that experts are more likely to express their opinions in a team (Stasser & Stewart, 1995).

Conflict can also have a direct effect on collaboration between team members. As indicated, collaboration is critical for team functioning. However, it is time-consuming and difficult. Because creative individuals are likely to have a working style that favors individual and independent work, working collaboratively is often difficult. In addition, collaboration is made easier by having a common language, a similar understanding of the tasks and goal of the team, and effective communication, yet interdisciplinary creative teams are less likely to have the common understandings that would facilitate effective communication (Cronin & Weingart, 2007; Jassawalla & Sashittal, 2006). Similarly, SMMs develop more quickly and effectively in homogeneous groups, suggesting that interdisciplinary teams will take longer or may not even develop SMMs.

Although the obstacles described previously are all challenging, they are not impossible. Interdisciplinary creative teams are becoming more common, and researchers must identify possible solutions to mitigate some of the difficulties associated with teamwork and specifically interdisciplinary creative teamwork. In the next section, we discuss what leaders can do to facilitate effective interdisciplinary creative teamwork.

LEADING CREATIVE INTERDISCIPLINARY TEAMS

One of the main reasons for the use of interdisciplinary teams for creative work such as science or R&D is the realization that the problems that we are attempting to solve are complex, and that the creative solution would require knowledge and expertise from multiple disciplines. Interdisciplinary creative teams bring unique advantages stemming from the diversity of knowledge of team members. However, in order to take advantage of these benefits, teams must overcome significant obstacles. Some of these barriers are common to all creative teams, but there are also unique obstacles that are associated with the interdisciplinary nature of the team. One way in which teams can overcome these obstacles is through the effective facilitation of the team leader. The team leader is in a unique position to help the team overcome these barriers and allow the team to capitalize on the advantages that are afforded by an interdisciplinary nature of the team. The model presented and the discussion in the previous section provides a guide for the discussion of what leaders can and should do to enhance creativity in interdisciplinary teams and to address the unique challenges that face such teams.

Leadership has been identified as a critical variable that influences creativity in individuals, teams and organizations (Shalley & Gilson, 2004; Stenmark, Shipman, & Mumford, 2011). Research suggests that leaders are in a unique position to support creativity through various mechanisms such as modeling creativity (Jaussi & Dionne, 2003), motivating employees (Amabile, 1983), providing goals and structure for the work (Reiter-Palmon & Illies, 2004), creating an environment that is conducive to creativity (Carmeli et al., 2010), and by serving as the team representative to the organization and outside the organization (Robledo, Peterson, & Mumford, 2010). However, the interest in leadership as a way to facilitate creative performance is relatively recent (Mumford & Hunter, 2005). Further, much of this research has viewed the leader as relatively passive. Finally, much of the work on the effect of leadership has focused on creativity at the individual level as opposed to the team level (Reiter-Palmon et al., 2011).

In this section, we provide some suggestions as to what leaders can do to overcome some of the obstacles facing interdisciplinary creative teams. The summary of important findings regarding each variable from the model depicted in Figure 9.1, as well as recommendations, are presented in Table 9.1. These recommendations are mainly based on research focusing on leaders and individual creativity or team processes that facilitate creative performance. The leadership recommendations can also be viewed as propositions for future research as only limited work has directly evaluated how leaders can facilitate creative performance in interdisciplinary teams.

One important issue that leaders must address is getting team members to share information and knowledge. Without this sharing, the advantage of the interdisciplinary team disappears. Knowledge sharing does not happen automatically, and team leaders can play an important role in facilitating the knowledge-sharing process. Specifically, studies have shown that empowering leadership, participative leadership, and leader consideration were all related to knowledge-sharing (de Vries, Bakker-Pieper, & Oostenveld, 2010; Locke, Alavi, & Wagner, 1997; Srivastava, Bartol, & Locke, 2006). These types of leadership styles encourage knowledge sharing by acknowledging the importance of the subordinate and the information he or she has and by asking team members to voice their opinions. Carmeli, Reiter-Palmon, and Ziv (2010) found that leaders who are attentive to employee needs, are open, and are accessible, foster employee creativity. Shin and Zhou (2007) showed that interdisciplinary teams were most creative when they had transformational leaders that facilitated effective knowledge sharing. The results of these studies suggest that leader behaviors that encourage knowledge sharing are critical for creativity. Specifically, those leader behaviors that are related to encouraging open communication, allowing employees to voice their opinions, and attending to employee needs are important.

Further, interdisciplinary team members may have difficulty in communication due to the differing problem constructions or representational gaps (Cronin & Weingart, 2007; Reiter-Palmon et al., 2008). As these are based on past experience and education, these differences may lead to misunderstanding and less communication, especially with individuals who are from different disciplines. In addition, experts within a team may find communication with similar others easier, forming subgroups within a team. Kratzer, Leenders and van Engelen (2004) have found that formation of subgroups in teams was inversely related to creativity and innovation within those teams. Leaders have an important role in reducing misunderstandings that may result from disciplinary differences. One important way in which leaders facilitate creativity in teams is by providing direction and goals for team members (Mumford & Hunter, 2005). In this way, leaders provide

Table 9.1 Recommendations for Leadership in Creative Interdisciplinary Teams

Term	*Effect*	*Leadership recommendation*
Diversity	Demographic diversity has little to no effect on team creativity and innovation. Functional diversity shows a positive relationship with team creativity and innovation.	Task-oriented leadership decreases the relationship between team-value diversity and team conflict, whereas person-oriented leadership increases this relationship. Leaders should focus team members on work-related issues and minimize the negative effects of personal differences through the acceptance of diversity.
Problem construction	More diverse PCs will result in more creative solutions if teams can use the diversity to construct a more comprehensive and cohesive problem construction.	Leaders should encourage limited task conflict and positive communication that would lead to integration of different problem constructions.
Creativity traits	Creative individuals are less likely to collaborate and work effectively in teams, resulting in less team creativity and innovation.	Leaders should channel information sharing to involve all members of the team in a way that is functional and beneficial as creative disciplinary experts are more likely to be assertive.
Team conflict—task or relational	Relational conflict has a negative effect on creativity. The relationship between task conflict and team creativity is more complex. Moderate levels of task conflict, especially early in project life, are beneficial. Too much conflict or too little conflict may be linked to lower creativity.	Leader behaviors focusing on facilitating open and honest communication and seeking to solve disagreements in positive ways such as consensus and inclusiveness are related to less relational conflict.Leaders should maintain an open, positive climate; manage conflict constructively; create environment where participants feel secure and free to communicate work-related issues; and not permit personal attacks or discussions of personal issues.

Collaboration	Effective collaboration has been found to facilitate team creativity and innovation.	Leaders should focus 011 creating effective communication opportunities; be a source of information; prevent forming disciplinary narrow subgroups; be boundary spanners Set collaboration norms; structure work such that it is more interdependent, stimulating and encourages closer collaboration.
Trust and psychological safety	Trust and psychological safety are related to team member willingness to share information and speak up, and reduced conflict.	Leaders should create an environment in which individuals feel confident that their opinions are heard, are willing to speak up even if their opinion is not popular or is the minority in the team, and be willing to suggest new ideas in the team is critical. Also, leaders who take steps to prevent relationship conflicts from occurring also promote trust and psychological safety among their team members.
Knowledge sharing	Information sharing among team members is related to team creativity and innovation. However, team members do not always share information, especially unique information.	Leadership styles like empowering leadership, participative leadership, and leader consideration have been found to encourage knowledge sharing. These leaders acknowledge the importance of subordinates and the information they have and ask team members to voice their opinion.Leaders need to model appropriate communication and knowledge sharing and facilitate knowledge sharing in a way that will not create conflict.
Shared mental models	SMMs are positively related to team creativity. However, SMM can also reduce creativity due to too much similarity and conformity.	Leaders can facilitate the development of SMM is through the process of team reflection. Participative leadership also encourages discussion and integration of information, leading to development of SMM.

a common ground and common language for team members. (Mumford, Connelly, & Gaddis, 2003; Sosik, Avolio, & Kahai, 1998).

Lovelace, Shapiro, and Weingart (2001) demonstrated that the nature of the communication within a team, whether collaborative or contentious, was critical. Collaborative communication was related to higher creativity and innovation, whereas contentious communication was related to lower creativity and innovation. Effective communication is therefore essential for effective knowledge sharing. Leaders have an important role in facilitating effective communication. Leaders themselves are a source of information for the team (Salas et al., 2005). Furthermore, leaders can provide more accurate and comprehensive information that would facilitate team adaptation to changes in the environment, or in the case of creative teams, can result in a better reflection of the information from all team members (Marks, Mathieu, & Zaccaro, 2001). In addition, leaders serve in a boundary-spanning role, that is, the link between the team and external sources of information in the organization. Communication with external sources has been found to be one of the better predictors of team creativity and innovation (Ancona & Cladwell, 1992; Hulsherger et al., 2009). Finally, leaders set expectations for appropriate interaction and communication patterns and norms (Salas et al., 2005)

However, it is important to remember that creative individuals may not require extra attention and encouragement to voice their opinions and share their knowledge. Creative individuals have been described as opinionated and confident, suggesting that team members will not have difficulty sharing information and voicing their opinions. Furthermore, in interdisciplinary teams, these individuals serve as the expert in a specific discipline. Studies evaluating knowledge-sharing find that experts are more likely to share information with the team (Franz & Larson, 2002; Stasser & Stewart, 1995). As a result, it is possible that in interdisciplinary creative teams, the difficulty that the team and the team leader encounter is not one of lack of sharing, but rather one of how to share information and voice opinions in a way that is effective and useful to the team members. As a result, it is important that leaders facilitate effective communication between team members. Leaders can do that by setting expectations for appropriate interaction and communication patterns and norms (Salas et al., 2005). Further, leaders are in the best position to intervene when communication between team members becomes difficult or contentious and to direct the communication in more positive channels.

As indicated by the model presented in the previous section (see Figure 9.1), trust and conflict are two important antecedents of information sharing. Trust and conflict may influence not only the degree of knowledge sharing—that is, how willing team members are to share the knowledge they have—but also how the knowledge is viewed and integrated by other team members. In interdisciplinary creative teams, conflict and low trust may be a more likely outcome of interactions as a result of the team makeup. As a result of the personalities of creative individuals, creative teams may suffer from less than optimal communication and interactions. Further exacerbating the difficult interaction pattern, interdisciplinary teams may suffer from misunderstandings and lack of clarity in communication due to the interdisciplinary nature of the team. These less than optimal and even difficult interaction patterns may result in increased conflict.

Conflicts can have detrimental effects on teams on various levels (Jehn & Bendersky, 2003). These effects can range from distracting team members from their tasks to undermining the team's ability to work together. Thus, maintaining a positive climate within the team and reducing conflict also constitutes an important aspect of the leadership role (Bass, 1990). Effective leadership involves managing the team such that conflict does not have a negative impact on team outcomes. Conflict management through effective leadership is even more relevant when teams are diverse in nature (Webber, 2002).

However, not all conflict is necessarily detrimental to team creativity. Task-related conflict has been suggested to be positively related to team outcomes, especially creativity. Task conflict allows team members to fully discuss differences in view points, to integrate them, and to understand underlying causes for these (Amason, Thompson, Hochwarter, & Harrison, 1995; Kurtzberg & Amabile, 2001; Mannix & Neale, 2005). However, others have argued that task conflict is not always beneficial. Jehn and Mannix (2001) suggested that task conflict breeds interpersonal conflict. In other words, high levels of task conflict may increase interpersonal conflict, which in turn, hinders creative performance. This indicates that how conflict is managed is particularly important.

Leaders can a have strong influence on team conflict by managing conflict and facilitating a positive tone of discussion (Wakefield, Leidner, & Garrison, 2008). Strategies that focus on integration of differences have been related to positive effects of conflict, whereas avoidance strategies have been linked to negative effects (Song, Dyer & Thieme, 2006). Similarly, Lovelace et al. (2001) indicated that open communication such that team members feel secure in voicing their opinions and disagreements results in positive effects for conflict. Finally, Weingart, Todorova, and Cronin (2010) have found that how conflict was managed could enlarge or reduce the divergence in how problems and issues are understood by diverse team members. When cooperative approaches to solve the conflict were used, team members developed problem constructions that took into account multiple points of view and were more complex. This in turn resulted in more innovative product development. On the other hand, task conflict can also result in divergent problem constructions that are not integrated and, therefore, result in less innovation.

Therefore, it is not surprising that leader behaviors focusing on facilitating open and honest communication and seeking to solve disagreements in positive ways such as consensus and inclusiveness were related to less relational conflict (Wakefield et al., 2008). Leaders can manage conflict effectively by creating an environment in which individuals feel secure and free to communicate and where negative forms of communication such as criticism that is not constructive or personal attacks are not permitted.

Contrary results were obtained in a recent study by Klein, Knight, Ziegert, Lim, and Saltz (2011), who found that task-oriented leadership attenuated the relationship between team diversity and team conflict, whereas person-oriented leadership increased this relationship. Specifically, Klein et al. speculated that the person-oriented leader would allow all team members to express their opinions and values equally. However, the expression of non-work-related values that were different between team members resulted in more conflict. Klein et al. did not differentiate between task- and relationship-oriented conflict. For leaders leading interdisciplinary creative teams, this finding is of particular importance. Although many studies suggest that encouraging individuals to speak up and creating an atmosphere of openness and acceptance of different opinions leads to increased creativity, the study by Klein et al. indicates that not all person-focused leadership has this effect. It is important to differentiate between allowing team members to openly and honestly discuss work-related issues as opposed to personal and non-work-related issues.

Creating an environment in which individuals feel confident that their opinions will be heard, are willing to speak up even if their opinion is not popular or are in a minority in the team, and being willing to suggest new ideas in the team, is critical. All these are characteristics of environments in which trust and the level of psychological safety are high. Trust and psychological safety are typically seen through norms for innovation, tolerance for risk and failure when innovation is not successful, and openness and willingness to try new ideas (West & Anderson, 1996). Research indicates that these norms are related to team creativity (Griffin, Neal, & Parker, 2007; Mathisen et al., 2008; Scott & Bruce, 1994). Furthermore, Hunter, Bedell, and Mumford (2007), in a meta-analysis of climate dimensions, identified positive collegial relationships as one of the best predictors of creativity and innovation.

As we have argued, creative individuals, and experts in particular, will be less likely to have difficulties or to be uncomfortable in expressing themselves, even under less than optimal conditions (for example, low psychological safety or high conflict). However, in this case, the role of the leader is not only to ensure that optimal conditions exist for individual team members to express themselves, but also to manage how these views are expressed and understood. By creating a climate that is supportive and safe, it is likely that discussions will be more productive. That is, team members are more likely to focus on the work-related issues and differences, are more likely to try to reach an understanding, work more collaboratively, and are less likely to engage in destructive communication and conflict. In the case of interdisciplinary teams, the emphasis of the leader's role shifts from encouraging expression of diverse views to managing how these views are expressed and how they are perceived.

As previously discussed, SMMs have been found to be effective predictors of team functioning and effectiveness, although the relationship with creativity and innovation is less conclusive (Cannon-Bowers et al., 1993; Kellermanns, et al. 2008; Pearce & Ensley, 2004). SMMs are not static, and are developed over time, through discussion of issues, shared knowledge, and learning from past mistakes and successes (Burke et al., 2006; Marks et al., 2002; West & Anderson, 1996). However, some research suggests that there is also the possibility that SMM in diverse teams will not develop over time, but rather teams will exhibit increased differentiation (Levesque, Wilson, & Wholey, 2001). Although we have already discussed how leaders can facilitate knowledge sharing, discussion, and effective communication, all of which will have an effect on SMMs, the direct relationship between leadership and SMM development is less clear (Dionne, Sayama, Hao, & Bush, 2010). Dionne et al. suggest that participative leadership in interdisciplinary teams, where leaders promote discussion and integration of information, is more likely to lead to the development of SMMs. Furthermore, this integration may lead to a more complete picture of the problem at hand, which would in turn lead to better team decision making and increased creativity.

Another way in which leaders can facilitate the development of SMMs is through the process of team reflection (West, 1996). Team reflexivity allows team members to evaluate past performance and to learn from it by engaging in discussion of strategies, approaches, specific tasks, and team performance. Research has shown that teams that engage in reflection are more creative (De Dreu, 2002; Muller, Herbig, & Petrovic, 2009; Tjosvold, Tang, & West, 2004) and that team reflexivity is particularly effective in diverse teams (De Dreu, 2002). However, teams typically do not engage in this process because it is time-consuming and can be emotionally difficult to team members. Moreover, the immediate benefit of reflection on what has happened is often viewed as negligible compared to a focus on future performance. Here, leaders can serve in a facilitator role by encouraging reflection and allowing time for reflection. In addition, leaders can ensure team reflection by modeling reflection and building team reflection into the normal routine of the team such as during team meetings.

Finally, collaboration is an important contributor to interdisciplinary team creativity. In interdisciplinary creative teams, collaboration, like other social processes, may not be easy to achieve due to the nature of creative individuals and the diversity of the team. Trust, as the model presented in Figure 9.1 suggests, is important in developing team collaboration. In addition, leaders who emphasize the importance of collaboration and facilitate the development of collaboration norms in a team are more likely to have a team that engages in collaborative behaviors (Mitchell, Boyle, & Nicholas, 2009; Taggar & Ellis, 2007). Collaboration norms may emphasize collaboration as an important aspect of team functioning, seeing collaboration as a team goal, and may be facilitated by formal aspects such as team rewards, as opposed to individual rewards (Mitchell et al., 2009). Finally, leaders may be able to facilitate collaboration by structuring the work in such a way that is more interdependent. Interdependent work will tend to force individual team members to work more closely together in order to accomplish the work (Jassawalla & Sahittal, 2006).

CONCLUSION

Interdisciplinary teams, including interdisciplinary creative teams, are becoming a common feature in many organizations today (Alves, Marques, Saur, & Marques, 2007). However, these teams face unique challenges. This chapter has outlined some of the barriers to creative performance that creative interdisciplinary teams face and has provided suggestions as to how leaders may address these challenges. Figure 9.1 provides a model that presents the important constructs that predict creativity in interdisciplinary teams, and the relationships between them. Table 9.1 provides a summary of the main factors or constructs discussed, the effect these factors have, and recommendations for leaders. Further, as the research on the relationship between leadership and creativity of interdisciplinary teams is limited, these recommendations can also serve as suggestions for future research.

One common theme that has emerged is that interdisciplinary creative teams must resolve an important paradox. Creative individuals by their very nature may not be willing or may lack the social skills that are required to work effectively in teams. On the other hand, interdisciplinary creative teams depend on the collaborative work of these creative individuals. Leadership is suggested as one possible way in which this paradox may be resolved. Specifically, leaders must find ways to reconcile the tension between the individualistic nature of the creative individual and the need for cooperation, collaboration, knowledge sharing, and the social processes required for creative team work (Hunter, Thoroughgood, Ligon, & Myer, 2011). Leaders may do this by the way they lead creative individuals in interdisciplinary creative teams.

Specifically, using more participative leadership techniques seems to facilitate the social processes required for effective teamwork. In addition, effective communication appears to be an important aspect that allows for constructive disagreements, task conflict, and open discussions and allows everyone to voice an opinion, with minimal negative effects on team creativity and innovation. There is a fine line between constructive criticism and negative communication. Leaders must provide their teams with guidance as to how to carry this type of communication most effectively.

REFERENCES

Alves, J., Marques, M. J., Saur, I., & Marques, P. (2007). Creativity and innovation management through multidisciplinary and multisectoral cooperation. *Creativity and Innovation Management*, *16*, 27–34.

Amabile, T. M. (1983). *The social psychology of creativity.* New York, NY: Springer-Verlag.

Amabile, T. M., & Gryskiewicz, N. (1989). The creative environment scales: Work environment inventory. *Creativity Research Journal*, *2*, 231–253.

Amason, A. C., Thompson, K. R., Hochwarter, W. A., & Harrison, A. W. (1995). Conflict: An important dimension in successful management teams. *Organizational Dynamics*, *24*, 20–35.

Ancona, D., & Caldwell, D. (1992). Demography and design: Predictors of new product team performance. *Organization Science*, *3*, 321–341. doi:10.1287/orsc.3.3.321

Baer, M. (2010). The strength-of-weak-ties perspective on creativity: A comprehensive examination and extension. *Journal of Applied Psychology*, *95*, 592–601. doi:10.1037/a0018761

Bantel, K. A., & Jackson, S. E. (1989). Top management and innovations in banking: Does the composition of the top team make a difference? *Strategic Management Journal*, *10*, 107–124.

Barczak, G., Lassk, F., & Mulki J. (2010). Antecedents of team creativity: An examination of team emotional intelligence, team trust and collaborative culture. *Creativity and Innovation Management*, *19*, 332–345.

Barron, F. B. & Harrington, D. M. (1981). Creativity, intelligence, and personality. *Annual Review of Psychology*, *32*, 439–476.

Bass, B. M. (1990). *Bass and Stogdill's handbook of leadership: Theory, research, and applications* (3rd ed.). New York, NY: Free Press.

Blansett, K. (2003). Creativity in science teams. In M. Beyerlein (Ed.), *Advances in interdisciplinary studies of work teams* (pp. 215–232). Bingley, England: Emerald.

Brophy D. R. (2006). A comparison of individual and group efforts to creatively solve contrasting types of problems. *Creativity Research Journal*, *18*, 293–315.

Bunderson, J. S., & Sutcliffe, K. M. (2002). Comparing alternative conceptualizations of functional diversity in management teams: Process and performance effects. *Academy of Management Journal*, *45*, 875–893.

Burke, C., Stagl, K., Salas, E., Pierce, L., & Kendall, D. (2006). Understanding team adaptation: A conceptual analysis and model. *Journal of Applied Psychology*, *91*, 1189–1207. doi:10.1037/0021-9010.91.6.1189

Burningham, C., & West, M. (1995). Individual, climate, and group interaction processes as predictors of work team innovation. *Small Group Research*, *26*, 106–117. doi:10.1177/1046496495261006

Cannon-Bowers, J., Salas, E., & Converse, S. (1993). Shared mental models in expert team decision making. In N.J. Castellan, Jr. (Ed.), *Individual and group decision making: Current issues* (pp. 221–246). Hillsdale, NJ: Erlbaum.

Carmeli, A., Reiter-Palmon, R., & Ziv, E. (2010). Inclusive leadership and employee creativity in the workplace: The mediating role of psychological safety. *Creativity Research Journal*, *22*, 250–260.

Carnevale, P., & Probst, T. (1998). Social values and social conflict in creative problem solving and categorization. *Journal of Personality and Social Psychology*, *74*, 1300–1309. doi:10.1037/0022-3514.74.5.1300

Crichton, M.T., Lauche, K. & Flin, R. (2005). Incident command skills in the management of an oil industry drilling incident: A case study. *Journal of Contingencies and Crisis Management*, *13*, 116–128.

Cronin, M.A., & Weingart, L.R. (2007). Representational gaps, information processing, and conflict in functionally diverse teams. *Academy of Management Review*, *32*, 761–774.

Cummings, J. (2004). Workgroups, structural diversity, and knowledge sharing in a global organization. *Management Science*, *50*, 352–364

Curseu, P.L., & Schruijer, S.G.L., (2010). Does conflict shatter trust or does trust obliterate conflict? Revisiting the relationships between team diversity, conflict, and trust. *Group Dynamics: Theory, Research, and Practice*, *14*, 66–79.

Damanpour, F. (1991). Organizational innovation: A meta-analysis of effects of determinants and moderators. *Academy of Management Journal*, *34*, 555–590. doi:10.2307/256406

De Dreu, C. (2006). When too little or too much hurts: Evidence for a curvilinear relationship between task conflict and innovation in teams. *Journal of Management*, *32*, 83–107. doi:10.1177/0149206305277795

De Dreu, C., Nijstad, B., Bechtoldt, M., & Baas, M. (2011). Group creativity and innovation: A motivated information processing perspective. *Psychology of Aesthetics, Creativity and the Arts*, *5*, 81–89

De Dreu, C. K. W. (2002). Team innovation and effectiveness: The importance of minority dissent and reflexivity. *European Journal of Work and Organizational Psychology*, *11*, 285–298.

De Dreu, C. K. W., & West, M. A. (2001). Minority dissent and team innovation: The importance of participation in decision making. *Journal of Applied Psychology*, *86*, 1191–1201.

de Vreede, G. J., Briggs, R. O., & Reiter-Palmon, R. (2010). Exploring asynchronous brainstorming in large groups: A field comparison of serial and parallel subgroups. *Human Factors: The Journal of Human Factors and Ergonomics Society*, *52*, 189–202.

de Vries, R. E., Bakker-Pieper, A., & Oostenveld, W. (2010). Leadership = communication? The relations of leaders' communication styles with leadership styles, knowledge sharing and leadership outcomes. *Journal of Business and Psychology*, *25*, 367–380. doi:10.1007/s10869–009–9140–2

Diehl, M., & Stroebe, W. (1987). Productivity loss in brainstorming groups: Toward the solution of a riddle. *Journal of Personality and Social Psychology*, *53*, 497–509. doi:10.1037/0022–3514.53.3.497

Dionne, S. D., Sayama, H., Hao, C., & Bush, B. (2010). The role of leadership in shared mental model convergence and team performance improvement: An agent-based computational model. *The Leadership Quarterly*, *21*, 1035–1049. doi:10.1016/j.leaqua.2010.10.007

Drach-Zahavy, A., & Somech, A. (2001). Understanding team innovation: The role of team processes and structures. *Group Dynamics: Theory, Research, and Practice*, *5*, 111–123. doi:10.1037/1089–2699.5.2.111

Edmondson, A. (1999). Psychological safety and learning behavior in work teams. *Administrative Science Quarterly*, *44*, 350–383.

Edmondson, A. C. (2004). Psychological safety, trust, and learning in organizations: A group-level lens. In R. M. Kramer & K. S. Cook (Eds.), *Trust and distrust in organizations: Dilemmas and approaches* (pp. 239–272). New York, NY: Russell Sage Foundation.

Ekvall, G. (1996). Organizational climate for creativity and innovation. *European Journal of Work and Organizational Psychology*, *5*, 105–123.

Ericsson, K. A. (1999). Creative expertise as superior reproducible performance: Innovative and flexible aspects of expert performance, *Psychological Inquiry*, *10*, 329–333.

Fay, D., Borrill, C., Amir, Z., Haward, R., & West, M. (2006). Getting the most out of multidisciplinary teams: A multi-sample study of team innovation in health care. *Journal of Occupational and Organizational Psychology*, *79*, 553–567. doi:10.1348/096317905X72128

Feist, G. J. (1998). A meta-analysis of personality in scientific and artistic creativity. *Personality and Social Psychology Review*, *2*, 290–309.

Fiore, S. M., Hoffman, R. R., & Salas, E. (2008). Learning and performance across disciplines: An epilogue for moving multidisciplinary research toward an interdisciplinary science of expertise. *Military Psychology*, *20*, 155–170.

Franz, T.M., & Larson, J. R (2002). The impact of experts on information sharing during group discussion. *Small Group Research*, *33*, 383–411

Galbraith, J. R. (2009). *Designing matrix organizations that actually work.* San Francisco, CA: Jossey-Bass

George, J., & Zhou, J. (2001). When openness to experience and conscientiousness are related to creative behavior: An interactional approach. *Journal of Applied Psychology*, *86*, 513–524. doi:10.1037/0021-9010.86.3.513

Gilson, L., & Shalley, C. (2004). A little creativity goes a long way: An examination of teams' engagement in creative processes. *Journal of Management*, *30*, 453–470. doi:10.1016/j.jm.2003.07.001

Gooch, J. (2005). The dynamics and challenges of interdisciplinary collaboration: A case study of "cortical depth of bench" in group proposal writing. *IEEE Transactions on Professional Communication*, *48*, 117–190.

Griffin, M. A., Neal, A., & Parker, S. K. (2007). A new model of work role performance: Positive behavior in uncertain and interdependent contexts. *Academy of Management Journal*, *50*, 327–334

Hennessey, B. A., & Amabile, T. M. (2010). Creativity. *Annual Review of Psychology*, *61*, 569–598.

Hoegl, M., Weinkauf, K., & Gemuenden, H. (2004). Interteam coordination, project commitment, and teamwork in multiteam R&D projects: A longitudinal study. *Organization Science, 15*, 38–55. doi:10.1287/orsc.1030.0053

Howell, J., & Shea, C. (2006). Effects of champion behavior, team potency, and external communication activities on predicting team performance. *Group & Organization Management*, *31*, 180–211. doi:10.1177/1059601104273067

Hulsheger, U., Anderson, N., & Salgado, J. (2009). Team-level predictors of innovation at work: A comprehensive meta-analysis spanning three decades of research. *Journal of Applied Psychology*, *94*, 1128–1145. doi:10.1037/a0015978

Hunter, S. T., Bedell, K. E., & Mumford, M. D. (2007). Climate for creativity: A quantitative review. *Creativity Research Journal*, *19*, 69–90.

Hunter, S. T., Thoroughgood, C. N., Myer, A. T., & Ligon, G. S. (2011). Paradoxes of leading innovative endeavors: Summary, solutions, and future directions. *Psychology of Aesthetics, Creativity, and the Arts*, *5*, 54–66. doi:10.1037/a001777

Jassawalla, A. R., & Sahittal, H. C. (2006). Collaboration in cross-functional product innovation teams. In M. M. Beyerlein, S. T. Beyerlein, & F. A. Kennedy (Eds.), *Innovation through collaboration* (pp.1–26). San Diego, CA: JAI Press.

Jaussi, K. S., & Dionne, S. D. (2003). Leading for creativity: The role of unconventional leader behavior. *Leadership Quarterly*, *14*, 475–498.

Jehn, K. (1995). A multimethod examination of the benefits and detriments of intragroup conflict. *Administrative Science Quarterly*, *40*, 256–282. doi:10.2307/2393638

Jehn, K. A. (1997). A qualitative analysis of conflict types and dimensions in organizational groups. *Administrative Science Quarterly*, *42*, 530–557.

Jehn, K. A., & Bendersky, C. (2003). Intragroup conflict in organizations: A contingency perspective on the conflict-outcome relationship. *Research in Organizational Behavior*, *25*, 189–244.

Jehn, K., & Mannix, E. (2001). The dynamic nature of conflict: A longitudinal study of intragroup conflict and group performance. *Academy of Management Journal*, *44*, 238–251. doi: 10.2307/3069453

Keller, R. T. (2001). Cross-functional project groups in research and new product development: Diversity, communications, job stress, and outcomes. *Academy of Management Journal*, *44*, 547–555.

Kellermanns, F., Floyd, S., Pearson, A., & Spencer, B. (2008). The contingent effect of constructive confrontation on the relationship between shared mental models and decision quality. *Journal of Organizational Behavior, 29*, 119–137. doi:10.1002/job.497

Klein, K. J., Knight, A. P., Ziegert, J. C., Lim, B., & Saltz, J. L. (2011). When team members' values differ: The moderating role of team leadership. *Organizational Behavior and Human Decision Processes, 114*, 25–36. doi:10.1016/j.obhdp.2010.08.004

Klimoski, R., & Mohammed, S. (1994). Team mental model: Construct or metaphor? *Journal of Management, 20*, 403–437.

Korner, M. (2010). Interprofessional teamwork in medical rehabilitation: A comparison of multidisciplinary and interdisciplinary team approach. *Clinical Rehabilitation, 24*, 745–755.

Kozlowski, S. J., & Bell, B. S. (2008). Team learning, development, and adaptation. In V. I. Sessa, & M. Fondon (Eds.), *Group learning* (pp. 15–45). Mahwah, NJ: Erlbaum.

Kratzer, J., Leenders, R., & van Engelen, J. (2004). Stimulating the potential: Creative performance and communication in innovation teams. *Creativity and Innovation Management, 13*, 63–71. doi:10.1111/j.1467-8691.2004.00294.x

Kratzer, J., Leenders, R., & van Engelen, J. (2006). Team Polarity and creative performance in innovation teams. *Creativity and Innovation Management, 15*, 96–104. doi:10.1111/j.1467-8691.2006.00372.x

Kurtzberg, T., & Amabile, T. (2001). From Guilford to creative synergy: Opening the black box of team-level creativity. *Creativity Research Journal, 13*, 285–294. doi:10.1207/S15326934CRJ1334_06

Levesque, L. L., Wilson, J. M., & Wholey, D. R. 2001. Cognitive divergence and shared mental models in software development project teams. *Journal of Organizational Behavior, 22*, 135–144.

Locke, E. A., Alavi, M., & Wagner, J. (1997). Participation in decision-making: An information exchange perspective. In G. Ferris (Ed.), *Research in personnel and human resources management* (pp. 293–331). Greenwich, CT: JAI Press.

Lovelace, K., Shapiro, D., & Weingart, L. (2001). Maximizing cross-functional new product teams' innovativeness and constraint adherence: A conflict communications perspective. *Academy of Management Journal, 44*, 779–793. doi:10.2307/3069415

Mannix, E., & Neale, M. (2005). What differences make a difference? The promise and reality of diverse teams in organizations. *Psychological Science in the Public Interest, 6*, 31–55. doi:10.1111/j.1529-1006.2005.00022.x

Marks, M. A., Mathieu, J. E., & Zaccaro, S. J. (2001). A temporally based framework and taxonomy of team processes. *Academy of Management Review, 26*, 356–376.

Marks, M., Sabella, M., Burke, C., & Zaccaro, S. (2002). The impact of cross-training on team effectiveness. *Journal of Applied Psychology, 87*, 3–13. doi:10.1037/0021-9010.87.1.3

Marta, S., Leritz, L. E., & Mumford, M. D. (2005). Leadership skills and the group performance: Situational demands, behavioral requirements, and planning. *The Leadership Quarterly, 16*, 97–120. doi:10.1016/j.leaqua.2004.04.004

Mathieu, J. E., Goodwin, G. F., Heffner, T. S., Salas, E., & Cannon-Bowers, J. A. (2000). The influence of shared mental models on team processes and performance. *Journal of Applied Psychology, 85*, 273–283.

Mathieu, J. E., Maynard, M. T., Rapp, T., & Gilson, L. (2008). Team effectiveness 1997–2007: A review of recent advancements and a glimpse into the future. *Journal of Management, 34*, 410–476.

Mathisen, G. E., Martinsen, Ø., & Einarsen, S. (2008). The relationship between creative personality composition, innovative team climate, and team innovativeness: An input—process—output perspective. *The Journal of Creative Behavior, 42*, 13–31.

McComb, S. A. (2007). Mental model convergence: The shift from being an individual to being a team member. In F. Dansereau & F. J. Yammarino (Eds.), *Multi-level issues in organizations and time* (Vol. 6, pp. 83–147). Amsterdam, the Netherlands: Elsevier.

Mesmer-Magnus, J., & DeChurch, L. (2009). Information sharing and team performance: A meta-analysis. *Journal of Applied Psychology, 94*, 535–546. doi:10.1037/a0013773

Milliken, F. J., Bartel, C. A., & Kurtzberg, T. (2003). Diversity and creativity in work groups: A dynamic perspective on the affective and cognitive processes that link diversity and performance. In P. Paulus & B. Nijstad (Eds.), *Group creativity* (pp. 32–62). New York: Oxford University Press.

Mitchell, R., Boyle, B., & Nicholas, S. (2009). The impact of goal structure in team knowledge creation. *Group Processes & Intergroup Relations, 12*, 639–651. doi:10.1177/1368430209340568

Mortensen, M., & Hinds, P. (2001). Conflict and shared identity in geographically distributed teams. *International Journal of Conflict Management, 12*, 212–238. doi:10.1108/eb022856

Mullen, B., Johnson, C., & Salas, E., (1991). Productivity loss in brainstorming groups: a meta-analytic integration. *Basic and Applied Social Psychology, 12*, 3–23.

Muller, A., Herbig, B., & Petrovic, K. (2009). The explication of implicit team knowledge and its supporting effect on team processes and technical innovations: An action regulation perspective on team reflexivity. *Small Group Research, 40*, 28–51. doi:10.1177/1046496408326574

Mumford, M., Reiter-Palmon, R., & Redmond, M. (1994). Problem construction and cognition: Applying problem representations in ill-defined domains. In M. Runco (Ed.), *Problem finding, problem solving, and creativity* (pp. 3–39). Norwood, NJ: Ablex.

Mumford, M. D., Connelly, S., & Gaddis, B. (2003). How creative leaders think: Experimental findings and cases. *The Leadership Quarterly, 14*(4–5), 411–432. doi:10.1016/S1048–9843(03)00045–6

Mumford, M.D., Feldman, J.M., Hein, M.B., & Nagao, D.J. (2001). Tradeoffs between ideas and structure: Individuals versus group performance in creative problem solving. *Journal of Creative Behavior, 35*, 1–23.

Mumford, M. D., & Hunter, S. T. (2005). Innovation in organizations: A multi-level perspective on creativity. In: F. J. Yammarino & F. Dansereau (Eds.), *Research in multi-level issues: Volume IV* (pp. 11–74). Oxford, England: Elsevier.

Nicholson, N., & West, M. (1988). *Managerial job change: Men and women in transition*. New York, NY: Cambridge University Press.

Nijstad, B. A., Diehl, M., & Stroebe, W. (2003). Cognitive stimulation and interference in idea generating groups. In P. B. Paulus & B. A. Nijstad (Eds.), *Group creativity: Innovation through collaboration* (pp. 137–159). New York, NY: Oxford University Press.

Oldham, G. R., & Cummings, A. (1996). Employee creativity: Personal and contextual factors at work. *Academy of Management Journal, 39*, 607–634.

Pearce, C., & Ensley, M. (2004). A reciprocal and longitudinal investigation of the innovation process: The central role of shared vision in product and process innovation teams (PPITs). *Journal of Organizational Behavior, 25*, 259–278. doi:10.1002/job.235

Perry-Smith, J., & Shalley, C. (2003). The social side of creativity: A static and dynamic social network perspective. *Academy of Management Review, 28*, 89–106.

Rank, J., Pace, V. L. & Frese, M. (2004). Three avenues for future research on creativity, innovation, and initiative. *Applied Psychology: An International Review, 53*, 518–528.

Reiter-Palmon, R., Herman, A. E., & Yammarino, F. J. (2008). Creativity and cognitive processes: Multi-Level linkages between individual and team cognition. In M. D. Mumford, S. T. Hunter, & K. E. Bedell-Evers (Eds.), *Multi-Level issues in creativity and innovation*. (pp. 203–267). Oxford, England: JAI Press.

Reiter-Palmon, R., & Illies, J. J. (2004). Leadership and creativity: Understanding leadership from a creative problem-solving perspective. *Leadership Quarterly, 15*, 55–77.

Reiter-Palmon, R., Mumford, M. D., & Threlfall, V. K. (1998). Solving everyday problems creatively: The role of problem construction and personality type. *Creativity Research Journal, 11*, 187–197.

Reiter-Palmon, R., & Robinson, E. (2009). Problem identification and construction: What do we know, what is the future? *Psychology of Aesthetics, Creativity, and the Arts, 3*, 43–47. doi:10.1037/a0014629

Reiter-Palmon, R., Wigert, B., & de Vreede, T. (2011). Team creativity and innovation: The effect of team composition, social processes and cognition. In *Handbook of organizational creativity* (pp. 295–326). Academic Press

Robledo, I. C., Peterson, D. R., & Mumford, M. D. (2011), Leadership of scientists and engineers: A three-vector model. *Journal of Organizational Behavior, 33*, 140–147. doi:10.1002/job.739

Rosen, A., & Callaly, T. (2005). Interdisciplinary teamwork and leadership: Issues for psychiatrists. *Australasian Psychiatry, 13*, 234–240.

Rosen, M. A., Bedwell, W. L., Wildman, J. L., Fritzsche, B. A., Salas, E., & Burke, C. S. (2011). Managing adaptive performance in teams: Guiding principles and behavioral markers for measurement. *Human Resource Management Review, 21*, 107–122.

Salas, E., Cooke, N., & Rosen, M. (2008). On teams, teamwork, and team performance: Discoveries and developments. *Human Factors, 50*, 540–547. doi:10.1518/001872008X288457

Salas, E., Nichols, D. R., & Driskell, J. E. (2007). Testing three team training strategies in intact teams. *Small Group Research, 38*, 471–488.

Salas, E., Sims, D., & Burke, C. (2005). Is there a "big five" in teamwork? *Small Group Research, 36*, 555–599. doi:10.1177/1046496405277134

Santanen, E. L., Briggs, R. O., & de Vreede, G. J. (2004). Causal relationship in creative problem solving: Comparing facilitation interventions for ideation. *Journal of Management Information Systems, 20*, 167–197.

Scott, S. G., & Bruce, R. A. (1994). Determinants of innovative behavior: A path model of individual innovation in the workplace. *Academy of Management Journal, 37*, 580–607.

Shalley, C. E., & Gilson, L. L. (2004). What leaders need to know: A review of social and contextual factors that can foster or hinder creativity, *The Leadership Quarterly, 15*, 33–53.

Shalley, C. E., Zhou, J., & Oldham, G. R. (2004). Effects of personal and contextual characteristics on creativity: Where should we go from here? *Journal of Management, 30*, 933–958.

Shin, S., & Zhou, J. (2007). When is educational specialization heterogeneity related to creativity in research and development teams? Transformational leadership as a moderator. *Journal of Applied Psychology, 92*, 1709–1721. doi:10.1037/0021-9010.92.6.1709

Silvia, P. J., Kaufman, J. C., Reiter-Palmon, R., & Wigert, B. (2011). Cantankerous creativity: Honesty-humility, agreeableness, and the HEXACO structure of creative achievement. *Personality and Individual Differences, 51*, 687–689.

Song, M., Dyer, B., & Thieme, R. J. (2006). Conflict management and innovation performance: An integrated contingency perspective. *Journal of the Academy of Marketing Science, 34*, 341–356.

Sosik, J. J., Avolio, B. J., & Kahai, S. S. (1998). Inspiring group creativity: Comparing anonymous and identified electronic brainstorming. *Small Group Research*, *29*(1), 3–31. doi:10.1177/1046496498291001

Srivastava, A., Bartol, K., & Locke, E. A. (2006). Empowering leadership in management teams: Effects on knowledge sharing, efficacy, and performance. *Academy of Management Journal*, *49*, 1239–1251.

Stasser, G. & Stewart, D. (1992). Discovery of hidden profiles by decision-making groups: Solving a problem versus making a judgment. *Journal of Personality and Social Psychology*, *63*, 426–434.

Stenmark, C. K., Shipman, A. S., & Mumford, M.D. (2011). Managing the innovative process: The dynamic role of leaders. *Psychology of Aesthetics, Creativity, and the Arts*, *5*, 67–80. doi:10.1037/a0018588

Stewart, D., & Stasser, G. (1995). Expert role assignment and information sampling during collective recall and decision making. *Journal of Personality and Social Psychology*, *69*, 619–628.

Taggar, S., & Ellis, R. (2007). The role of leaders in shaping formal team norms. *The Leadership Quarterly*, *18*, 105–120.

Tjosvold, D., Tang, M., & West, M. A. (2004). Reflexivity for team innovation in China: The contribution of goal interdependence. *Group and Organization Management*, *29*, 540–559.

Valacich, J.S., Dennis, A.R., & Connolly, T. (1994), Idea generation in computer-based groups: A new ending to an old story. *Organizational Behavior and Human Decision Processes*, *57*, 448–467.

Vincent, A. H., Decker, B. P., & Mumford, M. D. (2002). Divergent thinking, intelligence, and expertise: A test of alternative models. *Creativity Research Journal*, *14*, 163–178.

Wakefield, R.L., Leidner, D.E., & Garrison, G. (2008). A model of conflict, leadership and performance in virtual teams. *Information Systems Research*, *19*, 434–459.

Webber, S. S. (2002). Leadership and trust facilitating cross-functional team success. *Journal of Management Development*, *21*, 201–14.

Weingart, L. R, Todorova, G., & Cronin, M.A. (2010). Task conflict, problem-solving, and yielding: Effects on cognition and performance in functionally diverse innovation teams, *Negotiation and Conflict Management Research*, *3*, 312–337.

West, M., & Anderson, N. (1996). Innovation in top management teams. *Journal of Applied Psychology*, *81*, 680–693. doi:10.1037/0021-9010.81.6.680

West, M., & Richter, A. (2008). Climates and cultures for innovation and creativity at work. In J. Zhou & C. E. Shalley (Eds.), *Handbook of organizational creativity* (pp. 211–236). Hillsdale, NJ: Lawrence Erlbaum Associates.

West, M. A., Hirst, G., Richter, A., & Shipton, H. (2004). Twelve steps to heaven: Successfully managing change through developing innovative teams. *European Journal of Work & Organizational Psychology*, *13*, 269–299.

10 Leadership and Followership in Science and Technology

Michael E. Gorman and Philip S. Gorman

> I follow him to serve my turn upon him.
>
> Iago, *Othello*

Shakespeare's Iago was the good servant who enabled his leader's success until he realized there would be no reward—he would be taken for granted, and others would be preferred for promotion. Before the events in the play, Othello had been a great military leader, and Shakespeare makes it clear he depended on Iago—they had a close and trusting relationship. Othello promoted another, and the disillusioned Iago set in motion a campaign to convince the newly wedded Othello that his wife was unfaithful. For those who have not seen it, we will not give away the end of this famous play; suffice to say, it is a tragedy, for Othello, for Iago, and for others.

One lesson for leadership is that every leader depends on good followers—particularly good right hands who can be trusted. One lesson for followership is that it is better to raise a concern, to ask why, than to let it fester. This chapter outlines a set of heuristic principles and corollaries for ethical leadership and followership in science and technology. These principles are heuristic because they are guidelines, not rules.

The ethical guidelines and examples presented here are deliberately oriented towards creative leadership in science and technology, the kind that spurs innovations. A goal of the guidelines is to make sure the innovations are responsible, minimizing harm and maximizing benefit.

> *Principle 1:* Leaders and followers have to be able not only to listen to each other but to 'walk in each others' shoes'.

It is very important that leaders be able understand the situations their followers face—this makes it much easier for followers to give honest feedback. The best generals are those who spend time on the front lines as well as in headquarters, so they will understand what their followers are facing. The hazard of being close to the troops in war is the loss of commanders. King Karl XII of Sweden was shot in the autumn of 1718 while supervising his troops in the front lines as they built siege works toward a Norwegian fortress (Lindqvist, 2002).

Leaders who work closely with followers are not typically in danger in scientific and technological projects, unless there is a disaster involved, like the recent failure of nuclear power plants in Japan. It is important for a leader to listen, but he or she may not understand the follower's feedback unless he or she has tried to see the situation as a follower sees it. The same principle applies to followers—to communicate most effectively, the follower needs to see the leader's situation.

To illustrate we draw on examples from the experience of Michael Gorman,[1] who worked for two years as a program director at the US National Science Foundation (NSF), in a science, technology, and society program. The NSF is an independent federal agency created by Congress in 1950 to promote scientific progress, especially in the basic sciences, in ways that would advance national health, prosperity, and welfare (http://www.nsf.gov/about/). The 2010 budget was $6.9 billion. Award sizes range from $20 million Science and Technology Centers to $20,000 workshops and dissertation travel grants. There are interdisciplinary initiatives on topics like nanotechnology that span the foundation and basic funding for multiple fields of science, engineering, computer science, and social and behavioral sciences.

The NSF's leadership situation is much more ambiguous than the military one, in part because its mission is so different. The NSF was founded after World War II to support the kind of basic science research that turned out to be so critical in winning the war. The NSF is supposed to pay special attention to research that is creative and innovative (http://www.nsf.gov/about/transformative_research/).

Almost all NSF proposals are subjected to peer review, which can involve sending out proposals to reviewers and assembling teams of reviewers in a panel that argued about the relative merits of each proposal. Program directors (PDs) take leadership in seeking reviewers for proposals and identifying panelists, but the decisions about what to fund are based largely on the recommendations from these reviewers and panelists and have to be vetted with administrators who set the budgets and can micro-manage PD decisions. Many PDs are rotators who serve for 1 or 2 years and then go back to their home institutions. The administrators—including the NSF director—are also usually rotators, in part because full-time NSF employees are capped at a salary level below that of most administrators, who elect instead to be 'on loan' from their home institution at their full salary. (The 'loan' is specified in an Interagency Program Agreement, so those 'on loan' are called IPAs).

The NSF is therefore an unusual organization where both the 'troops' coming in and their 'generals' are newcomers to a complex bureaucratic culture that involves learning complex, elaborate procedures for doing tasks like recommending a proposal for funding. Here, then, is the perfect opportunity to learn together, to see each other's worlds. Unfortunately, in Gorman's experience in his part of the NSF, the new administrators spent little time trying to understand the PD's situation, though they did try to communicate what they were seeing and learning as they worked the upper echelons of the bureaucracy. Outside advisory committees constantly reminded NSF management that the PDs were overextended. Management was always sympathetic, then passed along more tasks for the PDs to do, typically ones directed at informing the upper levels of management about results and outcomes, but not about the processes used to achieve them.

The academic analogy would be a dean or provost who had never taught. Some administrators had been PDs, but Gorman's IPA supervisors had not. Every academic administrator should teach once in a while and should write an occasional grant proposal and publish an article. Gorman's experience suggested that NSF administrators who sign off on the decisions made by the PDs should handle one or two proposals in a cycle themselves, working with the PDs in the program to which the proposal was assigned. The PDs, in turn, should volunteer to help with tasks that allow them to work with administrators, to see what constraints and concerns they have to handle. Gorman's IPA supervisors were good at recruiting PDs to help with their administrative initiatives, but did not try to experience firsthand the proposal process they were administering.

This first leadership and followership principle are illustrated by other cases in this chapter. Other principles consistent with it are also be articulated. Each principle is supplemented by corollaries that make the advice in the general principle more specific.

> *Corollary 1:* Both leaders and followers have to honor each others' contributions by giving credit where it is due.

The great sociologist of science Robert K. Merton (1973) noted that credit is the coin of science. One of the best measures of credit is citations—how often (and by whom) a scientist's or engineer's articles are cited is an important factor in both reputation and promotion. A leader should make sure credit is given where credit is due across a research team or a laboratory or even an organization.

Order of authorship is important, with first generally being the most important. The leader is not always or even usually the first. In a large laboratory, the leader may have her name well down the list of authors, even at the end—and the main contribution may be having obtained the necessary resources and staff to support the research. But in every case, the leader must understand the work and be sure that it has been carried out properly. One solution to the problem of figuring out which coauthor did what is to require each author to indicate what her or his contribution was (see Eggert, 2011, for suggestions on how this approach could be applied to cognitive science). Coauthors must guarantee that their part of the work is sound and have an understanding of how their part of the research fits into the overall argument made in the paper.

Postdocs and students are particularly important because their positions are more vulnerable and they are earlier in their careers. For them, credit is not only coauthorship but also letters, contacts for jobs, and so on. Both postdocs and PhD students must be encouraged to publish and learn to take on the responsibilities of first authorship.

Corollary 2: Be a leader, not a boss.

Bosses give orders and expect obedience. Bosses do not care how a follower gets a result as long as she or he gets it. Leaders give orders, but listen to objections and concerns from their followers.

One of Gorman's NSF supervisors was a boss who leaned on him to get a set of critical proposals through the system before an arbitrary close-out date. The supervisor was right—the proposals had to be done. The reviews were in, and the NSF partners in this complex proposal had been contacted and had given e-mail over verbal approval, but the task of finding the appropriate budget codes and getting digital sign-offs for each partner remained. Gorman had tried to delegate these tasks to a more knowledgeable staff member who was supposed to help, but when the administrator applied the pressure, Gorman found the proposals had not moved. A new, 'improved' digital system for recommending proposals had just been installed, and everyone—even veteran staff—was on a learning curve.

The supervisor was right to order that the proposals be moved forward but wrong to show no interest in why the proposals were stuck in the system. Gorman provided feedback on the cause of the hold up, then set out to solve it, helped by another staff member. Together they completed the steps and sign-offs to fund a proposal with multiple partners across the NSF. Gorman mastered arcane elements of the software and process to avoid future problems—even though the time spent acquiring this information was wasted from an NSF standpoint because he was going to leave NSF in a year.

Leaders not only remind followers of priorities, but also help investigate and clear bottlenecks and look for opportunities to improve the system. Leaders are both charismatic and humble: They inspire both outsiders and followers, and they know how to follow Principle 1 by walking in the shoes of the follower. Organizations have to be designed to encourage leaders to manage, not just pass on orders.

Corollary 3: A lab director or business leader must be aware of the assumptions on which her or his mental models are founded, and encourage followers to look for results that suggest alternate mental models. This search for alternatives will promote creativity.

Bosses confuse their mental model of the world with reality and insist that followers conform to it. The Nobel Laureate Shockley founded Shockley Semiconductors (in 1956) to manufacture transistors. He hired top talent from a variety of places, including Gordon Moore and Rober Noyce. Shockley may have been a genius, but he was a boss instead of a leader;

he became more interested in having his employees pursue his own vision for a transistor than in taking full advantage of their talents and ideas. A group of eight, including Noyce and Moore, decided to leave Shockley and formed Fairchild Semiconductor. They succeeded in making a profit while Shockley's firm failed because he would not listen to any alternative to his core idea (Riordan & Hoddeson, 1997). He had a creative group and did not let them pursue their innovative ideas. Moore and Noyce went on to found Intel (Jackson, 1997).

Similarly, when an inventor is trying to prototype a new design, it makes sense to try to set up ideal conditions in which it will work to establish proof of concept. Alexander Graham Bell's mental model for the telephone was the human ear. He built a crude device that would translate sound waves into a fluctuating current that could be transmitted to the same sort of device that could act both as transmitter and receiver. His first transmissions of speech were barely discernible, but this proof of concept was sufficient for him to apply for a patent and continue research. He assumed refinements in technology would demonstrate the value of his design, and he was partly right and partly wrong—his ear mental model produced a great receiver but a poor transmitter (Gorman, 1995).

A leader should always be aware of the possibility that confirmation can turn into a bias. Canadian virologist Patrick Lee let one of his graduate students do a 'stupid' experiment that led to a surprising result. Lee's student wanted to cleave sialic acid off a retrovirus and make it inactive (Thagard, 2002). When the student compared the cleaved cells to a control, it was the latter that were inactive. Lee was sure the student had made an error, so he made him replicate the experiment several times—with the same result. Lee's mental model told him that the result had to be wrong, but he was willing to check and recheck until he decided the result was right and his assumptions had to be changed. This surprise sparked a program of research into what was going on in this control group of cells, one that eventually resulted in a publication in *Science* and a possible cure for cancer.

Kevin Dunbar saw a falsification bias in some of the molecular biology laboratories whose meetings he taped and analyzed (Dunbar & Fugelsang, 2005). The laboratory directors were more likely than were students and postdocs to question results that accorded with expectations.

Creativity can lead to confirmation bias in business as well. The idea of turning mortgages into securities was a creative one that allowed investors to buy bundles of mortgages. The ratings agencies, based on historical mortgage defaults, assigned these securities a high rating—one or two mortgages could default, but most would not.

The problem was that most of the mortgages bundled into these securities were made with no down payment or income verification to borrowers with very low credit ratings (McLean & Nocera, 2010, p. 135). Many of the borrowers who took these mortgages did not have enough income to make the payments. The ratings agencies displayed confirmation bias by assuming the past safety of mortgages would apply to these new securities and borrowers. Their confirmation bias was also an instance of a conflict of interest: They derived their income from the brokerage firms who had invented these new securities, and the firms threatened to move their business elsewhere if one of the agencies lowered its ratings. When the housing bubble burst and the borrowers walked away from the homes they had put no down payment on, the ratings agencies lowered their ratings—tantamount to warning that the ship might sink when water is flooding the engine room.

Confirmation bias was here coupled with conformity. The firms securitizing these mortgages were constantly looking at what other brokerage and mortgage companies were doing, and raced to capture higher shares of the expanding market. Those offered the loans saw all their peers doing it and gaining spending power; therefore, they did the same.

Savvy business leaders are more like the laboratory directors who mistrust the existing trend, who build in hedges and diversification to reduce risk, and who look for signals or trends that might disconfirm aspects of their mental models (Oetinger, 2010). These kinds of leaders will be more likely to spot a change in the business environment and encourage creative thinking about where new opportunities and challenges lie. The fact that other companies are following a trend may simply accelerate the system collapse.

Principle 2: Effective leadership and followership in organizations requires communication not only within but also across bureaucratic hierarchies and boundaries.[2]

Diane Vaughan (1996) studied the case of the *Challenger* shuttle disaster. The cause was failure of an O-ring that sealed the joints in the booster. Engineers like Roger Boisjoly had raised concerns about O-rings on previous launches, and Boisjoly was particularly concerned about the *Challenger* launch because temperatures were far colder than on any previous launch, and he was sure the rings showed more wear in colder weather, though the correlation was far from perfect. Vaughan argued that the NASA bureaucracy had reclassified the risk from the O-rings as acceptable. The conversation changed from proving the O-rings were safe to proving that they were unsafe. What originally would have been regarded as a deviation from best practices was now regarded as normal—hence, 'normalized deviance'.

In order to secure funding, NASA and its contractors and backers had portrayed the space shuttle as a safe, cost-effective way to boost cargo into space; indeed, it was often referred to as a 'space truck'.

The *Challenger* disaster could be viewed as a highly unusual event—the space truck had a rare accident preventable by design improvements and restrictions on launching at low temperature. It could equally be viewed as a sign that the shuttle was a high-risk experiment, not a reliable delivery vehicle. The O-rings were not the only persistent problem. After *Challenger*, the satellite business was transferred back to rockets, and the shuttle program turned more to science, including valuable repairs on the Hubble.

> The foam did it, the institution allowed it. (Columbia Accident Investigation Board member Duane Deal)[3]

From the first flight, the shuttles had shed debris from the foam tiles that encased much of the orbiter's surface, and these tiles had caused at least some damage to every shuttle flight. Originally, such damage was considered unacceptable, but as shuttle after shuttle flew despite foam debris, what was once seen as deviant became normal.

The shuttle *Columbia* launched on January 16, 2003. Eighty-two seconds into the launch, a chunk of insulating foam detached from a fuel tank and struck the left wing. This incident was not discovered until the morning of January 17, during a review of photographs of the launch. The photograph was blurry; it was hard to determine the size, shape, and momentum of the debris and exactly where it had struck.

Analysts classified this event as 'out of family', meaning it was deviant. NASA created a Debris Assessment Team (DAT) that reported directly to the Mission Management Team (MMT). The DAT asked if one of the astronauts could do a space walk to assess the damage, but never received an answer. The DAT asked that defense department satellites try to capture images of the wing. When the MMT found out about this request, they criticized the DAT for not submitting it through the usual channels and canceled it. According to Vaughan, "The MMT was supposed to operate in a decentralized way. That's not what happened. It was a very centralized operation with all the information flowing to the manager. It was very rule-oriented, very protocol-oriented".[4]

There were certainly arguments against enrolling the satellites or pursing another alternative, a space walk. Defense satellites are doing important surveillance work. Space walks are complicated and dangerous. Furthermore, even if damage were found, it was not clear what to do to rescue the shuttle. But the decision should have involved a larger trading zone embodying different expertise and perspectives—including the astronauts themselves, who may have wanted to know the extent of the damage. According to Kathy Thornton, a colleague of mine who was an astronaut, the astronauts knew that shuttle flights were risky. This particular group of astronauts on board the *Columbia* were told that foam had struck the wing but that it did not represent a threat. The commander and crew were not told any of the options for studying the risk, for example, satellite photos and even perhaps a space walk.

According to Richard Feynman, who served on the Presidential Commission evaluating the *Challenger* disaster,

> It appears that there are enormous differences of opinion as to the probability of failure with loss of vehicle and human life. The estimates range

> from 1 in 100 to 1 in 100,000. The higher figure comes from working engineers, and the very low figures from management. (Vaughan 1996, p. 274)

Feynman himself felt the probability was about 1 in 25; the *Challenger* and *Columbia* disasters suggest Feynman's estimate was close to the mark.

Communication across bureaucratic lines would have allowed working engineers who thought the probable failure was 1 mission in 100 to interact with not only managers who thought it was much more remote but also policy makers who assumed the system was safe.

TRADING ZONES

Trading zones emerge in locations (physical or virtual) where multiple cultures surmount apparently incommensurable linguistic barriers by developing common languages sufficient to get around barriers to collaboration—first a jargon, then a pidgin, and finally a full-scale creole. The beginning exchanges are simple, and require only a few common terms, but as the trades become more complex, the language has to evolve, too—and the end result, a creole, is a new combination of several older languages that can be taught to a new generation. These trading zones occur in scientific and technological collaborations, including ones that incorporate the humanities and social sciences (Gorman, 2010).

Bureaucratic procedures preserve order, and checks and balances, but for an organization to succeed, it also needs informal trading zones that cut across bureaucratic lines. To ensure communications across bureaucratic and expertise barriers, NASA could have set up trading zones in advance, or allowed them to flourish. After the *Columbia* foam strike and while the shuttle was still in orbit, members of this trading zone could have been assembled into a tiger team[5] to consider options. At least some team members would have to possess sufficient interactional expertise to appreciate insights from those with different training and backgrounds. Such a tiger team would also have to include managers responsible for the launch decision who would still have to make fast decisions but would know they could get the best knowledge in the organization mobilized to help them. Such a team could have also have been in touch with the astronauts themselves, sharing knowledge and involving them in decisions. Trading zones permit multiple stakeholders to coevolve solutions to problems (Gorman & Werhane, 2010). Participants in a trading zone come from different expertise communities and interact more or less as equals. Ideas, time, and resources are exchanged, typically in pursuit of a goal like building new kinds of giant telescopes (McCray, 2004).

Corollary 1: Leaders who facilitate trading zones act as catalysts

A catalyst leader is someone who takes the lead in creating and facilitating trading zones. Consider again the role of a program officer (PO) at the NSF. The PO is a catalyst for research in that she or he evaluates proposals that have been submitted from scholars and practitioners in a domain. The method for evaluation is peer review. The PO makes recommendations what to fund based on this peer feedback, and this decision is vetted by a division director. POs can collaborate to create solicitations for new types of proposal, but again the main evaluation mechanism is peer review. Therefore, the PO is not supposed to be a leader, but a catalyst who creates the conditions for the best research to emerge and prosper. Leadership in science and technology involves much more than giving orders and making decisions—it involves mobilizing talented professionals to work toward a common goal.

A catalyst leader can play an important role in bringing multiple disciplines and expertise together around a common goal. Gary Tabor has been the catalyst for the Freedom to Roam initiative that includes companies like Patagonia, nongovernmental organizations (NGOs) such as NatureServe (http://www.natureserve.org/), and people from government agencies such as the NSF (http://freedomtoroam.org/). The common goal is establishing wildlife corridors across North America, thereby providing a model that could be adapted for other parts of the world. Corridors for animal movement prevent species from being isolated in pockets too small to maintain genetic diversity. There are dozens of local strategies that can be used for creating such corridors, including wildlife bridges over highways, tunnels under railroads, access granted through private property, establishment of parks, and so on. If the accumulation of greenhouse gases leads to global warming, these corridors will provide routes for mass migration—and people will see the migrations in their backyards.

The role of the catalyst in an organization like this is to find individuals and organizations willing to participate and facilitate their engagement. When asked why they had joined, about a third of the members at a meeting I attended mentioned Tabor, highlighting his catalyst role. My thesis is that a catalyst leader has to have at least interactional expertise across multiple areas of the extended domain—has to be able to talk enough of the language of companies, scientists, landowners, hunters, and policy makers to convince them that they share a common goal, and must work together to achieve it. The catalyst leader also has to put the spotlight on others in the coalition, making sure they get credit—while keeping her or himself in the background. Remove the catalyst at the early stage, and the trading zone falls apart. If the catalyst leader is successful, eventually the trading zone stands up on its own.

> *Corollary 2:* In adaptive organizations, delegation involves autonomy and collaboration

Philip Gorman, the second author, works for a company, PROS Pricing, that on paper has a hierarchical management structure, but leadership responsibility is transferred downward: Leaders communicate goals to followers, who in turn have to transform those goals into subgoals and communicate them to others who will carry out actions. Consider the following example (see Figure 10.1).

A large global consulting company approached PROS at the executive level: A vice president (VP) at the consulting company whom we will call J contacted a VP at PROS whom we will call D. The consulting firm had been hired by consumer package goods companies that needed help with their pricing strategy. Here was a classic opportunity to create a trading zone: The consulting firm had the clients and the customers; PROS had expertise, experience, and software in pricing. PROS VP D needed an application that would help the consumer package goods company improve its pricing. He communicated this goal to Philip Gorman, who had to work with his counterpart R at the consulting firm.

Philip Gorman and R had to own the high-level management goal and collaborate to break the goals down to a more granular level so they could delegate. R suggested that consumer package goods companies needed to know how exactly customers respond to various kinds of promotions—discounts, special sales, flyers, and so on. If the application could predict whether a promotion would succeed or fail then it would help the consumer packaged goods company set more effective pricing.

R had access to the raw historical data on promotions. Real prices were needed in order to make this strategy work, PG subdivided this goal into two parts. One involved getting end-to-end sales data—when is a discounted product actually sold and at what price? This part PG delegated to C, who

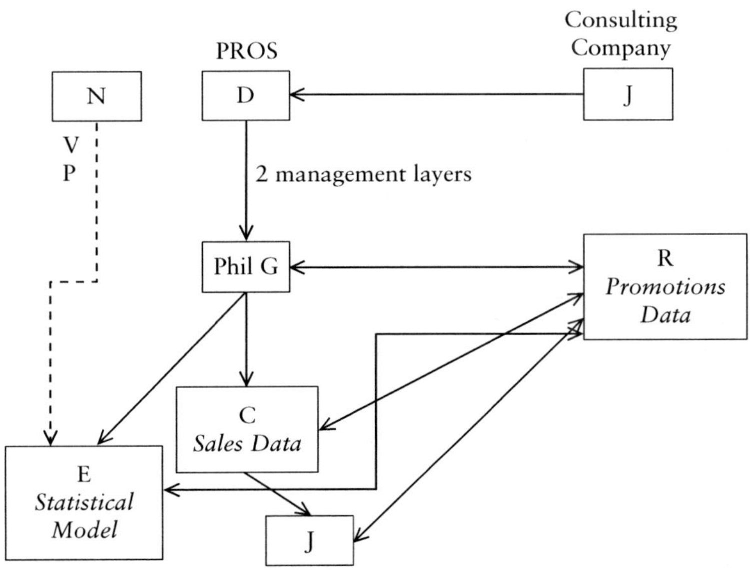

Figure 10.1 To come

had experience tracking a product from the manufacturing plant all the way to the customer. C in turn delegated number-crunching tasks to J. Both C and J had to collaborate with R to get the appropriate data and choose a model for tying a store sale to a manufacturing sale. The company R consulted for manufactured plastic bags[6] and sold them to major store chains. The manufacturer could give a particular store a discount on bags bought during a particular period—say 50% off over a holiday. The store could pass all or part of the discount onto the consumer. The data from the manufacturer and the stores were separate streams and had to be linked so the overall impact of a discount on sales could be assessed.

After the streams of data were linked, PG delegated development of a statistical model to E, with the goal of predicting promotion efficiency. E collaborated with others to run and tweak the model. E and C had to communicate regularly with Philip Gorman and R at the consulting firm, to make sure the models and analyses really addressed the goals.

PG and R were collaborated on setting goals and strategies and brainstorming who could follow. E and C were followers because they were delegated specific tasks that took advantage of their expertise. E and C did occasionally collaborate with others at their level to accomplish specific goals, and C occasionally collaborated with R to build the data model.

In terms of the management structure of PROS, Philip Gorman has two layers of management between himself and D. C, E, and Philip Gorman were on the same level, but E reported to N, not to D, so N had to be informed. Philip Gorman therefore led two people at his level on the project, including one technically in another part of the company. On other projects, there were occasions when Philip Gorman actually had several of his nominal bosses working for him on a project, so he could draw on their expertise. He was able to cross divisions of the company to recruit the right expertise. This kind of task-oriented, fluid leadership allows an organization to rapidly reconfigure to take advantage of new opportunities.

The PROS example involves delegation with autonomy, which is why the follower becomes a leader within constraints—a leader in defining how she or he will fulfill the goal, in constant communication and sometimes collaboration with those who assigned it and collaborating with others with relevant expertise. The farther down the delegation chain, the more detailed the goals, until finally the goal is specified almost at the task level. Only at the lowest level were there followers who were told to run specific analyses and did not need to understand the goals.

FLUID LEADERSHIP ACROSS ORGANIZATIONS IN A TRADING ZONE

Gorman worked with a group of participants from different federal agencies on the development of the 2001 strategic plan for the National Nanotechnology Initiative. All of the relevant the agencies would have to sign onto the relevant parts of the plan. Periodically, one of the agency representatives would wear her or his agency hat and block something, saying their agency probably would never agree to it, and often the participant checked with others at her or his agency. Depending on the response, consensus would be reached on new wording.

The PROS trading zones occurred across two organizations and across bureaucratic lines within the organizations. The strategic planning committee was a cross-organizational trading zone and included input from workshops that involved other stakeholders from the science, engineering, ethics, and business communities.

The Office of Science and Technology Policy at the White House (OSTP) was driving the effort, based on recommendations from the President's Council on Science and Technology (PCAST). There was a representative from OSTP present at virtually every meeting of the strategic planning group. The OSTP representative was not a manager like D was for Philip Gorman, who handed off a goal; the OSTP representative reminded the group of deadlines, of the kinds of documents that had to be produced and workshops that had to be conducted, and tried to keep the group on task without dictating the goals. The National Nanotechnology Coordinating Office (NNCO) was created not only to serve a catalytic role that complemented OSTP but also to provide the staff support for the strategic planning process. The NNCO's chief representative was also present at virtually every meeting; he pushed our group to go beyond the goals provided in earlier 3-year strategic plans and come up with objectives that the other agencies involved in the trading zone would be willing to implement.

The leader of the strategic planning group was from the Environmental Protection Agency, one of the participating agencies. She managed the meetings while participating actively in the discussions, providing language for goals and objectives. The leader could not order; she could not even manage in the traditional sense: She also had to act as a catalyst, keeping the group moving forward together.

So leadership in this committee was distributed and the goals were set by the participants, a very different situation from the PROS example, in which there were no formal committees and leadership set the goals.[7]

Gorman worked with a subgroup on goals related to the societal dimensions of nanotechnology, including environmental risks and benefits and legal and ethical dimensions. Within our subgroup, leadership shifted, based on who had time to shepherd the draft at a given stage. Decisions were made by consensus. Gorman was a new rotator at the NSF and knew only one person on the interagency group when he started, yet his suggestions were taken as seriously as anyone else's. Toward the end, a catalyst leader within the group emerged to shepherd the final stage; she kept the 'canonical document', but never used it to establish control; instead, she used the document to keep us focused on the outcome of the zone, down to the specifics of the language that would keep agencies engaged. At times, others would volunteer to take over the leadership roles—even Gorman did, on one section.

In the case of the NNI strategic plan, there were no followers, except perhaps NNCO staff who arranged for the meetings and sent out email reminders. In relatively equal trading zones, the follower–leader relationship is fluid—leadership shifts with the demands of the task and the expertise of those in the trading zone.[8] It was important for members of this group to do a little 'walking in each other's shoes'—to understand the constraints under which other agencies were operating and to try to create a set of goals each member of the group could sell to her or his bosses.

The PROS and NNI examples illustrate the way leadership can be distributed in different types of trading zones and emphasize the importance of the nature of the task. In the PROS case, a customer had a specific need that required a small, fluidly hierarchical trading zone across two companies. In the NNI case, government agencies had to form a trading zone around a strategic plan that they jointly developed in response to a mandate that such a plan be updated or developed every 3 years.

> *Principle 3:* Leaders and followers developing new technologies should think about the potential benefits and harms to segments of society.

The NNI deliberately incorporated consideration of societal impacts into its strategic planning. The NSF funded Centers for Nanotechnology in Society at Arizona State University (http://cns.asu.edu/) and the University of California at Santa Barbara (http://www.cns.ucsb.edu/) at about $13 million apiece over 10 years. The NSF has also funded numerous smaller projects, and ethical and societal concerns are also integrated into many of the nanoscience and nanotechnology projects, though the depth of the integration varies. One model is to outsource the ethics to a consultant or specialist. Another is to thoroughly integrate the ethics, putting social scientists and humanists into the laboratories (Fisher & Mahajan, 2010).

Milton Friedman famously argued that corporate leaders should not engage in social responsibility in part because they are not qualified: their job is to create profits for their shareholders (Friedman, 1996). Friedman's argument could be amended for scientists and engineers to say that they are not qualified to think about social responsibility—they should simply make discoveries and develop devices. For engineers, the motivation for creating new technologies is often the pursuit of 'technological sweetness' (Pacey, 1989)—the joy of creating something for its own sake, to show that it can be done. The novel *Frankenstein* is a story about a kind of biomedical engineer who pursues creating life to see if it can be done—without a thought for what he will do after he succeeds, or how society or his creation will respond.

The business counterargument to Friedman is that corporations are responsible not just to shareholders but also to stakeholders (Freeman, Harrison, & Wicks, 2007). To use a simple example, if a company makes a profit for its shareholders by utilizing the equivalent of slave labor somewhere else in the world, then it may be benefiting its shareholders in the short run, but the workers and the society they live in are stakeholders and they company should also address their concerns.[9] When it comes to issues like climate change, the entire planet becomes a stakeholder in companies that produce greenhouse gases or seek to mitigate them with new technologies.

This argument about stakeholders applies to both scientists and engineers. The NSF uses taxpayer funds to support science, and quite appropriately expects the scientists to address the societal impacts of their work.[10]

Engineers create technologies for stakeholders like those who use bridges, airplanes, the Internet, genetically modified crops, and so on. These stakeholders can become customers. Thinking about societal benefits means thinking about who would benefit from a technology, and that potentially leads to the discovery of new markets.

But how can engineers and scientists think about societal benefits if it is not part of their expertise?

> *Corollary 1:* One way to create technologies that benefit society—and to avoid harms—is to create a trading zone with those that have expertise in thinking about ethical and social concerns.

Scientists and engineers can form trading zones with ethicists and social scientists who make it their practice to think about societal benefits and harms. Here the acquisition of interactional expertise is critical—the ethicist or the social scientist has to work with the scientist to understand the potential uses for a technological breakthrough and to design alternatives and options. The dialogue between scientist, engineers, and ethicists can open up new alternatives that lead to better research from both a scientific and social standpoint (Fisher, 2007). To explore this idea, Gorman obtained an NSF grant to build a small trading zone that would insert societal concerns into a nanotechnology research project at the earliest or upstream stage. He worked with James Groves, a materials scientist; they co-advised a master's student who wanted an engineering degree but was willing to consider societal implications from the start (the grant funded the student). In order to trade, they had to develop a shared jargon. Instead of developing new terms, they shared meanings for existing ones. Gorman's understanding of a term like *directed self-assembly* was never as deep or as experiential as was Groves's and vice versa with respect to the concept of moral imagination.[11]

Each of us acquired some interactional expertise in the other's area. In Gorman's case, the interactional expertise complemented his knowledge gained from years of research into scientific reasoning (Gorman, 1992), thinking and analyzing how scientists deploy what they refer to as the scientific method—as it varies from field to field. He used this expertise to make suggestions about alternative research strategies and, as he gained interactional expertise from the patient explanations of his collaborators, was able to make those suggestions more specific to the project. By the end, Groves felt Gorman was enough of a partner in the design to deserve to be named on a patent proposal.

The team also had to develop a metaphoric language to talk about its goals. All three participants in this zone liked hiking, which is why this seemed a natural set of metaphors. Groves took the lead in creating the language. Distant mountains represented major global problems and opportunities, like human health, climate change, the prevalence of warfare, and so on. Closer foothills represented specific aspects of these problems, like the elimination of heart disease, or cancer, or providing more data on toxins introduced into the environment either as a form of biological warfare or as pollution. The graduate student was to build a bridge that could be used to reach the foothill. This bridge could connect to trails blazed by other researchers, or could suggest a new trail that ought to be blazed by work that extended beyond the master's thesis. The research topic selected by the first student incorporated nanotechnology, biotechnology, and cognitive reflection, three of the four NBIC technologies (the fourth is information technology), and included ethical considerations (Gorman, Groves, & Catalano 2004).

Patricia Werhane (1999), an expert on business and engineering ethics, suggested thinking about an alternate goal to the mountains—perhaps a village in the developing world, which would direct the project more toward Amartya Sen's emphasis on increasing affluence and education for women worldwide (Sen, 1999). In order to develop appropriate technologies, the engineers would have to be able to 'walk in the villager's shoes', seeing what they needed and understanding enough of their values to coevolve solutions to problems like providing safe drinking water (Gorman, Wardak, Fauss, & Swami, 2008).

> *Corollary 2:* Another, synergistic way to create technologies that benefit society is for scientists and engineers to engage in moral imagination

This ability to 'walk in the villager's shoes' is a key component of what is referred to as moral imagination, which is the ethics and values component of Principle 1. Moral imagination requires seeing one's own values as a perspective and being able to adopt the perspective of another (Werhane, 1999). So, for example, instead of assuming that the pursuit of basic science is guaranteed to provide societal benefits, the scientist must instead see that as a view, or an assumption, and be able to understand how others might regard science as the practice of an elite in the developing world to which other societies and cultures are forced to adapt. Sometimes, this kind of mutual understanding allows one party to persuade the other that she or he is wrong; sometimes both parties, or multiple parties, can move to a new view. Moral imagination is not relativism—it does not assume that all values are equally ethical. But moral imagination does begin with the premise that it is worth trying to understand someone else's view of the world.

Moral imagination is the equivalent of the scientific method applied to ethics, with empathy added. This process is science-like because one can consider values systems as hypotheses about how to live and about the exchange as a way of comparing hypotheses, then decide on ways to evaluate which one works better in a specific situation—or whether a new set of values need to be created to accommodate (say) a technological breakthrough that was not anticipated by traditional values systems. The process involves empathy, because one has to walk in another's shoes.

Moral imagination allows the concerns of stakeholders to be internalized by scientist and engineers, working not only with social scientists and ethicists but also with the stakeholders themselves. Catalyst leaders have to be able to exercise moral imagination—have to be able to see how a problem like wildlife corridors can be turned into an opportunity from the perspective of a landowner, and a hunter, and a company like Walmart.

Moral imagination blurs the leader–follower distinction, because stakeholders and engineers can coevolve new solutions to problems such as providing clean water or access to knowledge.[12] But there are situations where scientists and engineers have to take the lead in creating a better future.

Corollary 3: At least some scientists and engineers should become involved in policy and governance.

Deployment of technologies that benefit societies, and avoidance of ones that pose unacceptable risks, requires governance systems that are focused on the welfare of citizens, as opposed to just maintaining the power and position of an elite. Andrei Sakharov was proud of creating the Soviet hydrogen bomb until he realized that he had provided the weapon to a government that would not consult scientists or engineers or citizens on its use and did not care about the radioactive fallout from testing on Soviet soil (Gorman, 1998). Sakharov, a hero of the Soviet Union, changed from a follower in matters of policy to a leading dissident who was imprisoned until Gorbachev released him. Sakharov then joined the Soviet parliament and died while serving.

Civilizations in which the elite have focused on their own well-being lose the ability to adapt to changing circumstances (Diamond, 2005); witness the collapse of the Soviet Union. Governance systems have to include the capability to anticipate; here, science can be particularly useful at providing methods for capturing and analyzing data and for modeling future conditions. But scientists cannot just supply data; some need to become policy makers and all need to be involved in governance. In other words, scientists and engineers cannot just be followers in policy; they have to adopt leadership when appropriate.

CONCLUSION

The principles and corollaries outline in this article are meant to be the beginning of a discussion of how to promote creative leadership and followership in science and technology. Future empirical research should be directed at looking more closely at which kinds of leadership and followership are best suited to specific kinds of problems and contexts.

For example, on a Eureka problem where there really is one best solution, the main problem is to create a situation where this optimal idea or strategy is most likely to emerge and be recognized quickly (Gorman, 1992). To avoid confirmation bias, it is important to seek evidence that suggests the limitations of the idea or approach.

On a problem where there are pros and cons for different stakeholders and an urgent need to achieve a common goal, for example, sustaining a socio-technical-environmental system, trading zones are essential. Here the goal is a coevolutionary process that will lead to solutions that are better than the partners could have achieved separately.

Empirical work on these issues would have to involve triangulation among multiple methods (Plowright, Sokolow, Gorman, Daszak, & Foley, 2008), collecting longitudinal and cross-sectional data on evolving leadership and followership patterns. Methods could include coded transcripts (Dunbar, 1999), interviews (Hemlin, 2009), reflective diaries (Shrager, 2005) and could even involve the researchers as participant observers (Fisher & Mahajan, 2010). Research can also involve experimental simulations of leadership and followership situations, isolating and controlling variables (Gorman, 1992).

It is important to do this research not just for its own sake but also for the sake of our collective futures. We need creative leadership and followership to manage the process by which we are transforming the socio-technical-natural systems we inhabit (Allenby, 2005).

NOTES

1. The senior author will hereafter be referred to with his last name only; the junior author will be referred to with both names.
2. The account of the *Challenger* and *Columbia* disasters and their ethical implications is drawn from M. E. Gorman and Patricia H. Werhane (2010). Usingtrading zones to prevent normalized deviance in organizations.
3. Quoted in Bohmer, Edmondson, and Roberto (2004).
4. See Bohmer, Edmondson, and Roberto (2004, p. 10).
5. NASA rules actually called for creating a tiger team, but in this case, the bureaucratic rule that might have led to a trading zone was ignored.
6. The actual company did not manufacture plastic bags; this example is designed to protect the identity of the company.
7. One reviewer suggested that the main difference may have been between committee leadership and other types, but in fact, committees themselves vary widely in terms of hierarchy, rules for discussion, mission, and so on. A committee can be a way for members of a trading zone to do business, or a tool for management to communicate its imperatives to employees.
8. There can be very unequal trading zones where one or more parties dominate, and in those there will be followers (Collins, Evans, & Gorman, 2007).
9. In the age of NGOs, companies have learned that such stakeholders can also damage their profits by revealing information about worker conditions, envi-

ronmental damage, and so on—so the same actions can and often do benefit both shareholders and stakeholders.

10. See the special issue of *Social Epistemology* (Holbrook, 2009) devoted to the NSF's broader impacts criterion.
11. Directed or guided self-assembly refers to methods that create a pattern on a surface on which nanoparticles will assemble. Moral imagination refers to a process that involves comparing the mental models that underpin values and considering alternatives (Gorman, Werhane, & Swami, 2009).
12. IBM is a leader on a new initiative to create a science of services that emphasizes how providers and potential clients should co-evolve solutions to problems like providing drinking water, or managing socio-technical systems (Spohrer & Engelbart, 2004; Spohrer, McDavid, Maglio, & Cortada, 2006).

REFERENCES

Allenby, B. (2005). Technology at the global scale: Integrative cognitivism and earth systems engineering management. In M. E. Gorman et al. (Eds.), *Scientific and technological thinking* (pp. 303–344). Mahwah, NJ: Erlbaum.

Bohmer, R. M. J., Edmonson, A. C., & Roberto, M. A. (2004). *Columbia's Final Mission* (Business case no. 9–304–090). Cambridge, MA: Harvard Business School.

Collins, H., Evans, R., & Gorman, M. (2007). Trading zones and interactional expertise. *Studies in History and Philosophy of Science, 39*(1), 657–666.

Diamond, J. M. (2005). *Collapse: How societies choose to fail or succeed.* New York, NY: Viking.

Dunbar, K. (1999). How scientists build models: In vivo science as a window on the scientific mind. In L. Magnani, N. Nersessian, & P. Thagard (Eds.), *Model-based reasoning in scientific discovery* (pp. 85–100). New York, NY: Kluwer Academic Press.

Dunbar, K. N., & Fugelsang, J. A. (2005). Causal thinking in science: How science and students interpret the unexpected. In M. E. Gorman et al. (Eds.), *Scientific and technological thinking* (pp. 57–80). Mahwah, NJ: Erlbaum.

Eggert, L. D. (2011). Best practices for allocating appropriate credit and responsibility to authors of multi-authored articles. *Frontiers in Psychology,* V, doi: 10.3389/fpsyg.2011.00196

Fisher, E. (2007). Ethnographic invention: Probing the capacity of laboratory decisions. *NanoEthics, 1*(2), 155–165.

Fisher, E., & Mahajan, R. (2010). Embedding the humanities in engineering: Art, dialogue and a laboratory. In M. E. Gorman (Ed.), *Trading zones and interactional expertise: Creating new kinds of collaboration* (pp. 209–230). Cambridge, MA: MIT Press.

Freeman, R. E., Harrison, J. S., & Wicks, A. C. (2007). *Managing for stakeholders: Survival, reputation, and success.* New Haven, CT: Yale University Press.

Friedman, M. (1996). The social responsibility of business is to increase its profits. In T. Donaldson & P. H. Werhane (Eds.), *Ethical issues in business* (pp. 222–227). Upper Saddle River, NJ: Prentice Hall.

Gorman, M., Werhane, P., & Swami, N. (2009). Moral imagination, trading zones, and the role of the ethicist in nanotechnology. *Nanoethics, 3*, 185–195.

Gorman, M. E. (1992). *Simulating science: Heuristics, mental models and technoscientific thinking.* Bloomington: Indiana University Press.

Gorman, M. E. (1995). Confirmation, disconfirmation and invention: The case of Alexander Graham Bell and the telephone. *Thinking and Reasoning, 1*(1), 31–53.

Gorman, M. E. (1998). *Transforming nature: Ethics, invention and design.* Boston, MA: Kluwer Academic.

Gorman, M.E. (Ed.). (2010). *Trading zones and interactional expertise: Creating new kinds of collaboration*. Cambridge, MA: MIT Press.
Gorman, M. E., Groves, J. F., & Catalano, R. K. (2004). Societal dimensions of nanotechnology. *IEEE Technology and Society Magazine*, 29(55–64).
Gorman, M. E., Wardak, A., Fauss, E., & Swami, N. (2008). Title. In N. Savage et al. (Eds.), *A framework for using nanotechnology to improve water quality* (pp. 491–508). Norwich, NY: William Andrew, Applied Science Publishers.
Gorman, M. E., & Werhane, P. W. (2010). Using trading zones to prevent normalized deviance in organizations. In M. E. Gorman (Ed.), *Trading zones and interactional expertise: Creating new kinds of collaboration* (pp. 245–264). Cambridge, MA: MIT Press.
Hemlin, S. (2009). Creative knowledge environments. An interview study with group members and group leaders of university and industry R&D groups in biotechnology. *Creativity and Innovation Management, 18*(4), 278–285.
Jackson, T. (1997). *Inside Intel: Andy Grove and the rise of the world's most powerful chip company.* New York, NY: Dutton.
Lindqvist, H. (2002). *A history of Sweden*. Stockholm, Sweden: Norstedt Forlag.
McCray, P. (2004). *Giant telescopes*. Cambridge, MA: Harvard University Press.
McLean, B., & Nocera, J. (2010). *All the devils are here: The hidden history of the financial crisis*. New York, NY: Portfolio/Penguin.
Merton, R. K. (1973). *The sociology of science: Theoretical and empirical investigations*. Chicago, IL: University of Chicago Press.
Oetinger, B. v. (2010). Can trading zones and interactional expertise benefit business strategy? In M. E. Gorman (Ed.), *Trading zones and interactional expertise: Creating new kinds of collaboration* (pp. 231–242). Cambridge, MA: MIT Press.
Pacey, A. (1989). *The culture of technology.* Cambridge, MA: MIT Press.
Plowright, R. K., Sokolow, S. H., Gorman, M. E., Daszak, P., & Foley, J. E. (2008). Causal inference in disease ecology: Investigating ecological drivers of disease emergence. *Frontiers in Ecology and the Environment*, *6*, 420–429.
Riordan, M., & Hoddeson, L. (1997). *Crystal fire*. New York, NY: W.W. Norton.
Roberto, M. A., Bohner, R. M. J., Edmonson, A. C., & Ferlins, E. M. (YYYY.) Teaching Note to Columbia's Final Mission (Case HBS No. 5–305–033). Cambridge, MA: Harvard Business School.
Shrager, J. (2005). Diary of an insane cell mechanic. In M. E. Gorman et al. (Eds.), *Scientific and technological thinking* (pp. 119–136). Mahwah, NJ: Erlbaum.
Sen, A. (1999). *Development as freedom*. New York: Random House.
Spohrer, J. C., & Engelbart, D.C. (2004). Converging technologies for enhancing human performance: Science and business perspectives. In M. C. Roco & C. D. Montemagno (Eds.), *The coevolution of human potential and converging technologies* (pp. 50–82). New York, NY: The New York Academy of Sciences.
Spohrer, J. C., McDavid, D., Maglio, P. P., & Cortada, J. W. (2006). NBIC convergence and technology-business coevolution: Towards a services science to increase productivity capacity. In B. Bainbridge & M. C. Roco (Eds.), *Managing nano-bio-info-cogno innovations: Converging technologies in society* (pp. 227–253). Dordrecht, the Netherlands: Springer.
Thagard, P. (2002). Curing cancer? Patrick Lee's path to the reovirus treatment. *International Studies in Philosophy of Science*, *16*, 179–193.
Vaughan, D. (1996). *The Challenger launch decision*. Chicago, IL: University of Chicago Press.
Werhane, P. H. (1999). *Moral imagination and management decision making*. Oxford, England: Oxford University Press.

11 Creative Leadership

Its Meaning and Value for Science, Technology, and Innovation

Gerard J. Puccio, Marie Mance, and Jeffrey Zacko-Smith

Leadership has been a topic of intellectual discourse and systematic research for approximately 100 years; until the 1930s it was primarily the 'person' that was of concern and not what an individual or group of individuals did that mattered (Bass, 1990). During this time, conceptions of what constitutes effective leadership have evolved as a reflection of organizational life and broader societal trends. As we progress into the 21st century, a time that can only be described as 'change laden and innovation heavy', theories and models have begun to connect leadership directly to the ability to initiate and manage change. This evolution has occurred for good reason as the new millennium has ushered in unparalleled levels of complexity and dynamism in our organizational and personal relationships. Perhaps no sector has experienced greater levels of upheaval than in science and technology. For example, it has been reported that technological products undergo fundamental redesign every 6 to 12 months. These short product life cycles are emblematic of what some economists refer to as the age of innovation (Janszen, 2000). In keeping with this view, other economists have suggested that creativity now plays a more significant role in the workplace, communities, and the economy overall (Florida, 2002).

Creativity is often described as the catalyst to innovation, but creativity does not need to be left up to chance. For more than 50 years, the field of creativity studies has delved into the models, processes, and personal qualities that help us to understand the factors necessary to bring about original and valuable change that is important to leaders and leadership in all sectors. The purpose of this chapter is to explore the central role creativity, and specifically creative thinking, plays in leadership in the 21st century. To that end, our goal is to make a case for a new concept of leadership, called 'creative leadership', and to underscore the extent to which creative leadership serves as a catalyst to innovation. In this chapter we propose a definition of creative leadership, as well as a description of the attributes of effective creative leadership, touching on some of the skills that research has shown to be associated with creative leaders. We argue that given the pace of change in science and technology, organizations dependent on scientific breakthroughs and technological developments need creative leadership to survive and thrive in times of rampant change.

We begin this chapter by exploring the nature of creativity and why many claim that creative thinking is an essential skill for all in the 21st century. From our description of creativity we next move to evolving views of leadership and how recent descriptions of effective leadership embrace the concept of creativity. We then present our definition and argument for why creative leadership should be included in the current lexicon of leadership concepts. We close the chapter by exploring how to begin developing the skills of a creative leader.

UNPACKING THE CREATIVITY CONCEPT

Creativity as a 21st-Century Skill

Most authors who talk about life in the 21st century almost always begin their descriptions with some reference to the unprecedented pace of change. Generally the reference to the increased pace of change is presented as a passing statement; often little explanation is provided, as the author most likely assumes that the reader is fully aware of the heightened levels of change through his or her own daily life experiences. But what do the trends really show with regard to the supposed increase in change, and what data exist to clearly demonstrate that life is changing at an ever-faster rate. Here are several points that provide some evidence for the statement that life has increased at an exponential rate:

- In the last 200 years there have been a total of 25 inventions, both social and technical, that have qualitatively altered our lives, and it took more than 3,000 years before 1800 to achieve the same number of life-altering inventions (Henry, 2001).
- Product life cycles have become shorter and shorter. Manufactured products are subject to fundamental redesign every 5 to 10 years, whereas products in the high-tech world are subject to massive redesign every 6 to 12 months (Hunter & Schmidt, 1996; Williams & Yang, 1999).
- In the United States, students in school today are expected to change jobs more than 11 times between the ages of 18 and 42, and it is likely that they will hold more than 22 jobs before they retire from the workforce (Trilling & Fadel, 2009).
- In 1900 less than 10% of the United States population worked in what are considered 'creative' jobs, that is, work that requires higher order thinking and imagination. One hundred years later, this figure increased to 30% (Florida, 2002).
- Content on the World Wide Web doubles every 18 months, which has led to ubiquitous availability of information (Mehaffy, 2010). For example, in 2006 YouTube added 65,000 new videos every 24 hours;

today 2 billion videos are viewed per day. In 2008 Google reached a total of 1 trillion indexed webpages (Trilling & Fadel, 2009).
- In the 20th century the term *social networking* had no meaning; today Facebook has more than 750 million users who spend 700 billion minutes interfacing with this social networking site each month (McCracken, 2011).

The preceding data points are intended to provide concrete evidence of the accelerated pace of change in the 21st century, and although we may be certain that the pace of change has never been greater, what is equally true is that this level of change carries a higher degree of uncertainty, ambiguity, and volatility. As a pre-21st-century report about life in the new millennium, authored by the Hudson Institute (1997), noted,

> The pace of technological change in today's economy has never been greater. It will accelerate even further, in an exponential manner. Innovations in biotechnology, computing, telecommunications and their confluences will bring new products and services that are once marvelous and potentially frightening. And the "creative destruction" wrought by this technology on national economies, firms, and individual workers will be even more powerful in the twenty-first century than when economist Joseph Schumpeter coined the phrase fifty years ago. We cannot know what innovations will transform the global economy by 2020, any more than analysts in the mid-1970s could have foreseen the rise of the personal computer or the proliferation of satellite, fiber-optic, and wireless communications. (p. 2)

It would seem that what is certain about the future is that there will be greater levels of uncertainty. Indeed, a survey carried out by IBM showed that executive leaders around the world anticipate a more than 30% increase in the complexity associated with business environment (IBM, 2010). It is exactly this understanding of the future that has led to what appears to be a clear-cut conclusion among educational strategists; that creativity and innovation are among the most important skills students, and workers in general, must develop to be successful in the 21st century work world (National Center on Education and the Economy, 2008; Trilling & Fadel, 2009). Those who are concerned about how well we are preparing students to join this workplace have almost unanimously cited creativity and innovation as fundamental to professional success. As a book published by the National Center on Education and the Economy (2008) noted,

> For the past 25 years, we have optimized our organizations for efficiency and quality. Over the next quarter century, we must optimize our entire society for innovation . . . Creativity, innovation, and flexibility will not be the special province of an elite. It will be demanded of

> virtually everyone who is making a decent living, from graphic artists to assembly line workers, from insurance brokers to home builders. (p. 25)

Although the recent call for the importance of creativity in today's society is now more widely recognized, those in the field of creativity studies have been making this argument for more than 50 years (Guilford, 1950; Osborn, 1953). With respect to the overall importance of creativity in today's world, Trilling and Fadel (2009) said it best when they stated that

> As we journey deeper into the 21st century, creativity and innovation will become the brightest stars in the constellation of 21st century skills. New ideas, innovative products, novel services, and fresh solutions to local and global problems will increasingly power our emerging Innovation Age. (p. 104)

The complexities of the 21st century are not limited to the workplace or to the broader economy. Change, whether technical or social, cuts across all aspects of our lives, both professional and personal. Therefore, in our own work we have broadened the importance and relevance of creativity and creative thinking beyond a strictly workplace skill, and have referred to it as an essential life skill. As Puccio and Murdock described in 2001,

> Creative thinking is an essential life skill. It is a rational process that enables people to successfully produce novel and useful responses to open-ended challenges and opportunities. Creative thinking involves specific cognitive, metacognitive, and affective skills. Once internalized, these skills can be readily applied to all areas of life. (p. 70)

Perplexing situations that require imaginative responses are not limited to one narrow segment of life, whether one is concerned about a worker's ability to maintain a high standard of living in today's workplace (which is generally the reason provided by educational reformers for the importance of creative thinking as a 21st-century skill) or is focused on an organization's need to drive innovation in order to survive in the fast-changing economic landscape. Or perhaps one is the entrepreneur looking for the next business opportunity, the social entrepreneur who desires to create a sustained solution to a community challenge, an executive in a not-for-profit organization that is dealing with reduced levels of funding, an at-risk student who is struggling to maintain motivation in school, or a parent who is challenged by a teenage child. All of these life and work situations require creative thinking. And this is the exact reason why our vision at the International Center for Studies in Creativity, at Buffalo State College, is to "Ignite creativity around the world: Facilitating the recognition that creative thinking is a life skill". Although there has never been greater recognition of the importance of creative thinking, we are somewhat concerned that the main call for this skill is pinned primarily to professional and workplace success. Imagination knows no boundaries, and thus, creative thinking can and should be used to confront all complexities brought on by the 21st century, that is, personal, professional, social, political, and so on.

Defining Creativity

So what is creativity? How is it described and defined? Since the inception of the scientific study of creativity, which occurred in the middle of the 20th century, numerous definitions of this multifaceted phenomenon have been put forth. Although complete convergence on a single definition has not occurred, there is a description of creativity that has been adopted widely. Many of today's creativity scholars now define creativity as the ability to produce original ideas that serve some value or solve some problem (Kaufmann, 1988; MacKinnon, 1978; Talbot, 1997). This view counters the misconception that creativity is strictly about imagination, diversity in thought, or simply standing out as different. Rather, true creativity involves three ingredients, novelty, usefulness and realization. This definition draws a close conceptual relationship between creativity and problem solving. Kaufmann (1988) defined a problem as "a discrepancy between an existing situation and a desired state of affairs" (p. 88). And in categorizing types of problems, Kaufmann indicated that problems range along a continuum from 'well-structured' to 'ill-structured'. According to Kaufmann, three distinct features are indicative of ill-structured problems, namely, novelty, complexity, and ambiguity. Novelty refers to the lack of familiarity with the situation, the goal, or the challenge. Complexity relates to the number of pieces, variables, and elements that must be managed in order to resolve the situation. And ambiguity is present when there is a high degree of uncertainty as to which goal structures, solution paths, or outcomes are correct or most desirable. Well-structured problems do not necessarily require creative thinking; however, when a problem is novel, complex, and ambiguous, creative thinking is needed to discover a pathway that will close the gap between the dissatisfaction with the current state and the desired future state.

Since the mid-1950s, many scholars and practitioners have developed models that capture, in a systematic manner, the nature of the creative process. That is, how individuals and groups could apply their creative thinking in a way that would enhance the probability of producing novel and useful solutions to ill-structured problems. Through this line of work a number of creative process models have emerged that have been studied and applied over the years, such models as Creative Problem Solving (Osborn, 1953; Puccio, Mance, & Murdock, 2011), Synectics (Gordon, 1961), TRIZ (Altshuller, 2001), Design Thinking (Kelley & Littman, 2001), and de Bono's many techniques, such as his six thinking hats (de Bono, 1999) and lateral thinking (de Bono, 1977). These, and other models, were designed to make the creative process more deliberate, repeatable, and teachable, thereby improving the likelihood of a creative breakthrough rather than it occurring by mere chance.

It is beyond the scope of this chapter to provide in-depth descriptions of these creative-process models. That said, to understand how these deliberate creative processes work, we do believe it is valuable to have a minimal introduction to one of these models. For this purpose we briefly overview creative problem solving (CPS) because it is the model for which we devote much of our instruction and research efforts. For information about the other creative process models see the primary sources cited earlier or go to Puccio and Cabra (2010) for an overview and description of each of these creative-process models. CPS is a structured process designed to improve an individual's creative thinking skills and to enhance the probability that both individuals and teams can successfully resolve complex problems through the application of their imaginations.

The chapters in this section identified the challenge leaders face when attempting to foster collaboration among creative individuals who are accustomed to working alone. This is particularly pronounced for those who work in science and technology. Reiter-Palmon, de Vreede, and de Vreede (this volume) highlighted the fact that the use of interdisciplinary teams for the purpose of generating innovative breakthroughs is on the rise. Gorman and Gorman (in this volume) extended the importance of teamwork when they noted that collaborative engagement is not only important for the development of new and useful products, but is necessary for our collective future. When CPS becomes the shared creativity framework within a team, it provides the common language that enhances communication and coordination, similar to the benefits of shared mental models as noted by Reiter-Palmon et al.

Figure 11.1 provides a graphic depiction of the CPS framework. In the center of the model is a step called Assessing the Situation. This is referred to as the meta-step as all creative thinking efforts begin here, by developing an awareness of the situation at hand, whether it is a predicament that has presented itself or an opportunity that is being explored. Assessing the Situation has two objectives: (a) to develop a deep understanding of the present situation by identifying and describing relevant data and (b) to determine whether creative thinking is necessary and, if so, where to begin in the CPS process. If the problem solver determines that the present situation requires creative thinking, then he or she moves into the appropriate CPS step and then advances through the process as necessary. The remaining six steps are organized into three stages. If the problem solver needs to spend time further clarifying the situation, that is, either determining a vivid image of the desired end state (Formulating a Vision), or identifying specific challenges that must be addressed to achieve the future vision (Formulating the Challenges), then he or she will move into the Clarification stage of CPS. If the problem solver has a clear sense of the challenges, but does not yet have a solution in mind, then he or she would move into the Transformation stage, where tentative solutions would be generated (Exploring Ideas) and then

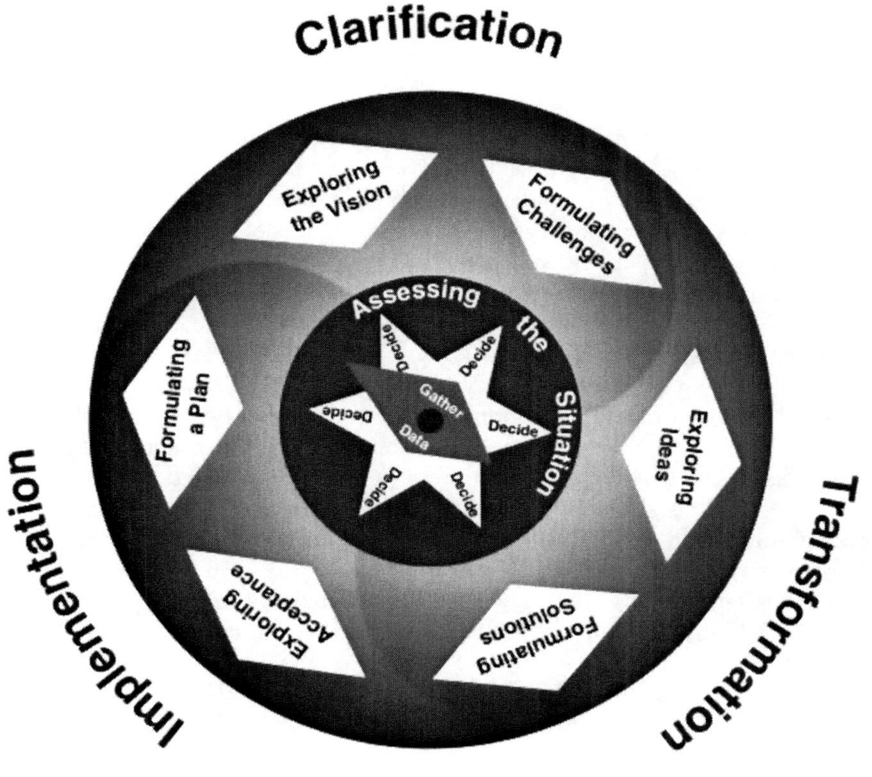

Figure 11.1 Creative problem solving: The Thinking Skills Model

developed into workable solutions (Formulating Solutions). Finally, when the problem solver is ready to bring a solution to fruition, he or she would work on the steps in the Implementation stage. Here he or she would begin with examining the context for sources that will either assist or resist the proposed solution (Exploring Acceptance) and then use this understanding to develop action steps that enhance the probability of having the solution adopted (Formulating a Plan).

The reader should note that a diamond shape depicts each step of the CPS process. This iconography represents the important balance between divergent and convergent thinking for all steps of the process. That is, each step begins with a search for many varied and original options (divergent thinking), followed by a careful culling and evaluation of the most promising alternatives (convergent thinking). The alternating movement between divergent and convergent thinking is the hallmark of the CPS process. CPS training is designed to help learners internalize this process model and to develop creative thinking skills, such as divergent (i.e., fluency, flexibility and originality in thought) and convergent thinking (i.e., screening, sorting and prioritizing). The CPS process subsumes a range of additional thinking skills and specific tools designed to carry out these skills. For a complete description see Puccio, Mance and Murdock (2011). We have spent more time digging into CPS, as later we make the case that leadership in the 21st century requires an ability to creatively solve complex problems; that is, leaders need to be creative problem solvers.

A number of empirical studies have closely examined the degree to which creative-process models, such as the CPS process just described, have been successful in improving individual creative performance (Puccio, Firestien, Coyle, & Masucci, 2006; Rose & Lin, 1984; Scott, Leritz, & Mumford, 2004a, 2004b; Torrance, 1972; Torrance & Presbury, 1984). Scott, Leritz, and Mumford (2004a, 2004b) carried out the most recent, and perhaps most robust, review of the effectiveness of creativity training programs. In one study, Scott et al. (2004b) reviewed 156 training programs to identify the most effective approaches for enhancing creative capacities. One of the conclusions drawn from this review was that creative process training programs that took a cognitive approach, such as CPS, were shown to be among the most effective. In examining the specific effects of creativity training, Scott et al. (2004a) reported that creativity-training programs were shown to significantly improve divergent thinking, problem solving, attitudes, and behavior. In summarizing their research findings regarding creativity training, these authors concluded that

> The results obtained in this study indicate that well-designed training can evidence substantial external validity. Creativity training contributed to divergent thinking, problem solving, performance, and attitudes and behavior for younger and older students and working adults, and for high achieving and more "run of the mill" students . . . Taken as a whole, these observations lead to a relatively unambiguous conclusion. Creativity training works. (p. 382)

Our review of the meaning of creativity has focused mainly on the creative process as it is applied to ill-structured problems. The process facet of creativity represents only one dimension of this complex phenomenon. Many scholars agree that creativity is made up of many facets, with four routinely cited in the literature. The creative process, as we have already described, which focuses on the stages of thought people engage in when solving problems in a creative manner. The next facet is the creative product, which refers to the tangible and intangible outcomes of creative thinking. Another major facet of creativity is the person, which relates to the study of individual attributes, attitudes, traits, and skills that contribute to creative behavior. The person facet has received the earliest and greatest attention with respect to creativity research. Early research focused on identifying the qualities that were useful in distinguishing highly creative individuals from their less creative counterparts (MacKinnon, 1962). Here the fundamental question was whether researchers could isolate the qualities that would be useful in predicting future creative performance. More recent research related to the creative person has moved away from strictly predicting the degree or amount of creativity and has explored the different forms and expressions of creativity (Basadur, Wakabayashi, & Graen, 1990; Kirton, 1976; Puccio, 1999). This approach, commonly referred to as 'style' of creativity, challenges the view that creative expression will come in only one form or manner. Finally, the fourth commonly identified facet of creativity is the environment. By environment we mean the physical and psychological conditions that either facilitate or inhibit creative behavior. Some of the more notable work carried out in this area has been conducted by West (1997); Anderson and West (1998); Ekvall (1996); Amabile, Burnside, and Gryskiewicz (1999); and, more specifically, in relationship to scientific and technological creativity, by Hemlin, Allwood, and Martin (2008).

Perhaps the first scholar to cite the four fundamental aspects of creativity just described, that is, process, product, person, and environment, was Rhodes (1961). When Rhodes originally described these four facets he noted that although these dimensions could be studied independently, in reality they interact and influence one another, creating a more holistic depiction of creativity. Csikszentmihalyi (1990) aptly described the importance of a holistic or interactionist approach to the study of creativity when he stated, "to study creativity by focusing on the individual alone is like trying to understand how an apple tree produces fruit by looking only at the tree and ignoring an apple, the sun and the soil that support its life" (p. 202). We hold a similar view of the importance of understanding creativity as an interaction of broader dimensions (see Murdock & Puccio, 1993), and therefore put forward a visual model with the purpose of depicting these interactions (Puccio et al., 2011). Adopting the four fundamental facets of creativity originally described by Rhodes, and supported by numerous other scholars (MacKinnon, 1978; Mooney, 1963; Stein, 1968), Figure 11.2 shows our interpretation of how these facets interact and influence one another within a system, such as an organization. To review models specifically related to organizational creativity see the work of Woodman, Sawyer, and Griffin (1993); Ford (1995); and Locke and Kirkpatrick (1995).

We present this systems model for a number of reasons. First, this systems model is designed to highlight the crucial factors that interact to generate creative outcomes. Second, this model is designed to show the relationships between the fundamental areas of creativity, that is, to move from a theoretical description of creativity to a practical one. Third, we wish to use this model to show the relationship between creativity and innovation. Finally, the model is presented to demonstrate the role of creative leadership within this system.

Creativity always occurs within some context, that is, home, classroom, organization, community, society, and so on. In each of these contexts you have an individual or set of individuals engaging in some thought process that is stimulated, to larger or lesser degrees, by their immediate environment (both psychological and physical). In these settings individuals bring to bear their individual differences, such as knowledge, imagination, experience, attitudes,

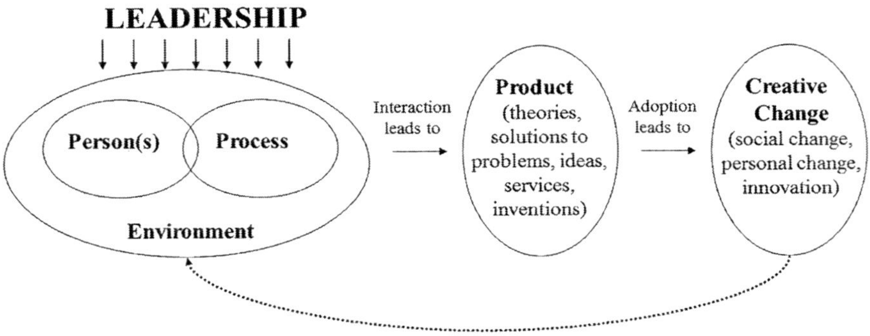

Figure 11.2 Creative change model: A systems approach

passions, and personality traits. Mixed with these individual characteristics is the process or methods individuals use to address needs, problems, and opportunities that require new thinking. Whether in a deliberate and structured manner, or by happenstance, creative thinking must be applied to resolve the complex problems that will natural arise in these various settings. Environment surrounds the person and process facet as the nature of the physical and psychological will either support individuals and facilitate creative thinking or limit individuals (i.e., their knowledge, imagination, passion, etc.) and undermine their efforts to engage in creative thinking. The positive interaction among these three elements leads to creative breakthrough, a tangible or intangible product. Creative products can take on many forms, including a new theory, a solution to a problem, a new service or manufactured product, an invention, and so on. As a holistic system, the facets of creativity are never left behind, and individuals must continue to employ creative thinking, and understand the environment, in order to have the creative product adopted. Sometimes the greatest creative effort does not come along with the breakthrough, but is in getting others to accept the proposed new product. When eventually adopted, creative change is realized. Like the creative product, creative change can come in many forms. When individuals adopt change, experiencing a shift in careers or adopting new lifestyles, we call this personal change. This view of creativity aligns with the self-actualized theories of creativity, in that through creativity we reach our highest potential (Maslow, 1954; Rogers, 1962). In other cases the change is designed to be more public, such as the pursuit of social change, as we have witnessed recently in the Middle East and Africa or historically as in the civil rights and women's rights movements.

Innovation can be thought of as another form of public change, that is, a creative change that occurs in the marketplace, the industry, or within an organization. Innovation can be defined as the adoption of a new and valuable idea (Amabile et al., 1999). These new and valuable ideas can result in new business models, organizational practices, products, or services. An important point here is that innovation and creativity are not synonymous or interchangeable terms; rather, they are interconnected whereby creativity (the interaction of person, process, and environment) will lead to innovation. For a more detailed description of the relationship between creativity and innovation, see Puccio and Cabra (2010).

As a system, there is an iterative effect caused by the adoption of creative change. The context will be impacted, to varying degrees, by the adoption of the creative change. Consider how digital music has had a radical impact on so many facets of the recording industry or how the digital books have changed the nature of book publishing and sales.

To the traditional creativity facets of person, process, environment, and product, we add a variable—creative leadership. It is our contention that the leadership behavior present in any given context, especially when it takes the form of creative leadership, is essential to this system, and, for purposes of this chapter, especially with reference to innovation as an organizational output. Our position with reference to the role of leadership within this system is reinforced by the Boston Consulting Group's 2007 survey of senior management that concluded (Andrew, Sirkin, Haanæs, & Michael, 2007):

> At the end of the day, a company's innovation efforts will succeed or fail depending on the quality of its leadership. A strong, committed leader will create an environment that supports innovation and drives it forward. (p. 26)

The extent to which the environment nurtures individuals and guides effective creative processes will be largely determined by the degree to which leadership operates in a creative manner. This underscores the fractal nature of the system, in that for leaders to have a positive influence on environments, people, and processes, they must personally possess the very same creativity principles and dimensions they wish to nurture in a particular setting. As Kouzes and Posner (1995) would say, leaders must model the way. The remainder of this chapter explores the concept of creative leadership, first by examining evolving views of leadership and then by specifically defining and delineating how we view creative leadership. We begin by locating the concept of creative leadership within the broader landscape of leadership theory and study. We then present a specific definition and description of creative leadership.

EVOLVING VIEWS OF LEADERSHIP: FROM GREAT MAN TO CREATIVE LEADERSHIP

When devising a method and model to help students both effectively understand leadership and develop skills associated with leadership effectiveness (skills such as facilitating group processes and engaging in effective communication and influence behaviors) Zacko-Smith (2010) outlined the

Three-T's Framework, which is based on the work of Gardiner (2006), who implicitly described leadership as existing distinctly along a continuum of understanding and practice. Starting with transactional leadership (Weber, 1947), which is technically and historically anchored in the very beginning of the formal study of the field itself, and is associated with productivity, efficiency and the exchange of work for reward (an assembly-line or a production-oriented mentality), the three parts of the framework build upon one another; it should be stated at the outset that no stopping point on this continuum is any better or any worse than any other, but rather all are important and useful in proper context (Hersey & Blanchard, 1977). The second part of the framework, transformational leadership (Burns, 1978), which was especially globally popular in the 1970s and 1980s, makes leadership much more relational, and is concerned with the empowerment of both the leader and constituents (the authors find the term *followers* disempowering and thus avoid its use); the level of relationship between the leader and constituents is greatly deepened. Finally, the last and most recent concept, one that has come to be known as transcendent leadership (Aldon, 2004; Gardiner, 2006; Larkin, 1995; Zacko-Smith, 2010), necessitates full, deep, and very authentic relationships between those involved in the leadership relationship, and requires that they all "transcended self into compassionate being and action," demonstrating a "leadership of service above self" (Gardiner 2006, p. 3). In the simplest terms, transactional leadership is about the individual (it is more autocratic), transformational leadership is about individuals in relationship (and is thus more democratic), and transcendent leadership is about the relationship's affect on the larger community (it is more holistic and laissez-faire); in fact, in the largest sense, it asks that all involved consider the effect their leadership relationship has on the rest of the world. An example of a transcendent leader would be Mohandas Gandhi; Henry Ford can serve to illustrate transactional leadership; and Steve Jobs of Apple Computers certainly serves as an example of the transformational leader.

As Zacko-Smith (2010) points out, virtually all leadership theory and practice, no matter the sector or situation, can fit into one or more parts of the Three-T's Framework, and many do, in fact, overlap distinctly. For example, the behavioral theory of leadership can be seen as falling under the second T, transformational leadership, because it stresses the impact of the relationship between leader and constituents, is more deeply interactive, and is meant to produce results that are about both personal and skill development (vs. basic task and/or organizational success, which would be a more transactional focus because it is not concerned with the person, per se). However, it can also be viewed through the lens of transactional leadership, or even transcendent leadership, depending on the level of relationship required and the context.

As illustrated by Figure 11.3, we would contend that creative leadership, specifically because it is highly relational, is grounded in process, utilizes diversity, requires self-reflection and is focused on accomplishing positive change (which is one aspect that differentiates it from change leadership in general), resides directly between the transformational and transcendent approaches (Zacko-Smith, 2010). This means that creative leadership is positioned very similarly to global, situational, relational, and even authentic leadership, among others. As will be shown, however, it introduces both a unique approach and a very actionable skill set that draws on decades of research on creativity, creative problem solving and innovation and that stresses not only the innate creative abilities of every one of us, but our innate ability to lead as well; like creativity, leadership is clearly something that can be learned (Bass, 1985; Conger, 1996), hence its interdisciplinary popularity as a subject and application within a variety of settings.

Creative Leadership, as we describe it above, is a new leadership paradigm, though one that, as with most leadership theories, is related to multiple other ideas and theories, as illustrated by Figure 11.3. In our rapidly changing and highly diverse world, skill in leading creatively is a crucial part of any leader's toolbox (Zacko-Smith, 2010); we should be ready to access it when necessary. Although we do not purport that creative leadership is universally applicable nor is even generally necessitated, as a paradigm and practice it offers leaders a distinct skill set, one that is well matched with contexts that require utilizing diversity (which is something that is all around

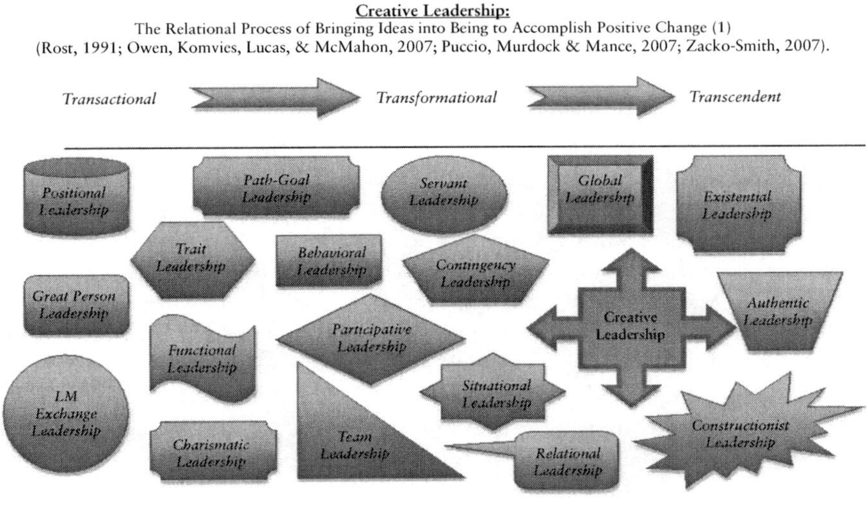

(1) Relational Leadership (Rl.) does not necessarily specify types of change outcomes, though Creative Leadership (Cl.) does (i.e. outcomes are positive. In addition leadership is viewed as a process since creativity generally requires space for emergence and incubation. CL is also relational in nature since such models are tied to knowledge of both self and other, and invite all stakeholders to become involved in processes, activities and decisions. Creative Leadership fils within the post industrial paradigms established by Roll (1991) and others, which expand understanding of transformation, revenue hierarchical structure (i.e. they invert the traditional "top down" leadership pyramid), stress influence and have "real purpose". Additionally. CL is considered "beginning transcendent" as it's skill set encourages intuitive and "higher order" decision making (multiple iterations. CL is also related to models that are non-essentialist and constructed, as discussed by Zocku-smith (2007), since it acknowledges the created nature of reality, the importance context, the need for flexibility and the usefulness of diversity (multiple perspectives). ***Model developed by Jeffrey Zacko-Smith, 2009***

Figure 11.3 Creative leadership's location within the field of leadership studies

us anyway), possessing a high tolerance for ambiguity (rapid change offers it as a by-product), and having the ability to initiate and manage change to meet personal, organizational, technological and, perhaps most important, mutually beneficial societal and social goals.

DEFINING CREATIVE LEADERSHIP

We offer creative leadership as a novel construct within the lexicon of leadership concepts and theories; where it fits is well illustrated by Figure 11.3. Leadership, as a primarily social and psychological phenomenon, and which by this we mean that it is fully relational (Komives, Lucas, & Mcmahon, 2006) will, and must, evolve with the times, and it would certainly seem that life in the 21st century calls for leaders to be creative in multiple arenas of their practice and to go further by nurturing the creativity of others (granting creative leadership its transformational status); this is a call applicable to every sector, from education to the sciences. As noted earlier, the exponential increase in change, which appears amplified in the sciences and technology, will result in greater levels of both ambiguity and chaos for organizations and communities. Therefore, for leaders to be most effective and successful, whether in positions of power or as self-directed/self-governing entities, they must be able to successfully manage change processes and resolve the complex problems generated by an almost constant evolution of ideas and needs; we believe success and effectiveness in a world of continuous flux has rendered creativity a core leadership competency. The new norm is constant change and thus the antidote is creative leadership. Heifetz, Grashow, and Linsky (2009) described the current environment in which leaders exist, a description that is even more applicable today:

> It would be profoundly reassuring to view the current economic crisis as simply another rough spell that we need to get through. Unfortunately, though today's mix of urgency, high stakes, and uncertainty, will continue as the norm even after the recession ends. Economies cannot erect a firewall against intensifying global competition, energy constraints, climate change, and political instability. The immediate crisis—which we will get through, with the help of policy makers' expert technical adjustments—merely sets the stage for sustained or even permanent crisis of serious and unfamiliar challenges. (p. 62)

And in response to this current reality, Heifetz et al. (2009) went on to describe how leadership must evolve toward what we think of and describe as creative leadership:

> In this context, leadership is an improvisational and experimental art. The skills that enabled most executives to reach their positions

> of command—analytical problem solving, crisp decision-making, the articulation of clear direction—can get in the way of success. Although these skills will at times still be appropriate, the adaptive phase of a crisis requires some new leadership practices. (p. 65)

A recent global survey of more than 1,500 CEOs further reinforces the creativity–leadership connection by underscoring the fact that leaders now recognize for themselves that success depends in large measure on their creativity; they acknowledge that creative thinking and problem solving are crucial to meeting organizational and societal goals and objectives (IBM, 2010). In this survey these executive-level leaders were asked to rank the most important leadership qualities needed over the next 5 years. Creativity was ranked number one with 60% of these executive leaders citing this skill as being of paramount importance for responding effectively to the current diverse and innovation laden business environment. As this report indicated,

> CEOs now realize that creativity trumps other leadership characteristics. Creative leaders are comfortable with ambiguity and experimentation. To connect with and inspire a new generation, they lead and interact in entirely new ways. (p. 23)

The discussion thus far makes the case that creativity is a core leadership competency, but our position goes a bit further and holds that the very nature of leadership has changed and thus has ushered in an entirely new form of leadership—creative leadership. Leadership in the 21st century has been fundamentally altered by the contexts it is embedded within. It is not about acquiring a few more tricks and skills, finding people with the proper traits and education, or having tools that can be applied and then shelved. Rather, we maintain that given the permanency of change, or crisis depending on one's perspective, the nature of leadership has been altered. And as we have argued elsewhere (Puccio et al., 2011), it is change that forms an inextricable bond between leadership and creativity; it is the glue that binds the two concepts strongly together. We look to leaders to drive change and/or to help us adapt to change, and creativity, as described earlier, can be defined as a process that leads directly to change; change, in fact, is both desired and serves as its end product by default. Therefore, creative leadership is about embodying change (i.e., that is an individual who does not just use creativity as a tool, but who also embodies the spirit of creativity in his or her leadership behavior). For instance, someone who constantly and consistently challenges the status quo, who persistently looks to future possibilities, who embraces the opportunities that come along with ambiguity and disruption, and who lives and works in a flexible way. With change, and particularly with creative change (i.e., the deliberate introduction of something that is both original and valuable), as the foundation on which we build our concept of creative leadership, we define this term in the following manner:

> The ability to deliberately engage one's imagination to define and guide a group towards a novel goal—a direction that is new for the group. As a consequence of bringing about this creative change, creative leaders have a profoundly positive influence on their context (i.e., workplace, community, school, family, etc.) and the individuals in that situation. (Puccio et al., 2011, p. 28)

This definition suggests that leaders in the 21st century make a conscious choice to search for new opportunities, invoking their own and others creative thinking in a deliberate manner. This definition also embodies the systems model of creativity described earlier, as leaders draw out and nurture others and the environments in which they work and live in so that creativity can flourish. Complex problems cannot, and should not, be solved by the leader alone. Thus, it becomes crucially important for leaders, both those in positions of authority and those who choose to make a difference through their influence, to deliberately engage in a process that draws on the creative thinking of others. When faced with complex problems and changing circumstances, diversity in perspectives, ideas, views, and opinions is necessary to ensure that a new and valuable approach is discovered.

To further expound on our view of creative leadership we present a figure that draws together three distinct but highly interconnected constructs—creativity, leadership, and innovation. The Venn diagram shown in Figure 11.4 illustrates how these three concepts interact and specifically highlights some of the points of intersection. Earlier we discussed the connection between creativity and leadership, so we do not delve any deeper into this area. We will, however, explore further where leadership intersects with creativity and with innovation. Our goal here is to highlight what we believe creative leaders do.

One concept that draws together leadership and creativity is creative problem solving. Earlier we described how creativity is a process for responding to ill-structured problems. Some now argue that a main capacity required to ensure leadership success in the 21st century is creative problem solving (Mumford, Zaccaro, Harding, Jacobs, & Flieshmann, 2000). Mumford and his colleagues posited a capacity theory of leadership that suggested that leadership performance is based squarely on an ability to resolve complex social problems. The qualities they used to define a complex social problem closely parallels the characteristics of ill-structured problems presented earlier. Specifically, complex social problems are marked by complexity (e.g., ill-defined problems with no single solution path), novelty (e.g., emerging situations for which past experience is not sufficient), and ambiguity (e.g., gaps in information). Mumford et al. (2000) suggested that such problems cannot be solved with routine thinking. Rather, creative thinking must be used to reshape and reform past knowledge and to discover or invent new ways forward. As such, these authors concluded that "The skills involved in creative problem solving influence leaders performance" (p. 17). It is for this

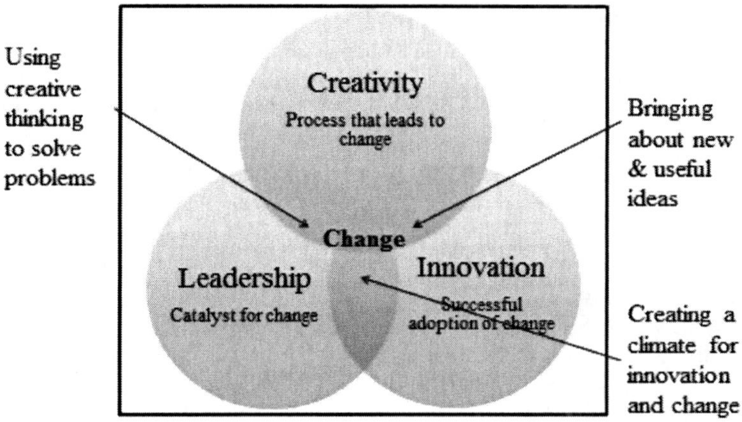

Figure 11.4 Understanding creative leadership through the intersections of creativity, leadership, and innovation

reason that we maintain that training in CPS does much to uplift leadership capabilities (Puccio et al., 2011).

Mumford and colleagues (e.g., Mumford, Baughman, Maher, Costanza, & Supinski, 1997; Mumford, Baughman, Supinski, & Maher, 1996; Mumford, Baughman, Threlfall, Supinski, & Costanza, 1996; Mumford, Supinski, Threlfall, & Baughman, 1996; Reiter-Palmon & Illies, 2004) have carried out research that has unpacked the cognitive abilities associated with creative problem solving, and Puccio et al. (2011) have outlined the thinking skills and strategies leaders can use to become more effective at resolving open-ended problems and, ultimately, to bring about change. Given the important relationship between leadership and creativity, especially with respect to organizational creativity, it is highly likely that leadership will continue to receive great attention by those interested in how creativity manifests itself in teams and organizations. Simply put, leadership behavior has emerged as the one of the most potent variables in predicting creativity in teams and organizations.

Where leadership and innovation intersect we wish to highlight the role leaders play in creating an environment that promotes innovation. Ekvall's (1983, 1991, 1996, 1997) research, for example, has illuminated the dimensions of psychological climate that promote a work atmosphere conducive to creative thinking and ultimately innovation. Ekvall, who has studied climate for the last 50 years, identified 10 dimensions of the creative climate, namely, dynamism, challenge, freedom, trust and openness, idea support, conflict, debate, idea time, playfulness/humor, and risk taking. Using his measure of the creative climate, the Creative Climate Questionnaire, Ekvall (1991, 1996, 1997) has demonstrated significant differences between organizations and units within organizations with respect to their creative output. In one of his most elaborate studies, Ekvall (1991) had independent researchers from a business school apply their model for assessing innovation to 27 different organizations. The application of the innovation criteria enabled the researchers to place each organization into one of three categories: innovative, average, or stagnant. Employees in these organizations also completed Ekvall's (1991) climate measure. Analysis showed that the most positive climates for creativity were found among those employees in organizations identified as being innovative.

More recent research has examined the nature of the relationship between leadership and the climate for creativity. Ekvall and others (Ekvall & Ryhammar, 1999; Sellegren, Ekvall & Tomson, 2008) have discovered that leadership acts as an intervening variable. That is, although leadership has a small direct effect on innovation, it does have a significant and profound effect on the climate for creativity. Thus, it is argued that for innovation output to be enhanced in work units and organizations, leaders should focus on creating a climate for creative thinking, and in turn, this creative climate will yield innovative outcomes. As noted earlier, there are other scholars who have delved into the nature of the work environment conducive to creativity and ultimately innovation. For a review and comparison of the creative environment dimensions proposed by different authors, see Puccio and Cabra (2010). For a critical review of the measures designed to assess the environment for creativity, see Mathisen and Einarsen (2004).

Through this figure we have attempted to underscore the relationships among creativity, innovation, and leadership. In particular, our goal was to underscore how leaders can adopt creative leadership practices. Here we focused on two important aspects: (a) the need for leaders to engage in creative problem solving and (b) the need for leaders to foster work climates that promote creative thinking and as a consequence produce greater degrees of innovative results. In the closing section of this chapter we further explore qualities and impact of creative leadership.

DEVELOPING CREATIVE LEADERSHIP SKILLS

So, how do you move from being a leader to being a creative leader? We have made the case that creative leaders are needed for the challenges of the 21st century; so what steps can one take to enhance one's own creativity skills and those of others in an organization? With respect to the latter point in this question, as Ligon et al. noted in their chapter in this volume, the impending wave of retirements among baby boomers puts great pressure on organizations to seriously explore succession planning for leaders of innovation. To explore issues related to developing creative leaders, we first describe a model for the development of creative leadership, Figure 11.5, (Puccio et al., 2011). Next, we present some practical suggestions that can be implemented immediately to enhance creativity. These suggestions are organized into two parts, using creativity to solve problems and creating a climate for change. These correspond to Figure 11.4 that illustrates the intersections of creativity and leadership (using creativity to solve problems) and leadership and innovation (creating a climate for creativity and change).

Creativity skills and behaviors can be learned, just like any other subject matter. This basic premise allows us to create an approach to growth and development that is both structured and deliberate. In the model, Figure 11.5, A Model for the Development of Creative Change Leaders, we describe four distinct levels of skill: Spectator, Student of Creativity, Skilled Facilitator, and Creative Change Leader. Beginning at the lowest level, the Spectator is someone who is unaware of the importance of creativity for leadership and does not take any specific actions to bring a creative approach to workplace or personal issues. To move to the next level, Student of Creativity, requires an awareness of the connection between creativity and leadership and a willingness to begin studying the creative process and aspects that influence creativity, such as climate and individual approaches to expressing creativity (psychological diversity). The third level, Skilled Facilitator, is described as a person who has gained some proficiency in the use of creative processes and tools and who practices them in real-life circumstances to deliberately bring creativity to bear in appropriate situations (i.e., ill-structured or complex problems). Finally, the Creative Change Leader has internalized the creativity skills and methods so that they are second nature and are a way of being. The integration of the creativity skills is such that it permeates the way this person interacts with others. This leader does not require a structured 'creative problem-solving meeting' to use creative approaches. This distinction between a Skilled Facilitator who uses creativity as a tool, and the Creative

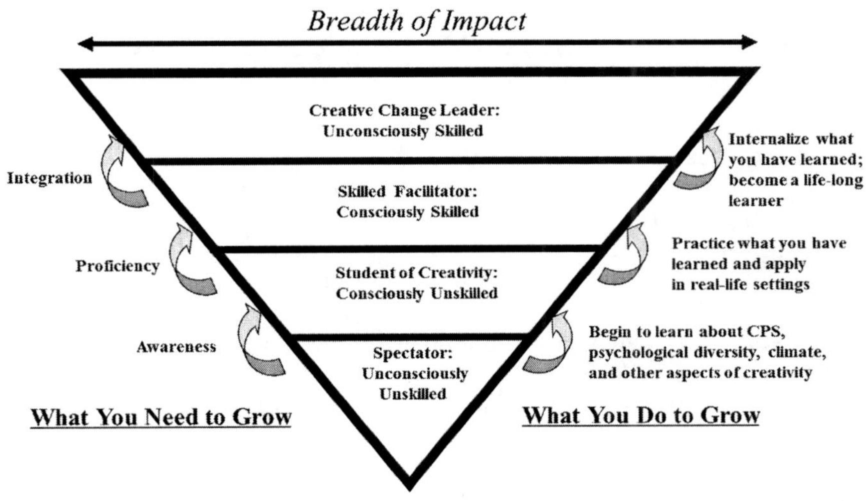

Figure 11.5 A model for the development of creative change leaders

Change Leader, who consistently lives and behaves in a creative manner, underscores the point we made earlier in our definition of creative leadership—that is, although some would suggest that creative problem solving is a tool or a core skill associated with effective leadership, our view of creative leadership goes further and suggests that it is about internalizing creativity. That is, creative leaders model the importance of creativity and innovation in regular conversations, budget discussions, promotion decisions, strategic planning, problem solving, and other organizational business.

We encourage leaders to evaluate where they fall on this scale, to consider a course of development to move from their current level, and, hopefully, to aspire to the highest level, the Creative Change Leader. Although there are a number of formal training programs and resources available to learn creative process, tools, and approaches, we now provide some suggestions that can be implemented immediately to enhance creative thinking in organizations.

Creativity and Leadership: Using Creative Thinking to Solve Problems

As we have stated, leaders need to solve complex, novel problems, yet most never received formal training in creative problem solving. In fact, most were never encouraged to deliberately use creative thinking in their careers. Now, as the need to lead change has become more critical, the need for creative thinking skills has never been more apparent. So, how can a leader use creative thinking to solve problems and encourage employees to do the same? In our limited space here, we have chosen to highlight some of the basic principles of creative thinking that a leader can implement immediately. These can be used with individuals or groups to stimulate creativity.

Creativity Principle: Separate and Balance Divergence and Convergence

One of the very basic principles of creative thinking is to separate the generation of ideas or options from their evaluation and selection. If you do nothing else, following this principle will increase your creative thinking. We are conditioned from a very young age to use critical thinking to evaluate the effectiveness of a concept, an idea, or an approach. Although this was very useful to our ancient ancestors—'I see a berry. Do I eat it or not?' and is still important in some situations—it is not helpful to creative thinking. In fact, the knee-jerk reaction of bringing up an idea and then quickly judging that it stifles creativity. We often compare this to trying to drive a car while having one foot on the accelerator and the other on the brake. If instead you first generate ideas (step on the gas), and then evaluate them (hit the brake), you will find that not only do you have more options to consider, that you are more likely to have more novelty.

The other point that we want to make here is that it is necessary, when seeking creative outcomes, to balance the use of divergence (i.e., generating many diverse and original options) and convergence (i.e., selecting and developing the most promising options). In fact, CPS is structured to reflect the equal importance of generating and evaluating, as symbolized by the diamond shapes in the CPS model. We often compare this balance to the process of breathing. Consider the natural rhythm of breathing in and breathing out. This is the same as the rhythm of divergence and convergence. They need to be balanced. Think of when you have experienced an overuse of divergence, perhaps a meeting during which many ideas are generated but no decisions are made and nothing comes of the ideas. On the other hand, an overuse of convergence might be reflected in a meeting during which the first idea generated becomes the solution without exploring other possibilities; that is, a decision or conclusion is reached so quickly that other more original or potent options are never even considered.

Creativity Principle: Defer Judgment

If you ask a group if they have ever used brainstorming they will invariably respond affirmatively. And if probed about the guidelines for brainstorming, a fairly consistent group response is 'Don't judge ideas'. Although technically true, a more precise statement would be 'Defer judgment'.

Creative thinking is enhanced when the flow is allowed to continue without the interruption of judgment, either positive or negative. Returning to the car analogy, consider how much more quickly you get to your destination by driving the highway, as compared to taking the local streets with stop-and-go traffic.

Creative thinking requires judgment at some point, however when generating options it is necessary to delay the judgment phase until later.

Creativity Principle: Use Affirmative Judgment

Managing judgment is an essential skill in creative thinking. As noted earlier, first we have to defer judgment to get a wide range of options. Once we have generated this large number of novel options, how do we evaluate them in a creative way so as not to lose the novelty? There may be some very strange and unusual options that can easily fall off the radar screen with heavy-handed judgment. It is important to recognize that few ideas or new concepts come out fully developed. At the initial stages of evaluation it would be easy to discount crazy ideas, that is, the development of Post-it Notes, which came from a glue that did not work, or the development of Viagara, which resulted from the unusual side effects of blood pressure medication. Using affirmative judgment is as simple as looking at what is right with an idea rather than focusing on what is wrong. This leaves the door open for the really novel, perhaps crazy but intriguing idea, to be further developed into a workable solution.

Creativity Principle: Allow Incubation Time

'Let me think about it . . .' is a phrase that we may hear from our leaders. Sometimes it is a convenient and indirect way of saying no. But, if used deliberately, extra time to consider an idea, or a problem, can bring new insights and a fresh perspective on the situation. Perhaps a more accurate statement of incubation time would be 'Let me *not* think about it . . .' because incubation means taking a break from the idea, or problem to give your brain time to unconsciously work on it. Think, for example, of a time when you were tongue-tied, caught off guard by a sharp comment. Then, later that same day, a witty response jumps into your head. Why did you not think of it at the time, when you were with the person? Sometimes your brain needs a little time to think about it, whether you are aware of what is happening or not.

So, use this creativity principle deliberately in your work life. Perhaps an invitation at the end of a meeting for everyone to get back to you with any incubated thoughts on the subject. You might adopt an operating structure that is used in some formal meetings—new business is brought up and discussed at one meeting and then decided at the next—giving people time to incubate. If time is of the essence, even delaying a decision until the end of the day or taking a short break in the meeting before deciding will give a little breathing room to allow for incubation.

Innovation and Leadership: Creating a Climate for Creativity and Change

Earlier, we discussed the impact that leadership has on people, processes, and environment (see Figure 11.2). The leader sets the tone, creating an atmosphere in which others either feel safe to use their creativity or do not. So, what can you do, as a creative leader to model creativity behaviors and create an environment that is safe for everyone to contribute their ideas, thoughts, and insights so that innovation and change can occur?

We can draw from Ekvall's research (1983, 1991, 1996) here to guide us. One of the first things you can do is to become aware of your current climate. Is it conducive to creativity? Using Ekvall's dimensions of the creative climate, outlined earlier, we have posed a series of questions for you to use in your evaluation.

1. Do individuals have time to generate and develop new ideas? (Idea time)
2. How are new ideas or suggestions treated? Are they valued? Given reasonable consideration? (Idea support)

3. Does the workplace allow for fun? Do people joke and keep things light? (Playfulness/humor)
4. Is the work that people are doing stimulating? Does it require them to think in new and different ways? (Challenge)
5. Is there autonomy so that individuals can use their own judgment? (Freedom)
6. Is there a general feeling that others are supportive and have your best interest at heart? Can people speak freely without fear of reprisal? (Trust and openness)
7. Is there a sense of movement, action, and activity? (Dynamism and liveliness)
8. Do people engage others in conversations that represent opposing opinions? (Debate)
9. Are individuals encouraged to try new, unproven ideas or follow their hunches into the unknown? (Risk taking)
10. Is there an absence of interpersonal dissent, gossip, backbiting, and personal attack? (Lack of conflict)

Once you have evaluated your current climate, make a deliberate effort to make changes in the areas you believe are most needed. Some suggestions include the following:

- Make a point of involving others in resolving problems, particularly those closest to the problem.
- Consider setting up project timelines to reflect the need for idea time for new or different outcomes to emerge.
- Provide feedback using an affirmative, developmental approach.
- Encourage debate and alternative opinions.
- Allow for failure; recognize how it contributes to the learning.
- Avoid micromanaging and controlling—allow individuals to have latitude within their roles. (Puccio et al., 2011)

CLOSING THOUGHTS: CREATIVE LEADERSHIP SKILLS AS RELATED TO SCIENCE, TECHNOLOGY, AND INNOVATION

The general list of creative leadership skills presented in the previous section is certainly not exhaustive, and you may already be engaging in some of them. In our closing section, we will explore creative leadership skills specifically linked to science, technology, and innovation. Earlier we noted that creative leaders, if they wish to promote innovation, must also embody creativity themselves. After evaluating innovation in organizations Andrew and Sirkin (2006) offered leadership recommendations that reflect this belief. These authors suggested that "effective leaders of innovation (and, in fact, anyone involved in the process) tend to possess a set of qualities and skills that may be less necessary for leaders of other types of efforts" (p. 195). The qualities and skills they suggest are distinctive to leaders who promote innovation were: tolerance for ambiguity, the ability to assess and be comfortable with risk, the ability to quickly and effectively assess an individual, the ability to balance passion and objectivity, and the ability to change. It is our contention that many of the items included in the leadership recommendations offered by Andrew and Sirkin relate directly to qualities and abilities that have been long associated with creativity. And, in particular, the final ability, that is, the ability to change, underscores our contention that leaders serve as catalysts for change and aligns with our description of creative leadership. Note how their description of the 'ability to change' corresponds with the systems model of creativity we presented earlier in Figure 11.2:

> Sometimes the required change is on an organizational level. You realize that people, processes, and structures that used to be successful no longer are. They need to be changed, and someone needs to have the insight to see the required change, an understanding of how to make the change happen, the courage to attempt it, and the determination to see it through. (Andrew & Sirkin, 2006, p. 197)

Andrew and Sirkin's quote highlights the fact that creative leadership is a needed catalyst for change in organizations. Given the pace of change in the areas of science and technology, perhaps leaders in these sectors must be even more vigilant than are those in other industries in continuously questioning the status quo. For example, Colvin's (2008) description of how Motorola's inability to keep up with change undermined their success in the mobile phone industry highlights the rise and fall of organizations in an era of increased innovation and globalization:

> Motorola-heroic and innovative pioneer of the cell phone at first, then scorned failure when it didn't jump on digital phones, then reborn champion when it created the sleek RAZR, then goat once again when it couldn't produce a successor, and finally a casualty of competition with its decision to unload its cell phone business completely. (p. 146)

Nokia provides a more recent example of a once proud innovative company in the high-tech sector that has fallen on difficult times. A business article in the April 11 issue of *Time* magazine chronicled the challenges now facing this high-tech giant (Faris, 2011). Since 2007 Nokia's share of the smartphone market has plummeted from 51% to 27%. As Faris noted, "Even though Nokia was the first firm to introduce a smart phone, in the 1990's, its complicated interface never caught on, and the company stood in shock as Apple redefined the market with the iPhone's touchscreen and app store" (p. 2). And even though Nokia has spent $4 billion on research and development, up until recently much of this investment has been focused on the refinement of an operating system dedicated to making calls not on accessing mobile data. The market has turned quickly, but will Nokia be able to make a change? We would contend that in large part this will be determined by the organization's leadership.

Hemlin and Olsson (2011) published a recent study that specifically examined the leadership behaviors of leaders in both universities and industry that proved to promote creativity. Using a critical incident approach to examine real-life examples of creativity, these researchers identified six leadership behaviors across both university and industry leaders; that is, these researchers found little difference in the kinds of leadership behaviors that leaders used to stimulate creativity in the private and public sector. The six sets of behaviors were as follows: provide expertise (e.g., generation of new research ideas, alternative approaches and new perspectives), coordinate group research (e.g., leading open-ended discussion, suspending judgment, initiate future oriented discussion), assign tasks (e.g., delegation of tasks, demonstrating trust, etc.), support group conditions (e.g., reward group members, show enthusiasm, demonstrate care, etc.), enhance external contacts (e.g., encourage travel, foster collaboration with others outside organization, etc.), and promote independence (e.g., provide autonomy, solicit opinions of group members, etc.).

We believe that the real-life leadership behaviors displayed by those in university and industry labs in the Hemlin and Olsson (2011) study reflect the spirit of creative leadership outlined in this chapter. Embedded in these behaviors are the general creative behaviors related to creative process skills and creative environment we described earlier.

To close, we think of creativity as an applied science. Therefore, we encourage the reader to closely consider the degree to which he or she embodies aspects of creative leadership. Becoming a creative leader requires awareness that new behaviors are needed to facilitate creative outcomes. Learning creativity skills, such as those outlined in this chapter, is similar to learning any new skill—taking small steps, trying on new behaviors, and practicing. Hopefully some of the suggestions we have made will start you on your journey.

REFERENCES

Aldon, L. J. (2004). *Transcendent leadership and the evolution of consciousness.* London, England: Lightening Source.

Altshuller, G. (2001). *And suddenly the inventor appeared: TRIZ, the theory of inventive problem solving* (L. Shulyak, Trans.). Worcester, MA: Technical Innovation Center.

Amabile, T. M., Burnside, R. M., & Gryskiewicz, S. S. (1999). *User's manual for assessing the climate for creativity: A survey from the center for creative leadership.* Greensboro, NC: Center for Creative Leadership.

Anderson, N. R., & West, M. A. (1998). Measuring climate for work group innovation: Development and validation of the team climate inventory. *Journal of Organizational Behavior, 19*, 235–258.

Andrew, J. P., & Sirkin, H. L. (2006). *Payback: Reaping the rewards of innovation.* Boston, MA: Harvard Business School Press.

Andrew, J. P., Sirkin, H. L., Haanæs, K., & Michael, D. C. (2007). *Innovation 2007: A BCG senior management survey* (BCG Report). Boston, MA: Boston Consulting Group.

Basadur, M., Wakabayashi, M., & Graen, G. B. (1990). Individual problem-solving styles and attitudes toward divergent thinking before and after training. *Creativity Research Journal, 3*, 22–32.

Bass, B. M. (1985). *Leadership and performance beyond expectations.* New York, NY: Free Press.

Burns, J. M. (1978). *Leadership.* New York, NY: Harper & Row.

Colvin, G. (2008). *Talent is overrated: What really separates world-class performers from everybody else.* London, England: Portfolio.

Conger, J. A. (1996, Winter). Can we really train leadership? *Strategy & Business*, pp. 52–65.

Csikszentmihalyi, M. (1990). The domain of creativity. In M. A. Runco & R. S. Albert (Eds.). *Theories of creativity* (pp. 190–212). Beverly Hills, CA: Sage.

de Bono, E. (1977). *Lateral thinking.* Harmondsworth, England: Penguin Books.

de Bono, E. (1999). *Six thinking hats* (Rev. ed.). Boston, MA: Little, Brown.

Ekvall, G. (1983). *Climate, structure and innovativeness in organizations: A theoretical framework and an experiment* (Report 1). Stockholm, Sweden: FAradet.

Ekvall, G. (1991). The organizational culture of idea-management: A creative climate for the management of ideas. In J. Henry & D. Walker (Eds.), *Managing innovation* (pp. 73–79). London, England: Sage.

Ekvall, G. (1996). Organizational climate for creativity and innovation. *European Journal of Work and Organizational Psychology, 5*, 105–123.

Ekvall, G. (1997). Organizational conditions and levels of creativity. *Creativity and Innovation Management, 6*, 195–205.

Ekvall, G., & Ryhammar, L. (1999). The creative climate: Its determinants and effects at a Swedish university. *Creativity Research Journal, 12*, 303–310.

Faris, S. (2011, April 14th). Nokia's lifeline? *Time Magazine,* Business, pp. 1–2.

Florida, R. (2002). *The rise of the creative class: And how it is transforming work, leisure, community, and everyday life.* New York, NY: Basic Books.

Ford, C. M. (1995). Creativity is a mystery: Clues from the investigators' notebooks. In C. M. Ford & D. A. Gioia (Eds.), *Creative action in organizations: Ivory tower visions & real world voices* (pp. 12–49). Thousand Oaks, CA: Sage.

Gardiner, J. J. (2006). Transactional, transformational and transcendent leadership: Metaphors mapping the evolution of the theory and practice of governance. *Leadership Review, 6*, 62–76.

Gordon, W. J. J. (1961). *Synectics.* New York, NY: Harper & Row.

Guilford, J. P. (1950). Creativity. *American Psychologist, 5*, 444–454.

Heifetz, R., Grashow, A., & Linsky, M. (2009). Leadership in a (permanent) crisis. *Harvard Business Review, 87*(7/8), 62–69.

Hemlin, S., Allwood, C. M., & Martin, B. R. (2008). Creative knowledge environments. *Creativity Research Journal, 20*, 196–210.

Hemlin, S., & Olsson, L. (2011). Creativity-stimulating leadership: A critical incident study of leaders' influence on creativity in research groups. *Creativity and Innovation Management, 20*, 49–58.

Henry, J. (2001). *Creativity and perception in management.* London, England: Sage.

Hersey, P., & Blanchard, K. H. (1977). *Management of organizational behavior: Utilizing human resources.* Upper Saddle River, NJ: Prentice Hall.

Hudson Institute (1997). *Workforce 2020: Work and workers in the 21st century.* Indianapolis, IN: Hudson Institute.

Hunter, J. E., & Schmidt, F. L. (1996). Intelligence and job performance: Economic and social implications. *Psychology, Public Policy, and Law*, *2*, 447–472.

IBM. (2010). *Capitalising on complexity: Insights from the global chief executive officer (CEO) study.* Portsmouth: England: IBM United Kingdom Limited.

IBM Gobal CEO Study (2006). *Expanding the innovation horizon,* Somers, NY: The IBM Corporation.

Janszen, F. (2000). *The age of innovation: Making business creativity a competence, not a coincidence.* London, England: Prentice Hall.

Kaufmann, G. (1988). Problem solving and creativity. In K. Grønhaug & G. Kaufmann (Eds.), *Innovation: A cross-disciplinary perspective* (pp. 87–137). Oslo, Norway: Norwegian University Press.

Kelley, T., & Littman, J. (2001). *The art of innovation.* New York, NY: Doubleday.

Kirton, M. J. (1976). Adaptors and innovators: A description and measure. *Journal of Applied Psychology*, *61*, 622–629.

Komives, S.R., Lucas, N., & Mcmahon, T.R. (2006). *Exploring leadership: For college student's who want to make a difference.* San Francisco, CA: Jossey-Bass.

Kouzes, J. M., & Posner, B. Z. (1995). *The leadership challenge: How to keep getting extraordinary things done in organizations.* San Francisco, CA: Jossey-Bass.

Larkin, D. (1995). *Beyond self to compassionate healer: Transcendent leadership* (Unpublished doctoral dissertation). Seattle University, Seattle, Washington. Retrieved from http://books.google.com/books?id=jkbBNwAACAAJ&dq=inauthor:%22Diana+K.+Larkin%22&hl=en&sa=X&ei=GfQYUaSAI-He0QH6loCwAw&ved=0CC0Q6AEwAA

Locke, E. A., & Kirkpatrick, S. A. (1995). Promoting creativity in organizations. In C. M. Ford & D. A. Gioia (Eds.), *Creative action in organizations: Ivory tower visions & real world voices* (pp. 115–120). Thousand Oaks, CA: Sage.

MacKinnon, D. W. (1962). The nature and nurture of creative talent. *The American Psychologist*, *17*, 484–495.

MacKinnon, D. W. (1978). *In search of human effectiveness.* Buffalo, NY: Creative Education Foundation

Maslow, A. H. (1954). *Motivation and personality.* New York, NY: Harper & Row.

Mathisen, G. E., & Einarsen, S. (2004). A review of instruments assessing creative and innovative environments within organizations. *Creativity Research Journal*, *16*, 119–140.

McCracken, H. (2011, July 25). Social showdown. Google+ has a shot at giving Facebook serious competition. *Time*, p. 55.

Mehaffy, G. L. (2010). The red balloon project: Re-imagining undergraduate education. Draft paper delivered as part of the American Association of State Colleges and Universities presentation. A version of paper to appear in *Teacher – Scholar: The Journal of State Comprehensive University.* For information on this project go to: http://www.aascu.org/programs/RedBalloonProject/

Mooney, R. L. (1963). A conceptual model for integrating four approaches to the identification of creative talent. In C. Taylor & F. Barron (Eds.), *Scientific creativity: Its recognition and development* (pp. 331–340). New York, NY: Wiley.

Mumford, M. D., Baughman, W. A., Maher, M. A., Costanza, D. P., & Supinski, E. P. (1997). Process-based measures of creative problem-solving skills: IV. Category combination. *Creativity Research Journal*, *10*, 59–71.

Mumford, M. D., Baughman, W. A., Supinski, E. P., & Maher, M. A., (1996). Process-based measures of creative problem-solving skills: II. Information encoding. *Creativity Research Journal*, *9*, 77–88.

Mumford, M. D., Baughman, W. A., Threlfall, K.V., Supinski, E. P., & Costanza, D. P. (1996). Process-based measures of creative problem-solving skills: I. Problem construction. *Creativity Research Journal*, *9*, 63–76.

Mumford, M. D., Supinski, E. P., Threlfall, K.V., & Baughman, W. A. (1996). Process-based measures of creative problem-solving skills: III. Category selection. *Creativity Research Journal*, *9*, 395–406.
Mumford, M. D., Zaccaro, S. J., Harding, F. D., Jacobs, T. O., & Fleishman, E. A. (2000). Leadership skills for a changing world: Solving complex problems. *The Leadership Quarterly*, *11*, 11–35.
Murdock, M. C., & Puccio, G. J. (1993). A contextual organizer for conducting creativity research. In S. G. Isaksen, M. C. Murdock, R. L. Firestien, & D. J. Treffinger (Eds.), *Understanding and recognizing creativity: The emergence of a discipline* (pp. 249–280). Norwood, NJ: Ablex.
National Center on Education and the Economy. (2008). *Tough choices or tough times: The report on the new commission on the skills of the American workforce.* San Francisco, CA: Wiley.
Osborn, A. F. (1953). *Applied imagination: Principles and procedures of creative problem-solving.* New York, NY: Scribner's Sons.
Puccio, G. J. (1999). Creative problem solving preferences: Their identification and implications. *Creativity and Innovation Management*, *8*, 171–178.
Puccio, G. J., & Cabra, J. F. (2010). Organizational creativity: A systems perspective. In J. Kaufmann & R. J. Sternberg (Eds.), *The Cambridge handbook of creativity* (pp. 145–173). New York, NY: Cambridge University Press.
Puccio, G. J., Firestien, R. L., Coyle, C., & Masucci, C. (2006). A review of the effectiveness of CPS training: A focus on workplace issues. *Creativity and Innovation Management*, *15*, 19–33.
Puccio, G. J., Mance, M., & Murdock, M. C. (2011). *Creative leadership: Skills that drive change* (2nd ed.). Thousand Oaks, CA: Sage.
Puccio, G. J., & Murdock, M. C. (2001). Creative thinking: An essential life skill. In A. Costa (Ed.), *Developing minds: A source book for teaching thinking* (pp. 67–71). Alexandria, VA: ASCD.
Reiter-Palmon, R., & Illies, J. J. (2004). Leadership and creativity: Understanding leadership from a creative problem-solving perspective. *The Leadership Quarterly*, *15*, 55–77.
Rhodes, M. (1961, April). An analysis of creativity. *Phi Delta Kappan*, pp. 305–310.
Rogers, C. R. (1962). Toward a theory of creativity. In S. J. Parnes & H. F. Harding, (Eds.), *A source book for creative thinking* (pp. 63–72). New York, NY: Scribners.
Rose, L. H., & Lin, H.-T. (1984). A meta-analysis of long-term creativity training programs. *The Journal of Creative Behavior*, *18*(1), 11–22.
Scott, G., Leritz, L. E., & Mumford, M. (2004a). The effectiveness of creativity training: A quantitative review. *Creativity Research Journal*, *16*(4), 361–388.
Scott, G. M., Leritz, L. E., & Mumford, M. D. (2004b). Types of creativity training: Approaches and their effectiveness. *The Journal of Creative Behavior*, *38*, 149–179.
Sellgren, S. F., Ekvall, G., & Tomson, G. (2008). Leadership behaviour of nurse managers in relation to job satisfaction and work climate. *Journal of Nursing Management*, *16*, 578–587.
Stein, M. I., (1968). Creativity. In E. F. Boragatta & W. W. Lambert (Eds.), *Handbook of personality theory and research* (pp. 900–942). Chicago, IL: Rand McNally.
Talbot, R. J. (1997). Taking style on board. *Creativity and Innovation Management*, *6*, 177–184.
Torrance, E. P. (1972). Can we teach children to think creatively? *The Journal of Creative Behavior*, *6*, 114–143.
Torrance, E. P., & Presbury, J. (1984). The criteria of success used in 242 recent experimental studies of creativity. *The Creative Child and Adult Quarterly*, *9*, 38–243.
Trilling, B., & Fadel, C. (2009). *21st Century skills: Learning for life in our times.* San Francisco, CA: Jossey-Bass.

Weber, M. (1947). *The theory of social and economic organization.* (A.M. Henderson & T. Parsons. Trans.). New York, NY: The Free Press.
West, M. A. (1997). *Developing creativity in organizations.* Leicester, England: British Psychological Society.
Williams, W. M., & Yang, L. T. (1999). Organizational creativity. In R. J. Sternberg (Ed.), *Handbook of creativity* (pp. 226–250). Cambridge, England: Cambridge University Press.
Woodman, R. W., Sawyer, J. E., & Griffin, R. W. (1993). Toward a theory of organizational creativity. *Academy of Management Review, 18*(2), 293–321.
Zacko-Smith, J.D. (2010). The three-T's framework for leadership education. *Academic Exchange Quarterly, 14*(2), 119–123.

12 Conclusion

Sven Hemlin, Carl Martin Allwood, Ben R. Martin, and Michael D. Mumford

This concluding chapter sets out to synthesize and analyze the main findings to emerge from the preceding chapters about creativity and leadership in the realm of science, technology, and innovation (STI). A second aim is to draw conclusions about the specific implications of the key issues in this synthesis with regard to the management and leadership of those engaged in science, technology, and innovation. A third and final aim is to attempt to identify new directions in creativity and leadership that might prove particularly fruitful for future research.

MAIN FINDINGS

Organizations and the organizational environments that have an impact on creative thinking and behavior through the efforts of leaders and their coworkers constitute the setting for much creative work. Moreover, the complexity of the work tasks involved and the ill-defined nature of the problems posed influence how creativity may best be led, particularly at the work-group level. Mumford, Peterson, and Robledo (this volume) focus on the expertise, responsiveness and planning capacities of leaders, the evaluative feedback they provide to coworkers, and their attempts to generate a positive work climate, all of which are essential for creative efforts in science and engineering. In addition to these demands on the leader, Denti and Hemlin (this volume) stress that members of work groups need creative role models and clear expectations with regard to what constitutes creative performance. These functions can, and indeed should, be provided and facilitated by leaders. It is, of course, also necessary that the members of work groups possess creative abilities, self-efficacy, and the intrinsic motivation required for creative endeavor. However, very creative work-group members will perhaps not need all aspects of the leadership influence described above as much as slightly less creative individuals (Tierney, Farmer, & Graen, 1999). Hunter, Cushenbery, Ginter, and Fairchild (this volume) suggest that creativity and innovation models can be changed and improved by new technologies and that leaders have an important role to play in their successful implementation.

It is essential that leaders are creative and visionary with respect to the technologies they introduce, promote, and support.

How do the best research leaders carry out such tasks, and what are their characteristics? The three empirical studies in the second section of the book each deal with these questions, but they employ rather different materials and techniques. One answer is that creative leaders need 'deep' domain expertise. It is not sufficient to be a very skillful manager if one does not possess the necessary degree of domain expertise to be a creative leader of others working in that domain (Simon, 1965). However, such expertise alone is not sufficient to lead and inspire others to creative performance. The leader should also by their own example and management act to inspire the work group's members to continual perseverance at their tasks. In addition, for research groups to become creative, the leader must commit him or herself to providing the necessary resources, taking part in and coordinating the groups' work, and engaging in extensive networking. In addition, some authors in this book note that the career stage of the leader is likely to influence creative leadership in various ways. For example, Verbree, van der Weijden, and van den Besselaar (Chapter 5, this volume) found in an empirical study of Dutch research group leaders in life sciences that senior leaders were focusing on securing grants for research, whereas less experienced leaders emphasized profiling of their research; that is, in a sense, they were creativity seekers. For this reason, it is important that human relations managers take full account of the wider situation of the leader of an innovative research group. For example, it may be important to consider not only the professional motives but also the more personal motivations driving a leader.

Very frequently, creative and innovative research and development efforts involve interdisciplinarity. This issue is thoroughly dealt with by Reiter-Palmon, de Vreede, and de Vreede, and also by Gorman and Gorman in their respective chapters (both in this volume). As was previously mentioned by Mumford, Peterson, and Robledo (this volume), complex and ill-defined problems demand creative efforts to be solved, and in an innovation context a *multidisciplinary* approach may not suffice. Rather, the demands for more integrated *interdisciplinary*[1] approaches are constantly growing and indeed a fully integrated *transdisciplinary* approach that moves beyond the original disciplinary structure is often needed (see, e.g., Reiter-Palmon, de Vreede, & de Vreede, this volume; Stokols et al., 2010).

A key point made by Reiter-Palmon et al. (this volume) is that both the composition and climate of teams are crucial in interdisciplinary creative endeavors. Much research testifies to the conclusion that the presence of members with different perspectives is an essential ingredient for success in creative innovation contexts. This point is elaborated more specifically in the following. As noted by several authors in this book and elsewhere, it is one of a leader's more important tasks to put together teams with a fruitful mix of complementary competencies that are needed to address the specific tasks at hand. The responsibility of leaders operating in an interdisciplinary context to ensure that the group interacts in a constructive way is a difficult and challenging one. As described in the chapter by Gorman and Gorman (this volume), part of the challenge for leaders to be successful in an interdisciplinary context is that they must be able to establish how best to support the meeting places—or 'trading zones'—where concepts and approaches that are discipline specific are 'translated' or at least explained to other group members. Previous research on interdisciplinary research shows that one difficulty and even danger in this context is that the demands for performance that quickly yields concrete results may not give the interdisciplinary group sufficient time to develop common goals, a suitable and commonly accepted theoretical language, and a well-integrated theoretical framework and problem formulation. Furthermore, Defila and Di Giullo (2010) argue that this will not happen automatically even when group members are well motivated. Group members have to be properly managed. For this to come about, leaders often need specific training. In general, in order for interdisciplinary groups to cope with potential difficulties arising from disciplinary differences, they should be trained in communicating about similarities and differences between disciplines with respect to different approaches to concepts, theories, and research methods. This is a major challenge for a leader to accomplish.[2]

The chapter by Gorman and Gorman also deals with another crucial issue, namely that the sustainability of results, solutions, and innovative products is likely to be a required characteristic of future creativity. This basically ethical viewpoint on creativity is something that both leaders and their coworkers should take into consideration, and which may be a fruitful topic for discussions within the group especially in the early stages of projects.

Another, less often discussed, issue in research and development (R&D) is how creative leaders can be replaced after they retire or move on to another job. This important problem is analyzed by Ligon, Demborski, Mapp, and Zongrone (this volume), who examine the issue of succession planning for both top and operational-level leaders in STI. A final application-oriented chapter about creative leadership is that by Puccio, Mance and Zacko-Smith (this volume), who stress that leaders not only need creativity to become better leaders but also, and very importantly, that this skill can be honed by training. More specifically, they suggest how leaders may develop their cognitive problem-solving capacity and how a team climate promoting creativity may be enhanced by leaders.

In summary, a number of important aspects of creativity and leadership in STI have been examined in this volume. They include the following:

- Research group leaders and creative scientific performance
- The leadership of creative work in interdisciplinary projects
- Different stages in the careers of leaders, and their effects
- Leadership in the different phases of projects
- Leadership training for creativity
- Environmental influences on leaders

- Leadership cognition and motivation
- Creativity and technological development
- A leader's understanding of how technology can be used to enhance creativity
- Leaders in engineering and technology
- Sustainability and creativity
- Shared leadership

Why are these issues important? Research group leaders are active at the operational level of scientific work; in other words, they are engaged in doing what they can to foster the creative performance of individuals and the group. Second, many scientific and technological projects are interdisciplinary by nature. To take just one topical example, environmental problems demand collaborative efforts of teams composed of a wide range of disciplinary specialties, such as meteorology, biology, psychology, and sociology. Third, the individual expertise, cognitive skills, and problem-solving capacities of leaders are crucial for evaluating the ideas of others and therefore for stimulating creative progress in STI. Fourth, the leader's own motivations are important since leaders act as role models and 'energizers' for their coworkers. If leaders do not show a sufficient degree of passion for what they are doing, their coworkers may tend to be less creative. Moreover, it has long been recognized that leaders need to ensure that work teams have a suitable composition as well as shaping the team climate. Fifth, leaders need training in creative problem solving, although they may be experts and problem solvers in a field, leaders do not always have creative skills. Sixth, leaders are influenced by their environment (their teams, the organization in which they work, and other colleagues round the world), making it important for leaders to be open to new experiences, but simultaneously able to shut out extraneous 'noise' to focus on the problems at hand. One example of the important influence that leaders can bring to bear in creative work is how they work to adopt and introduce creativity-enhancing technologies. Seventh, organizations need to pay attention to human resource issues with creative leaders. Such individuals are not easily substituted, and their specific needs in the different phases of their careers should be fully taken into account.

However, there are certain other issues in the field of creativity and leadership on STI that are rather less well dealt with in the chapters of this book. Among these are the following, some of which are further discussed later:

- Coworkers (or 'followers') and their characteristics
- Leadership in innovation-related work (as opposed to scientific or technological work)
- Organizational and top-management leaders

It is worth briefly considering why these issues on creativity and leadership are less well covered in the volume. First, leaders clearly do not work in a vacuum but lead others, that is, their co-workers. Here, we should note that the term *followers* is sometimes preferred in this connection, perhaps partly reflecting a cultural difference in research terminology between Europe (where the term *coworkers* is often preferred) and the US (where the term *followers* is more common).[3] However, the term *followers* might be taken to imply that the leader acts largely on his or her own, without the need for effective communication with and coordination of the other members of the research group, an idea that goes against much of the spirit of the chapters in this book, and indeed of much current research in this area in general. Hence, for this reason, we prefer in this chapter to use the term *coworkers* (even though it is slightly more awkward). Clearly, the coworkers are vital for creativity to occur within groups or organizations. Yet in spite of this, the perspective of the coworkers has not traditionally been given as much attention in leadership research, the focus instead having been primarily on the leaders and their perspective. However, the perspective of the coworkers is important to study, not just with regard to the question of how to achieve an effective team climate, but also in an effort to explain why some leaders are more successful than others. This is considered in the chapter by Denti and Hemlin (this volume), which amongst other things deals with the leader–member exchange (LMX) theory, where the importance of the reciprocal relations between leaders and coworkers is explicitly recognized.

Second, innovation leadership and innovation management in business or industry are less well dealt with in this book. This is partly explained by the fact that there is already an extensive literature on innovation management and on innovation more generally (e.g., Isaksen & Tidd, 2006), including numerous studies in the industrial economics and organizational studies literature. Nevertheless, many of the issues covered in this book are relevant to the innovation literature. For example, ideation is increasingly important in industries striving to be innovative (Björk, 2011), and the facilitation of new ideas being adopted, diffused, and accepted in organizations is clearly of interest for psychologists and others to investigate.

Third, top management and senior leaders in creative organizations tend to be rather less studied empirically,[4] as was pointed out by Ligon and her coauthors (this volume). There is currently a gap here in the literature on creativity and leadership that needs to be filled.

PRACTICAL IMPLICATIONS

Implications for Leadership

Leaders in science, technology, and innovation differ from other leaders in more general work domains in a number of crucial respects. First, leaders in STI need to have expertise in at least one domain, where they possess deep knowledge typically gained over a period of about 10 years or so (Ericsson & Charness, 1994). The knowledge that leaders have thus gained is essential in enabling them to evaluate and provide constructive feedback on the ideas and work of their coworkers. Secondly, moreover, in the arena of STI, leaders have to have the necessary creative skills, in particular to facilitate work processes in a creative direction. Thirdly, STI leaders need to be strongly motivated in their work in order to ensure success in their own creative efforts but also to provide a role model for their co-workers who likewise aspire to be creative. A motivated leader makes it clear that they expect their coworkers to strive continually towards creative performance. Motivation is also what drives leaders to join in the work of others in the team and indeed to become a team member themselves. Fourthly, and more generally, leaders in STI are perhaps rather less formally 'managerial' than are leaders in other domains. However, as stressed earlier, they should possess the skills needed to put together creative teams, to shape and maintain a good team climate, and to have effective time- and work-planning abilities, which demand management skills (Hemlin, 2006). Moreover, it is no easy task for organizations to replace leaders in STI because of the long learning period required to become an expert, so well-prepared succession planning is essential. Finally, it would seem that STI leaders tend to favor new technologies that may be used to facilitate creativity and to shape new forms of creativity (see, e.g., Nielsen, 2012, for a discussion of how Internet can be used to boost collective intelligence). If we look back, it is evident that technological advances such as the computer and its applications have had a tremendous importance in STI.

Implications for Creativity

Authors such as Puccio, Mance, and Zacko-Smith (this volume) argue that leaders need to be creative problem solvers. As is also clear from the chapters in this book, this adage is valid for many more aspects of the wider situation in which innovation occurs in an organization than simply the research and development tasks at hand. Other aspects such as the composition of the group, communication within the group, visionary use of technology, and career planning for individual members all require an element of creativity and creative problem solving.

Over the years, the interest of researchers from various parts of psychology such as the psychological judgment and decision-making research tradition has grown stronger with respect to organizational issues and how human thinking and decision making relate to the organizational context. There appears to be considerable potential in trying to integrate this research with some of the issues dealt with in the present volume, especially with respect to creativity. Two important stages in the creative process are reviewed in Chapter 1 in this volume, namely, idea generation and idea evaluation. We should hasten to add that these two phases are likely to overlap to a considerable extent. Hence, what is said in the following about the generation of creative ideas may also have relevance for the evaluation of ideas, and *vice versa*.

As we noted in the introductory chapter to this volume (Chapter 1), previous research on idea generation has strongly suggested that bringing to bear a range of perspectives in the research context tends to promote the generation of potentially creative ideas (Campbell, 1960; Simonton, 2003, 2011). Likewise, extensive research has shown that researchers with a broad range of interests and activities and with a broad experience base in their upbringing and professional life are often the most creative researchers (e.g., Feist, 1998, 2006; Simonton, 2003). Research on decision making in cognitive and social psychology likewise attests to the fruitfulness of bringing together many perspectives. For example, the idea of the 'wisdom of the crowd' is not only relevant as an example of how such an approach can be used to enhance the quality of an output (see Hunter, Cushenbery, Ginther, & Fairchild, this volume). The notion can also be generalized to the composition and work of research groups, especially to ensure that the group has the requisite degree of heterogeneity. For example, in research on group decision-making, it has been found that the presence of an outspoken minority among group members may improve the group's creativity (Wilke & Kaplan, 2001). However, such heterogeneity in groups also represents a challenge for the leader, because he or she has to constantly monitor the situation in order to ensure that fundamental differences in assumptions and personalities between group members do not lead to nonconstructive conflicts (Allwood & Bärmark, 1999). Researchers have argued that a certain moderate level of conflict might stimulate creativity and innovation, in comparison to groups with a high or a low level of conflict, and that a high degree of homogeneity in a group is normally associated with a low level of conflict (Selart & Schei, 2011). A further example of the usefulness of combining many perspectives in one group relates to the process of brainstorming. Although this technique may not have fulfilled its promise in its original form, more recent research has found that, when members of a brainstorming group are initially are given time to produce and reflect on their own ideas, it is more fruitful to discuss in a later stage members' ideas in a group context where the level of criticism is at first turned down (Dewett, 2003; Paulus & Yang, 2000). We argue that it is important for leaders to be well acquainted with such findings from the research literature if they are to organize brainstorming sessions in an optimal way. These examples illustrate the value of creativity skills training for leaders.

The utility of the multiperspective approach is also reflected in research cited in Heath, Larrick, and Klayman (1998). They provide examples of situations where individual group members were only able to come up with a small fraction of the total number of ideas generated by the group as a whole. Similarly, but at a more general level, Larrick (2009) argued that broadening the framework for the conceptualization of the decision and for the various activities involved in the decision-making process, will frequently lead to better decisions. Research from a variety of different areas has concluded that taking an average of the group members' individual conclusions will usually lead to a better overall outcome compared with the conclusion of an individual picked at random from the group. Seen from a statistical perspective, an average is of value since it will eliminate some of the random error in an individual's conclusion. On a smaller scale, the Dialectical Bootstrapping Model (DBM) proposed by Herzog and Hertwig (2009) illustrates the same phenomenon. It may be seen as the application of the 'wisdom of the crowd' idea to the individual level. The DBM, both theoretically and empirically, shows that when two numerical estimates by the same individual generated from somewhat different aspects of the person's background knowledge are averaged, the result is generally better than randomly picking one of the two estimates or randomly picking an answer from a group participant. The reason is that combining different estimates tends to cancel out random errors and in certain circumstances even some systematic errors. However, an important precondition for the DBM to work is that the two or more estimates differ from one another. Thus, a further contributing factor explaining the utility of many perspectives in a research group is that a suitable mix of perspectives is present within the group. Here, one of the more delicate aspects of a leader's tasks is to put together research groups with a suitable profile of competencies and perspectives. In this context, an important challenge for future research is to identify principles for how an optimal combination of different perspectives can be selected for groups in different research contexts.

The second important part of the creative process mentioned above is the evaluation and narrowing down of suggestions for creative ideas. For example, suggestions on how to develop a project will each be associated with various types of risks, such as the possibility that the research idea or product to be developed will not ultimately prove successful. This aspect is becoming all the more important given that, as Puccio, Mance and Zacko-Smith (this volume) note, the pace of change is now very rapid, implying a higher degree of uncertainty, ambiguity, and volatility. One needs to generate a broad set of alternative scenarios to capture the range of possible future outcomes, and the bringing together of many perspectives is often essential in accomplishing this. In this context, one of the leader's tasks is to facilitate the process whereby the risks of different options are evaluated so that realistic conclusions can be drawn. It is also well known that groups and group discussions are associated with a higher level of risk taking—compared with individuals (Prabhu, 2011). It is therefore important that leaders are fully aware of this phenomenon and have the ability to constructively balance such situations. In addition, when planning research groups and who will lead them, it is relevant to bear in mind that creative persons may often be more prepared to take risks than may less creative persons, that younger persons tend to take more risks than do older persons, and that men may (still) have a tendency to take more risks than will women (see the research reviewed in Byrnes, Miller, & Schafer, 1999; Prabhu, 2011).

At a more general level, Heath, Larrick, and Klayman (1998) suggested that organizations can provide the facilities for undertaking 'cognitive repairs' of individual shortcomings in thinking, judgment and decision-making processes. An organization can (and often does) provide individuals and groups with norms and routines to help check for known types of shortcomings in their judgment and decision processes, when hypotheses are generated, when information is collected, and when conclusions are drawn. For example, banks are frequently using software for credit assessments to assist in difficult judgments and decisions (the so-called transactional lending technology). In general, and applied to the context of this volume, taking responsibility for the development of the organization's repertoire of cognitive repair, for example, for such general human tendencies as 'confirmation bias' (a tendency to narrowly seek and evaluate evidence in favor of, rather than against, one's own opinions), would seem to be a task for senior research leaders. Finally, as noted by Puccio et al. (this volume), and given all the various demands on leaders noted in this chapter, the issue of how effective training can be provided for leaders to help them to carry out their wide-ranging tasks deserves appreciably greater attention.

NEW DIRECTIONS

According to the stereotypical view of scientists and engineers, creative achievement in scientific and technical fields is ultimately attributable to individuals working alone, individuals who must overcome almost insurmountable obstacles. This image reflects a somewhat romantic view of scientists and engineers, which, however evocative, simply does not reflect the realities of creative work in science and engineering today. Creative work in the sciences or in engineering is inherently collaborative and highly complex, and may involve hundreds, if not thousands, of collaborators. In one way or another, all of the scholars contributing to the present volume recognize this point. Moreover, it is a point we must bear firmly in mind whenever we carry out research, attempt to understand, or formulate policies in the hope of improving the creativity and innovativeness of scientists and engineers.

Although this point is of considerable importance, our foregoing observations also point to another crucial conclusion. When people need to work together to perform complex tasks, leadership becomes a key influence on performance. Indeed, the evidence provided by Mumford, Peterson, and Robledo (this volume) and Mumford, Scott, Gaddis, and Strange (2002) indicates that effective leadership may, in fact, be *the* most important determinant of the success of creative endeavors in the fields of science and engineering. Given the substantial impact of the innovations produced by scientists and engineers on our world, products ranging from nuclear reactors to the internet, it is critical that we improve the leadership of scientists and engineers if we seek to produce ever more innovative new products. In fact, the present volume has much to say about the nature of effective leadership in the fields of science and engineering. In particular, leaders must define realistic technical missions (Mumford et al., this volume), they must motivate their coworkers (Denti & Hemlin, this volume), they must establish viable teams and a climate promoting creative work (Reiter-Palmon et al., this volume), and they must think creatively (Puccio et al., this volume).

Although the conclusions emerging from this volume are of a certain importance in their own right, the various studies brought together collectively point to some potentially important directions for future research. One key direction is evident from the chapter by Mumford, Peterson, and Robledo (this volume), who argue that the leadership of creative and innovative efforts requires substantial adaptive planning under conditions where plans might unfold over a period of 20 years or more. Unfortunately, little is known about the key capacities required for successful long-term adaptive planning nor about how leaders attempt to execute these plans in a dynamic environment (Byrne, Mumford, Barrett, & Vessey, 2009). This gap needs to be addressed over coming years.

As important as this 'creative planning' may be, it is clear from various chapters in the present volume that the leaders of creative efforts cannot just plan. Leaders must also be able to 'sell' their plans to organizations, governments, and others in the field (Howell & Boies, 2004). They must be able to staff the projects flowing from these plans – selling their projects to scientists and engineers (Ligon et al., this volume). And, they must create a demanding, albeit relatively safe, environment in which scientists and engineers can freely pursue their work (Reiter-Palmon et al., this volume). What should be recognized here is that few other forms of work make such substantial conceptual, social, and organizational demands. As a result, there is a need for research examining how leaders of creative efforts manage the multiple, and often apparently contradictory, demands on them.

Our foregoing observations echo the strong *cognitive* emphasis evident in the chapters by Mumford, Peterson, and Robledo (this volume) and by Puccio, Mance, and Zacko-Smith (this volume). In contrast, studies of leadership have traditionally focused more on *motivation*, in particular on how leaders should motivate their coworkers (Bass & Bass, 2008; Yukl, 2010). The chapter by Hemlin and Denti (this volume) very much follows in this tradition, examining the mechanisms by which creative leaders motivate their coworkers. This divergence in approach is noteworthy because it points to two further potentially important directions for future research. First, we need studies expressly contrasting and integrating cognitive and motivational models of leadership with respect to the carrying out of creative work. These studies should seek to take into account the likely interactions among different cognitive and motivational actions taken by the leaders of creative efforts. Second, these studies should examine cognitive and motivational effects on leader performance at different levels of analysis and under different boundary conditions (Rosing, Frese, & Bausch, 2011).

With regard to these interactions between cognition and motivation, another potential direction for future research is suggested by Reiter-Palmon, de Vreede, and de Vreede (this volume). They provide a compelling argument that creative teams possess shared mental models—models that serve to direct cognition, to structure interactions, and to motivate scientists and engineers. However, we need more research on the key characteristics of viable mental models among teams of scientists and engineers as well as the mechanisms by which leaders contribute to the formulation and dissemination of viable mental models for creative work. It may be fruitful for such research to take into consideration results from the research area naturalistic decision making (see, e.g., Elliott, 2005). Moreover, we need studies examining the impact of shared mental models for arriving at a creative climate, for planning activities, and for the evaluation (both internal and external evaluation) of creative work. Such studies should consider not only how to facilitate for shared models and how this is beneficial for the group's work, but also how such common understanding should best be balanced with optimal variation in the team members' knowledge.

In addition, as elaborated earlier, more research is needed on how leaders and coworkers together can integrate the various perspectives and knowledge of the different group members in such a way that 'the wisdom of the crowd' is reformulated into the best possible conclusions or outputs of the group. It is the leader's task to monitor and facilitate this process, bringing it to a successful conclusion.

Perhaps the most clear-cut conclusion to emerge from the chapters presented in this volume is that the leadership of scientists and engineers engaged in creative work is unusually demanding. This point is of no little importance because in organizations we often tend to make an assumption that contradicts the complex task of leading scientists and engineers. We assume that those leading scientists and engineers are 'smart' individuals so they can easily take care of themselves and their teams. In this regard, the chapter by Ligon, Dembroski, Mapp, and Zongrone (this volume) is of some importance, serving to remind us that systematic developmental programs may be critical in helping prepare people to lead scientists and engineers.

However, development is a complex, multifaceted process. We need far more research on the exact nature of the developmental interventions likely to prove of value in helping scientists and engineers acquire the skills they will need to lead creative efforts successfully. For example, rotational assignments are often held to be valuable for those who will be asked to lead scientists and engineers, but when should these rotational assignments occur, where should people be rotated to, and for how long? Similarly, scientists and engineers appear to acquire leadership skills through 'hands-on' experience (Vessey, Barrett, & Mumford, 2011). This observation, however, raises the question of what types of project assignments are likely to prove most beneficial in developing leadership skills among scientists and engineers—small high-risk projects or larger more structured projects? Similarly, peers exert profound effects on scientists and engineers (Mumford et al., 2005). Unfortunately, we know little about the kind of peer exchanges that are most beneficial in developing leadership talent, nor how individuals should best be prepared to profit from these experiences.

In all developmental interventions, timing is critical. Indeed, the chapter by Eubanks, Palanski, Swart, Hammond, and Oguntebi (this volume) stresses this very point, suggesting that providing training in time-management skills may be especially valuable at the point when scientists and engineers move into more senior leadership positions. This observation, however, reminds us that effective developmental studies must take training or interventions into account vis-à-vis the demands that will be made on a leader at that particular stage in their career. Some years ago, studies of scientific and engineering performance often employed a career process framework (Zuckerman, 1977). Unfortunately, career development models rarely explicitly examined how careers unfolded for those asked to lead scientists and engineers. Undoubtedly, there would be substantial value in studies examining the careers of more and less successful leaders of creative efforts in the sciences and engineering.

Moreover, when we consider career processes it is important to bear in mind that the leadership of scientists and engineers occurs in an environment, a scientific or technical field (Csikszentmihalyi, 1999) drawing upon certain technologies (Hunter, Cushenbery, Ginther, & Fairchild, this volume) and operating under certain policies (Verbree et al., Chapter 6, this volume). The environmental factors may be powerful influences shaping the course of the careers of those asked to lead scientists and engineers. However, we know far too little about how these environmental variables influence a person's capacity for effective leadership and the consequent behavior of the scientists and engineers they are asked to lead. We need studies examining trade-offs in the effects of differing policies on the behavior of leaders—for example, will a focus on research productivity result in less concern about group climate and the motivation and wellbeing of staff? Can peer exchange systems be created on the internet to facilitate the development of scientific leadership? And how do leaders appraise, and transmit information about, environmental change through social networks? How do leaders act when they anticipate or encounter what they see as irrelevant resistance in the organization to new ideas? There would be value in obtaining answers to these and other related questions.

Traditionally, we have construed the work of scientists and engineers as being essentially technical and largely value-free work. However, as Gorman and Gorman (this volume) remind us, the leadership of scientists and engineers has profound ethical implications. In fact, recent research suggests that ethical decision-making is positively related to creativity in the social, biological, and health sciences (Mumford et al., 2010). What we do not know, however, is how leaders take into account ethical issues as they formulate project plans, sell their projects, and manage their scientists and engineers. We would hope that future research will be initiated to examine how the leaders of scientists and engineers set about appraising relevant ethical issues and then using these ethical issues to help guide their efforts in directing scientists and engineers.

CONCLUSION

We live in a rapidly developing and complex society in which science and technology are major driving forces. New technologies such as social media are connecting people in new ways and into smaller or larger networks based on various issues. This phenomenon may have an impact also on creativity (and its leadership) in STI. Scientific breakthroughs, which may help to solve global problems such as environmental pollution are urgently needed and are dependent on creative solutions in interdisciplinary research programs. As has been argued in this book, we firmly believe that leadership in STI is essential for creative outcomes in the contemporary world. However, leaders must take on new functions and get appropriate skills to make it possible for individuals and teams to become more creative. Indeed, as this book shows we need a creative leadership.

NOTES

1. According to Klein (2010), multidisciplinarity involves merely drawing upon and juxtaposing knowledge, methods, perspectives or whatever from two or more disciplines, while interdisciplinarity entails an element of linking, blending and integrating those various inputs. Transdisciplinarity, in contrast, involves a higher-level and more fundamental transformation of those inputs, so that the research then transcends the original disciplinary boundaries (see also Martin, 2011, p. 457).
2. For further research on the promises and challenges of interdisciplinary research, see, for example, Aagaard-Hansen (2007), Allwood and Bärmark (1999), Bärmark and Wallén (1986), Hollingsworth and Hollingsworth (2000), Stokols et al. (2010), and Weingart (2000).
3. This usage among researchers may also, to some extent, reflect a cultural difference between the US and Scandinavian countries with respect to how the relationship between leaders and the other group members is viewed. In the US and other countries, leaders are often viewed as more directive and as acting independently of the coworkers. In contrast, leaders in Scandinavia are to a larger extent expected to listen and coordinate although they are also expected to manage.
4. This may simply reflect that it is generally harder for academic researchers to negotiate the necessary access to top managers in order to study them.

REFERENCES

Aagaard-Hansen, J. (2007). The challenges of cross-disciplinary research. *Social Epistemology*, *21*(4), 425–438.

Allwood, C. M., & Bärmark, J. (1999). The role of research problems in the process of research. *Social Epistemology*, *13*, 59–83.

Bärmark, J. & Wallén, G. (1986). The interaction of cognitive and social factors in steering a large scale interdisciplinary project. In D. E. Chubin, A. L. Porter, F. A. Rossini, & T. Connolly (Eds.), *Interdisciplinary analysis and research: theory and practice of problem-focused research and development* (pp. 229–240). Mount Airy, MD: Lomond.

Bass, B. M., & Bass, R. (2008). *The Bass handbook of leadership*. New York, NY: Free Press.

Björk, J. (2011). Analyzing and realizing collective ideation in firms (PhD. diss.). Göteborg: Chalmers University of Technology.

Byrne, C. L., Mumford, M .D., Barrett, J. D., & Vessey, W. B. (2009). Examining the leaders of creative efforts: What do they do and what do they think about? *Creativity and Innovation Management*, *18*, 311–324.

Byrnes, J., Miller, D., & Schafer, W. (1999). Gender differences in risk- taking: A meta-analysis. *Psychological Bulletin*, *125*, 367–383.

Campbell, D. T. (1960). Blind variation and selective retention in creative thought as in other knowledge processes. *Psychological Review*, *67*, 380–400.

Csikszentmihaly, M. (1999). Implications of a systems perspective for the study of creativity. In R. T. Sternberg (Ed.), *Handbook of creativity* (pp. 312–338). Cambridge, England: Cambridge University Press.

Defila, R., & Di Giullo, A. (2010). Managing consensus in interdisciplinary teams. In R. Frodeman, J.T. Klein, & C. Mitcham (Eds.), *The Oxford handbook of interdisciplinarity* (pp. 482–485). Oxford, England: Oxford University Press.

Dewett, T. (2003). Understanding the relationship between information technology and creativity in organizations. *Creativity Research Journal*, *15*, 167–182.

Elliott, T. (2005). *Expert decision-making in naturalistic environments: A summary of research* (Report DSTO-GD-0429). Edinburgh, South Australia: Australian Government, Department of Defence, Defence Science and Technology Organization, Land Operations Division Systems Sciences Laboratory.

Ericsson, K. A., & Charness, N. (1994). Expert performance: Its structure and acquisition. *American Psychologist*, *49*, 725–747.

Feist, G.J. (1998). A meta-analysis of personality in scientific and artistic creativity. *Personality and Social Psychology Review*, *2*, 290–309.

Feist, G.J. (2006). How development and personality influence scientific thought, interest and achievement. *Review of General Psychology*, *10*, 163–182.

Heath, C., Larrick, R. P., & Klayman, J. (1998). Cognitive repairs: How organizations compensate for the shortcomings of individual learners. *Research in Organizational Behavior*, *20*, 1–37.

Hemlin, S. (2006). Managing creativity in academic research. Could creative action and management be reconciled in research? *Science Studies*, *19*(1), 83–92.

Herzog, S.M., & Hertwig, R. (2009). The wisdom of many in one mind. Improving individual judgments with dialectical bootstrapping. *Psychological Science*, *20*, 231–237.

Hollingsworth, J.R., & Hollingsworth, E.J. (2000). Major discoveries and biomedical research organizations: Perspectives on interdisciplinarity, nurturing leadership, and integrated structure and culture. In P. Weingart & N. Stehr (Eds.), *Practicing interdisciplinarity* (pp. 215–244). Toronto, Ontario, Canada: University of Toronto Press.

Howell, J. M., & Boies, K. (2004). Champions of technological innovation: The influences of contextual knowledge, role orientation, idea generation, and idea promotion on champion emergence. *The Leadership Quarterly, 15,* 123–143.

Isaksen, S. G., & Tidd, J. (2006). *Meeting the innovation challenge: Leadership for transformation and growth.* Chichester, England: John Wiley & Sons.

Klein, J.T. (2010). A taxonomy of interdisciplinarity. In R. Frodeman, J.T. Klein, & C. Mitcham (Eds.), *The Oxford handbook of interdisciplinarity* (pp.15–30). Oxford, England: Oxford University Press.

Larrick, R. P. (2009). Broaden the decision frame to make effective decisions. In E. A. Locke (Ed.), *Handbook of principles of organizational behavior* (2nd ed., pp. 461–480). Chichester, England: Wiley & Sons.

Martin, B.R. (2011). What can bibliometrics tell us about changes in the mode of knowledge production? *Prometheus, 29,* 455–479.

Mumford, M. D., Connelly, M. S., Scott, G. M., Espejo, J., Sohl, L., Hunter, S. T., & Bedell, K.E. (2005). Career experiences and scientific performance: A study of social, physical, life, and health sciences. *Creativity Research Journal, 17,* 105–129.

Mumford, M. D., Scott, G. M., Gaddis, B. H., & Strange, J. M. (2002). Leading creative people: Orchestrating expertise and relationships. *The Leadership Quarterly, 13,* 705-750.

Mumford, M.D., Waples, E.P., Antes, A.L., Brown, R.P., Connelly, M.S., Murphy, S.T., & Devenport, L.R. (2010). Creativity and ethics: The relationship of creative and ethical problem-solving. *Creativity Research Journal, 22,* 74–89.

Nielsen, M. (2012). *Reinventing discovery. The new era of networked science.* Princeton, NJ: Princeton University Press.

Paulus, P. B., & Yang, H.-C. (2000). Idea generation in groups: A basis for creativity in organizations. *Organizational Behavior and Human Decision Processes, 82,* 76–87.

Prabhu, V.P. (2011). Risk-taking. In M.A. Runco, & S.R. Pritzker (Eds.), *Encyclopedia of creativity* (Vol. 2, pp. 319–323). New York, NY: Academic Press.

Rosing, K., Frese, M., & Bausch, A. (2011). Explaining the heterogeneity between the leadership-innovation relationship: Ambidextrous leadership. *The Leadership Quarterly, 22,* 956–974.

Selart, M., & Schei, V. (2011). Organizational culture. In M.A. Runco & S.R. Pritzker (Eds.), *Encyclopedia of creativity* (Vol. 2, pp. 193–196). New York, NY: Academic Press.

Simon, H. A. (1965). Understanding creativity. *Carnegie Review, 1964/1965*(2) [unpaginated].

Simonton, D. K. (2003). Scientific creativity as constrained stochastic behavior: The integration of product, process, and person perspectives. *Psychological Bulletin, 129,* 475–494.

Simonton, D. K. (2011). Creativity and discovery as blind variation: Campbell's (1960) BVSR model after half the century mark. *Review of General Psychology, 15,* 158–174.

Stokols, D., Hall, K.L., Moser, R.P., Feng, A.X., Misra, S., & Taylor, B.K. (2010). Cross-disciplinary team science initiatives: research, training, and translation. In R. Frodeman, J.T. Klein, & C. Mitcham (Eds.), *The Oxford handbook of interdisciplinarity* (pp. 471–493). Oxford, England: Oxford University Press.

Tierney, P., Farmer, S. M., & Graen, G. B. (1999). An examination of leadership and employee creativity: The relevance of traits and relationships. *Personnel Psychology, 52,* 591–620.

Vessey, W. B., Barrett, J., & Mumford, M. D. (2011). Leader cognition under conditions of threat: "Just the facts." *The Leadership Quarterly, 22,* 710–728.

Weingart, P. (2000). Interdisciplinarity: The paradoxical discourse. In P. Weingart & N. Stehr (Eds.), *Practicing interdisciplinarity* (pp. 25–41). Toronto, Ontario, Canada: University of Toronto Press.

Wilke, H., & Kaplan, M. (2001). Task creativity and social creativity in decision-making groups. In C. M. Allwood & M. Selart (Eds.), *Decision making: Social and creative dimensions* (pp. 35–51). Dordrecht, the Netherlands: Kluwer Academic Publishers.

Yukl, G. (2010). *Leadership in organizations*. Englewood Cliffs, NJ: Prentice Hall.

Zuckerman, H. (1977). *Scientific elite: Nobel laureates in the United States*. New York, NY: Free Press.

Contributors

Carl Martin Allwood, PhD, is a professor of psychology at the department of psychology, University of Gothenburg and previously at Lund University (1998–2008), both in Sweden. His research mainly concerns social cognitive psychology (including judgment, especially different forms of metacognition, and decision making as it occurs in real life), forensic psychology, culture-oriented psychology, and research psychology. He has edited several books, for example, on creative knowledge environments (with S. Hemlin and B. Martin, 2004) and has published more than 70 articles in international scientific journals.

Lily Cushenbery is a doctoral student in Industrial and Organizational Psychology at the Pennsylvania State University. Her research areas include leader error recovery and malevolent creativity. Lily also does research with the International Center for the Study of Terrorism and has managed a number of research studies and applied projects. She has been an author for several articles and book chapters in the areas of leadership and innovation.

Kate T. Dembroski works in the Human Resource Leadership Development Program at Lockheed Martin Corporation. Prior to joining Lockheed Martin, Kate received her MS in Human Resource Development at Villanova University.

Leif Denti is a doctoral student at the University of Gothenburg, Department of Psychology. His research interests include leadership, creativity, and innovation.

Dr. Dawn L. Eubanks is an Associate Professor of Behavioral Science and Strategy in Warwick Business School at The University of Warwick. She holds a PhD and an MS in Industrial Organizational Psychology from The University of Oklahoma. Prior to that she earned an MS. in Applied Psychology and Quantitative Methods from University of Baltimore and a BA in Psychology from Illinois State University. Dawn's research on the themes of leadership and creativity and innovation has resulted in publications appearing in journals such as *The Leadership Quarterly, Creativity Research Journal,* the *Journal of Applied Social Psychology,* and *Human Resource Management Review.* In addition to her publications, she has earned an EPSRC grant in excess of £1 million and several other grants totaling in excess of £240,000. Dawn is on the editorial boards of *The Leadership Quarterly* and the *Journal of Occupational and Organizational Psychology.*

Joshua Fairchild is a doctoral student in Industrial/Organizational Psychology at The Pennsylvania State University. His research focuses on leadership, creativity, and technology in the workplace. He has collaborated with colleagues in a number of design and technology disciplines, and his work has been published in a variety of academic outlets.

Nicole Ginther is a doctoral student in Penn State's Industrial/Organizational Psychology program. Her research interests are primarily related to personnel selection and include the study of technology for leadership and organization improvement. Nicole recently completed her master's thesis on polychronicity and its role in personnel selection.

Michael E. Gorman earned an MS (1978) and a PhD (1981) in Social Psychology at the University of New Hampshire. He is a Professor in the Department of Science, Technology & Society at the University of Virginia, where he teaches courses on ethics, invention, psychology of science and communication. Currently he is working as a Program Director in the Science, Technology & Society program at the National Science Foundation (NSF). His research interests include experimental simulations of science, described in *Simulating Science* (Indiana University Press, 1992), and cognition, invention, and ethics, described in *Transforming Nature* (Kluwer Academic Press, 1998). With support from the NSF, he conducted a multiyear cognitive study of the invention of the telephone whose results appeared in *Social Studies of Science* and *Thinking and Reasoning.* NSF supported his work with Patricia Werhane on case studies that combined ethics, invention, and design, described in *Ethical and Environmental Challenges to Engineering* (Prentice Hall, 2000). NSF also supported work that led to his edited volumes on *Scientific and Technological Thinking* (Lawrence Erlbaum Associates, 2005), and *Trading Zones and Interactional Expertise: Creating New Kinds of Collaboration* (MIT Press, 2010). He is a member of the editorial boards of the *Journal of Psychology of Science and Technology* and *TopiCS in Cognitive Science*—for the latter, he edited a special issue on "Cognition in Science and Technology". He did a 2-year rotation as an NSF STS program director. His current research is in the kind of interdisciplinary trading zones that will be needed for scientists, engineers, and other stakeholders to collaborate on the development of new technologies.

Philip Gorman is a member of PROS Professional Services team. He specializes in advanced analytics reporting configuration, complex data models, and science integration with other reporting and pricing work flows. He has implemented PROS software for customers in multiple industries, including medical distribution, medical devices, delivery services, auto parts and heavy equipment, high-tech manufacturing, and industrial manufacturing. He earned a Bachelor of Arts degree in Chemistry and Economics at Rice University, where he was a Trustee Distinguished Scholar.

Dr. Michelle Hammond is a lecturer in organizational behavior at the University of Limerick, Ireland. She is also course director for the MSc in Work and Organisational psychology. Michelle's research focuses on leadership, work–life balance, and creativity and innovation. In addition to authoring *An Integrative Theory of Leader Development: Connecting Adult Development, Identity, and Expertise,* she has published in journal such as *The Psychology of Aesthetics, Creativity, and the Arts,* the *Journal of Vocational Behavior,* and *Human Resource Management Review.*

Sven Hemlin, PhD, is a professor of psychology. He is affiliated with the Department of Psychology and Gothenburg Research Institute (GRI), School of Business, Economics and Law, University of Gothenburg. At GRI he is the Director for the Leadership, Innovation and Co-workership program. He is also a Visiting Professor of Social Psychology at the Department of Technology and Society, University of Skövde. He was a Visiting Research Fellow at SPRU, University of Sussex, Visiting Professor at Copenhagen Business School and University of Tampere. He is doing research and teaches in Organizational Psychology and in Science, Technology and Innovation Studies. He has published in several scholarly journals such as *Leadership Quarterly, Creativity Research Journal, Creativity and Innovation Management,* the *International Journal of Innovation Management, Scientometrics, Science, Technology & Human Values and Journal of Managerial Finance.* Currently, he is involved in research on creative leadership (with Lisa Olsson), innovation management (with Leif Denti) and bank credit decision making (with C.-C. Trönnberg and Magnus Jansson).

Samuel T. Hunter is an Assistant Professor of Industrial and Organizational Psychology at The Pennsylvania State University. His primary areas of research include leadership and innovation management. Within these areas, Dr. Hunter has published numerous journal articles, books, and book chapters. Dr. Hunter has also received funding for his work from agencies such as the National Science Foundation and the Office of Naval Research.

Ginamarie S. Ligon is an Assistant Professor at University of Nebraska at Omaha in the College of Business Administration. Her primary research areas are the identification and development of high-level talent. Prior to joining UNO, she worked at Villanova University and at Psychology Associates Management Consulting Firm.

Marie Mance is the Associate Director of the International Center for Studies in Creativity. She has coauthored several articles on creative leadership, most recently *Creative Leadership: Welcome to the 21st Century,* with Jeffrey D. Zacko-Smith and Gerard J. Puccio. The second edition of her book with Gerard J. Puccio and Mary C. Murdock, *Creative Leadership: Skills that Drive Change,* was released in 2011. Marie has traveled internationally to present creativity and creative problem solving courses and workshops in Nigeria, Singapore, and South Africa. She has also designed and delivered workshops in creativity, strategic planning, and other change initiatives for a number of organizations in the public and private sector. Marie is a past President of the Niagara Frontier chapter of ASTD (American Society for Training and Development) and a Colleague of the Creative Education Foundation. She holds an MS in Creativity and MEd in Counseling/Student Personnel.

Robyn C. Mapp works in human resources at Gamesa, which is a wind energy organization headquartered in Philadelphia. Prior to joining Gamesa, she earned her MS in Human Resource Development at Villanova University. She has also held positions at IMS Health, Federated Department Stores, Sesame Place, and AmerisourceBergen Corporation.

Ben R. Martin is Professor of Science and Technology Policy Studies in SPRU (Science and Technology Policy Research), University of Sussex. Between 1996 and 2004, he served as SPRU's Director. He is currently also an Associate Fellow at the Centre for Science and Policy (CSaP), and Senior Visiting Fellow at the Centre for Business Research, Judge Business School, both at the University of Cambridge. His research for the last 35 years has focused on science policy. He helped pioneer techniques for evaluating scientific laboratories, research programs, and national scientific performance and for conducting 'technology foresight'. He served on the Steering Group for the UK Foresight Program. In recent years, he has conducted research on the benefits from government funding of basic research, the changing nature and role of the university, the impact of the UK Research Assessment Exercise, creative knowledge environments, and the evolution of the field of science policy and innovation studies. He served as Deputy Chair of the EU High-Level Expert Group that put forward the rationale for establishing a European Research Council in order to pursue 'frontier research', and as Chair of an ESF panel reviewing the possibilities for creating a research output database for the social sciences and humanities. He also served as specialist adviser to the House of Lords Select Committee on Science and Technology. Since 2004, he has been Editor of *Research Policy.* In 1997, he was awarded the de Solla Price Medal for Science Studies. He has a BA (first class honors) in Natural Sciences from the University of Cambridge, and an MSc in the Structure and Organisation of Science and Technology from the University of Manchester.

Dr. Michael D. Mumford is the George Lynn Cross Distinguished Research Professor of psychology at the University of Oklahoma where he directs the Center for Applied Social Research. He received his doctoral degree from the University of Georgia in 1983 in the fields of organizational psychology and psychometrics. Dr. Mumford is a fellow of the American Psychological Association (Divisions 3, 5, 10, 14), the Society for Organizational Psychology, and the American Psychological Society. He has written more than 270 articles on creativity, leadership, planning, and ethics. He has served as senior editor of *The Leadership Quarterly* and is on the editorial boards of the *Creativity Research Journal*, the *Journal of Creative Behavior*, *IEEE Transactions on Engineering Management*, the *International Journal of Creativity and Problem Solving*, and *Ethics and Behavior*. Dr. Mumford has served as principal investigator on grants totaling more than $30 million from the National Science Foundation, the National Institutes of Health, the Department of Defense, the Department of State, and the Department of Labor. He is a recipient of the Society for Organizational Psychology's M. Scott Myer award for applied research in the workplace.

Dr. Joy Oguntebi is an Assistant Professor of Management at the Rochester Institute of Technology's Saunders College of Business. She received her PhD from the University of Michigan in Industrial & Operations Engineering, with a concentration in engineering management. Her research interests include organizational learning, knowledge management, team effectiveness, global teams, and cultural intelligence. Her research work has been published in a selection of outlets including the Decision Sciences Journal of Innovative Education, and she has presented her research at several domestic and international conferences.

Michael E. Palanski is an Assistant Professor in Management, teaching Organizational Behavior, Leadership, and Business Ethics for both undergraduate and graduate students. He holds a PhD in Organizational Behavior and Leadership from the State University of New York at Binghamton. He also holds an MA in Theological Studies from Covenant Theological Seminary and a BS in Business Economics from Grove City College. Palanski's research interests include leadership, ethics, and levels of analysis issues. He has published multiple articles in journals such as *The Leadership Quarterly* and the *Journal of Business Ethics*. Prior to becoming a professor, he was a retail product manager and an online banking specialist.

David Peterson is a doctoral student in the Organizational Psychology program at the University of Oklahoma. His research interests include creativity, innovation, and leadership.

Gerard J. Puccio is Department Chair and Professor at the International Center for Studies in Creativity, Buffalo State; a unique academic department that offers the world's first Master of Science degree in creativity. Gerard has written more than 50 articles, chapters, and books. His most recent book, coauthored with his colleagues Marie Mance and Mary Murdock, is titled *Creative Leadership: Skills that Drive Change* (2nd ed.). In recognition of his outstanding work as a scholar, Dr. Puccio received the State University of New York Chancellor's Recognition Award for Research Excellence and the President's Medal for Scholarship and Creativity. Dr. Puccio is an accomplished speaker and consultant; he has worked with major corporations, universities, and numerous school districts. He has delivered creativity workshops and presentations around the world, in such countries as France, England, Spain, Italy, Mexico, Tanzania, Hong Kong, Singapore, the Dominican Republic, and Canada. Gerard holds a PhD in organizational psychology from the University of Manchester, England.

Dr. Roni Reiter-Palmon is the Isaacson Professor of I/O Psychology at the University of Nebraska at Omaha. She is also the Director of Research for the Center for Collaboration Science. Her research focuses on cognitive and personality variables that facilitate or hinder creativity at the individual and team level, as well as social process in teams that facilitate or hinder creativity. She is the Associate Editor of Journal of Creative Behavior. Her work has appeared in leading journals in the field such as *Leadership Quarterly*, *Creativity Research Journal*, the *Journal of Applied Psychology*, and *Psychology of Aesthetics, Creativity and the Arts*.

Peter van den Besselaar is a professor of organisation studies at the Faculty of Social Sciences and the Network Institute, VU University Amsterdam. Previously, he was a professor of communication studies and an associate professor of social informatics (both at the University of Amsterdam), and director of the Netherlands Social Science Data Archive. His research focuses on social and organizational aspects of information and communication technology, and on the organization and dynamics of science and innovation. His publications have been published in, among others, *Scientometrics*, *Higher Education Policy*, the *Journal of Informetrics*, the *Journal of the American Society of Information Science and Technology*, *Research Policy*, *Communications of the ACM*, and *Lecture Notes in Computer Science*. In his current research, he tries to develop and apply e-science approaches to science and technology studies.

Issac Robledo is a doctoral student in the Organizational Psychology program at the University of Oklahoma. His research interests include creativity, innovation, and leadership.

Juani Swart is the Professor of Human Capital Management, Head of Organization Studies, and Director of the Work and Employment Research Centre at University of Bath. Juani specializes in Knowledge management and the management of Knowledge workers. She is Head of the Organization Studies group and Director of the Work and Employment Research Centre (WERC), which is focused on Human Resource Management in Professional Services Firms. She is known for her work on the transfer of human capital into intellectual capital, thereby linking the intellectual capital, HRM, and Performance debates. Her current research is focused on the links between employee attitudes (commitment) and behaviors such as knowledge sharing, creative engagement, and innovation. Juani conducts her research in cross-boundary contexts wherein which multiple stakeholders have an impact on the strategic impact of HRM practices. She has published widely in the area of people management in knowledge intensive firms, intellectual capital structures, and systems approaches to knowledge.

Inge van der Weijden is a researcher at the Centre for Science and Technology Studies at the Leiden University. Her main research interests are situated in the area of academic leadership and management, scientific careers, and performance of scientists and the groups they are working in.

Dr. Gert-Jan de Vreede is the Frederic W. Kayser Distinguished Professor of Information Systems at the University of Nebraska at Omaha where he is the Managing Director of the Center for Collaboration Science. He is also affiliated with the Faculty of Technology, Policy and Management of Delft University of Technology. He has published more than 200 peer-reviewed scholarly works. He facilitated hundreds of workshops in the United States, Europe, and Africa in areas such as strategy development, Delphi studies, operational risk management, product development, expert consultations, process redesign, and gaming simulations. In addition, he has trained hundreds of people in general and domain-specific facilitation skills.

Maaike Verbree is a researcher at the research group Marketing, Market Research and Innovation and lecturer of research methods, statistics and consumer behavior at the HU University of Applied Sciences Utrecht. Her current research interests include market communication, consumer behavior, and science policy. Her dissertation focused on academic leadership of research groups.

Triparna de Vreede is a PhD. Student in Industrial/Organizational Psychology at the University of Nebraska at Omaha. She also is a research associate at the university's Center for Collaboration Science. She is a trained facilitator of thinkLets-based Group Support Systems workshops. Her research focuses on the psychological foundations of thinkLet-based collaboration processes, cognitive processes of creativity, technology and work practice transition in groups, and creativity in groups.

Jeffrey David Zacko-Smith, MPA, EdD is Assistant Professor of Leadership and Creative Studies, and Coordinator of the undergraduate Leadership Minor Program. He holds a Doctorate in Educational Leadership and a Master of Public Administration (MPA) Degree from Seattle University (2008 & 2004). Jeffrey's research, writing and activism focuses on multiple areas, including creative leadership, authentic leadership, transcendent leadership, diversity leadership, social constructionism, and issues of social justice. Jeffrey has presented his writing and research at numerous national and international conferences; has published in the *Leadership Review* (2007), *Academic Exchange* (2010), and in *Multicultural Education* (2010); authored a book chapter and a monograph; is a member of the International Leadership Association (ILA); and is a trainer with the National Coalition Building Institute (NCBI). Jeffrey welcomes correspondence at smithjd@buffalostate.edu.

Bianca M. Zongrone is a doctoral student at University of Buffalo in the Management Department. Prior to this work toward a PhD in Organizational Behavior, she obtained her MS in Human Resource Development at Villanova University, and she a BS in Business and Management at Rensselaer Polytechnic Institute.

Index

academic leadership *see* leadership
age and creative discovery 185
AVE *see* average variance extracted
average variance extracted (AVE) 200

BIM *see* building information modeling
brainstorming 322
breadth of knowledge 188
building information modeling (BIM) 92

CAD (Computer-Aided Design) 92–3, 94, 104
charismatic leadership *see* leadership
climate 12, 16–17, 19–20, 71, 316–20, 325, 327; creative climate 326; group climate 327; team climate 317–20; work climate 316; *see also* team
cloud computing 91–2
cognition 11, 14–16
cognitive 318–19, 322, 325–6; ability, 70; capacity 190; *see also* leadership
cognitive problem solving style 70; repairs 324; *see also* problem-solving
collaboration 244–5, 251–2, 254–6, 259–60
collaborative 326, 331
commitment *see* research involvement
complex (ity) 323–4, 328, 331, 333
Computer-Aided Design *see* CAD
concept generator (software) 90
concept selection 88; role of technology in 88–9
conceptual combination 89–90; role of technology in 90
confirmation bias 272–3, 284
conservation 69–70
coordination vs autonomy 114, 120, 134
co-workers 320–1, 332 *see also* follower
creating a climate for creativity and change 308
creative change model: a systems approach 295–6
creative climate questionnaire 303–4
creative environment 295
creative knowledge environment (CKE), 7, 23, 73, 187, 205
creative leader 17, 20; defined 300, 302; leadership 15–16, 297; leadership related to other leadership theories 298–9; problem solving 1, 6, 8–9, 12; problem-solving model (CPS) 9, 20; self-efficacy 10
creative leadership skills related to science technology and innovation 309–11
creative person 294
creative problem solver 321
creative problem solving 292
creative process 74, 294; creative process models 82, 291; early models 82; eight stage model 83; integration with multilevel frameworks 83–5
creative product 294
creative role model 316
creativity 1–4, 6–20, 58, 121, 123–5, 184, 316–22, 324, 328; big C 7; definition of 2, 83, 291; eminent 184; everyday 184; facets of 294–5; leadership, change connection 301; leadership, innovation connection 302–3; new forms of 321; small *c* 7;

training programs effectiveness 294; traits 245, 247, 251–4, 256, 258, 260; a 21st century skill 288
crowd funding *see* crowd sourcing
crowd-sourced funding *see* crowd sourcing
crowd sourcing 90–1, 95, 104–5

decision making 321–3, 326, 328
development 211, 212, 214, 215, 221, 227, 228
divergent and convergent thinking 13, 70, 188, 293–4; separate and balance 306
diversity 70

early career success 188–9; predictors of 188–9
Ekvall, G. 303–4
electronic databases 87–8
empowerment 69, 71
ethical 318, 327–8
ethics 279–80, 282
evaluate 321, 323–4
expertise 5, 243, 247–9, 251–4, 256, 258, 316–17, 319–20; domain 317

feedback 86, 316, 321; from products 95–6; user 86–7, 93–4, 96, 105
follower 268–71, 273, 275, 277–8, 280, 283–4, 319–20; *see also* co-workers
formal leader *see* leader
four P's (person, process, product, and press) 3
funding 115–16, 132, 150, 153, 164, 169, 171–73

generations 151–3, 162–4
goal 244, 250–3
goal setting 76
group composition 163, 177
group management 120–3, 134

heterogeneity 70, 322
heuristic algorithms 88–9
heuristics *see* heuristic algorithms
hierarchical linear modeling (HLM) 71
historiometric approach 191
human capital 115–16, 132

idea 319, 320, 321–3, 327; creative 323; evaluation 93, 321; generation 11, 90, 321–2; implementation 94, 321–2; new 327; role of technology in 90–5
ideation 320
Implementation phase 6
informal leader *see* leader
information gathering 87; role of technology in 87–8
innovation 58, 83; age 290; closed 3; definition of 296; distinction from creativity 83; exploitative 3; exploratory 3; incremental 3; need for 289; open 3; radical 3; relationship to creativity 297; role of leaders in 297; social 4
innovative 2
innovative capacity 2
innovative efforts 4
interaction effects 171–3
interdisciplinary 317–19, 328; teams 16, 240–4, 247–9, 251–4, 256–60; versus multidisciplinary teams 240; *see also* team
intrinsic motivation 10, 17, 68–9, 71
invention 1, 3, 20
inventor 1, 272

job requirements 215

kickstarter.com 95
Kirton Adaption-Innovation Instrument (KAI) 70
knowledge sharing 247–9, 251, 253, 255

late career success 189–90
leader 2, 4–5, 12–20; departing 157, 164–5, 169, 171; formal 4; informal 4; mature leaders 165–9; senior 216, 221–2, 225; starting 157–8, 164–7, 173; support 65, 71
leader-member exchange (LMX) 16, 67, 71, 320
leadership 1–20, 58, 252–60, 268–71, 73, 76, 78–80, 83–4; academic 114, 118–20, 133, 150–1, 157, 175–6; charismatic 13–14; and the climate for creativity 304; cognitive 5; creative 15, 20; culture 4;

development for creativity 304; empowering, 69; flexible 17; four critical dimensions of 13; implications for 5; motivational 5; participative 19; role 4, 18; scientific and engineering 19; skills 189–90; skills and qualities distinctive to promoting innovation 310; styles 5; theories 297; three t's framework of leadership theories 298; transactional 66, 298; transcendent 298; transformational 13–14, 16–17, 20, 66, 71, 298; using creativity principles 307–8
life cycle 154–6, 165–9
LMX *see* leader-member exchange

management 316–17, 319–21, 327; time 327
manager 4–5, 19
managerial 4
managerial culture 4
mediator variable 67–8
mental model 71–3, 85, 326; shared 249–52, 255, 258–9
model for development of creative change leaders 305–6
modeling 252, 255, 259
moderator variable 67–8
monitoring 95; role of technology in 95–6
motivation 316–17, 319, 321, 325–7
motivational leadership *see* leadership
multilevel consideration 186

National Science Foundation (NSF) 69–71, 76, 79–81, 85
need for achievement 76
need for autonomy 76
Netflix 88–9, 91, 104
network(ing) 162, 165, 173
network management 123, 135–6
Nobel Prize laureates 187, 198, 203
norms 250, 255–6, 258–9

partial least squares (PLS) 200
performance indicators *see* research performance
planning 316, 323, 325–6; adaptive planning 325; career planning 321; creative planning 325; succession planning 318, 321; time- and work-planning 321
PLS *see* partial least squares
proactivity 76
problem construction 85–6, 243–4, 253–4, 257; role of technology in 86–7
problem identification *see* problem construction
problems 316–19, 328; complex 317; ill-defined 317; types of 291, 302
problem-solving 318–19, 321; cognitive 318; creative 319, 321
process loss 241
productivity loss *see* process loss
psychological safety *see* trust
Pygmalion effect 65

R&D *see* research and development
rapid prototyping 93, 104
reflection 255, 259
reputation 155, 167, 176
research and development (R&D) 5–7, 21; definition of 6–7
research involvement 118, 133
research performance 124–8
resource strategy 114, 115–18, 132, 137
rewards 118, 122, 134, 165; reward systems 76
Rhodes, M. 295
risk 323–4, 327; risk taking 323
role model 316, 319, 321

science, technology and innovation (STI) 1–2, 6–7
self-efficacy 10, 17, 68, 71
senior leaders *see* leaders
shared information 248
shared mental models *see* mental model
simulation 94, 104
social innovation *see* innovation
starting leaders *see* leader
STI *see* science, technology and innovation
strategy formation 215, 229, 231–2
sustainability 318–19

team: communication 241, 244–5, 247–60; conflict 244–8, 251, 254–8; creativity 241, 243, 245–51,

254–60; demographic 242–3, 254; diversity 245, 250–1, 254, 257–9; functional diversity 243–4, 248, 254; relational 246, 254, 262; task 244, 246–7, 254, 257
team composition *see* team, diversity
technology 68–9, 72, 76, 79, 81, 83; adoption 97; challenges of 101–2; incentives for 99–100; leader influence on 97–100; modeling 98–9; promoting 97–8; role of vision in 100–1
time allocation 133, 155–6, 162–3, 169–71, 173
trading zone 74–80, 84, 318
transactional leadership *see* leadership
transformational leadership *see* leadership
trust 245–7, 251, 255–6, 258–9
try new things 189
types of problems *see* problems

Web 87
Web 2.0 *see* Web